MEASURES AND MEN

MEASURES AND MEN

· · ·

WITOLD KULA

· · ·

TRANSLATED BY

R. SZRETER

PRINCETON UNIVERSITY PRESS

Publication of this book has been aided by
grants from the Alfred Jurzykowski Foundation, Inc., and the Louis A. Robb
Fund of Princeton University Press

CONTENTS

292020

CONTENTS

TRANSLATOR'S NOTE

The author of the present work, professor of the University of Warsaw and the doyen of Polish economic historians, is a founder-member and a past President of the International Economic History Association. The only book of his to have appeared so far in English translation (by L. Garner, from the Italian edition) is *An Economic Theory of the Feudal System* (London, 1976). The English version of his magnum opus, *Problems and Methods of Economic History*, completed by the present translator, has yet to find a publisher. Chapter 13 of the latter work was devoted to historical metrology, and it was out of Professor Kula's fascination with the social significance of weights and measures that, in due course, the present book was born.

Of the four parts into which *Measures and Men* is divided, the first and last are general or international in character and world-wide in scope, the second deals with developments in Poland, and the third is devoted to France as the country that gave the world the metric system. In translating Part Two in particular, a determined effort was made generally to translate Polish terms into English, but some, for which no satisfactory equivalents could be found (e.g. the common unit of arable land, the *łan*), can be looked up by the reader in the short glossary that follows this note. Geographical names have been retained in their original Polish, except a few where a well-established English version exists, e.g. Warsaw for Warszawa. Lastly, apart from chapter 20, I have shortened a number of footnotes and omitted a score or so, all of which alterations were accepted by Professor Kula. I wish also to offer my thanks to him for answering promptly and lucidly a number of queries that arose in the course of translation.

My thanks for help are due also to several historians on the campus of Birmingham University, namely Mr. R. J. Knecht, Dr. J. T. Łukowski and Professor R.E.F. Smith. My two major debts of gratitude, however, are to Professor Ronald E. Zupko of Marquette University, Milwaukee, a scholar of unrivalled knowledge in the field of historical metrology, for his prompt and generous reading of the entire typescript and for making many pertinent suggestions which freed the translation from a number of errors, and to my wife, Dulcie M. Szreter, whose birthplace as well as academic training render her English doubly better than mine, and whose careful reading of the text and resultant criticisms (my frequently scant

vii

appreciation of them at the time notwithstanding) led to countless improvements in style and clarity. Finally, I wish to acknowledge the intervention of Professor Bert F. Hoselitz of the University of Chicago, whose acquaintance with my earlier translation of *Problems and Methods* led him, at an early stage of the proceedings, to clarify the relationship between that book and the present one to the Princeton University Press, while the latter—as represented throughout by Edward Tenner and at the copyediting stage by Elizabeth Powers—has been businesslike, positive, and courteous in its dealing with me.

R. S.
*University of Birmingham,
England*

GLOSSARY

folwark	demesne farm, within a larger manorial estate, generally worked by unpaid compulsory labor of serfs (the corvée)
garniec	Polish dry or liquid measure of approx. 4 liters (3.771 from 1764, 4.1 from 1819), often translated as "gallon"
lustracja	(pl. *lustracje*; from Latin "lustratio") surveys/inventories of the royal domains, first instituted in the 1560s
łan	(pronounced "won") the commonest but far from standardized Polish unit of measurement of agricultural land, the Chełmno *łan* of some 170,000 square meters and the Franconian *łan* of some 226,000 to 253,000 square meters being the most widely applied
morga	(also *morg* or *mórg*) Polish unit of land measurement—5,600 square meters
osep	(also *sep*) basic corn rent levied by feudal landowners
Referendaria Koronna	the Crown Referendary's court, established in the sixteenth century to hear appeals from peasants (serfs) on the royal domains against decisions by tenants or *starostas*
Sejm	(also *Seym*) before the partitions, the assembly of Polish nobles, often translated as "Diet" (The local *sejmik* or *seymik* is often translated as "dietine.")
starosta	an office with several meanings; here, usually, an intendant of the royal domains
strickle, striked	a term applied to a measure or container that is filled with grain or some other dry good and then made level with the rim by applying a straight piece of wood or "strickle" (The amount is known as a "striked" measure—some writers prefer "stricken" or "struck"—as opposed to "heaped.")
szlachta	(sing. *szlachcic*) Polish landed nobility, very numerous and including many smallholders or even-rent-payers (some writers prefer "gentry"), with the "magnates," owning huge tracts of land, at the other extreme

wojewoda   (English "voivode") provincial governor whose area of authority (voivodeship, *województwo*) approximated an English county

włóka   (often transliterated as *vloka*) a unit of land measurement, approximately 168,000 square meters

PART ONE

· 1 ·

THE REPRESENTATIONAL* AND FUNCTIONAL CHARACTER OF PAST MEASURES

"Who invented measures?" The puzzled reader may well suppose that measures, like the wheel or fire, are an anonymous invention, one that cannot be ascribed to any particular individual. But there he would be mistaken. A source so prestigious that I would not dare question its correctness tells us that the inventor of weights and measures was—Cain! This wicked son of Adam and Eve, having killed his brother Abel, went on to commit many other sins, and "he was"—so Flavius Josephus tells us—"the author of weights and measures, an innovation that changed a world of innocent and noble simplicity, in which people had hitherto lived without such systems, into one forever filled with dishonesty."[1] Our source is indeed worthy of respect, and reliable in telling us how, in the simple reasoning of the Biblical tradition, the notion of measure is associated with cheating; how it symbolizes the loss of primeval happiness, and how it derives directly from original sin. (This Biblical tradition was undoubtedly transmitted faithfully to posterity by Josephus who, although he collaborated with the Romans, did come from a priestly family in Jerusalem.) As we go on to consider a selection of fragments from the social history of weights and measures, we shall indeed come to see the extent to which that history is replete with injustice and dramatic struggles.

It is a matter of common knowledge that old measures bearing the same names can signify vastly different magnitudes, depending on the time, the place, and the substance measured (*ratione loci, ratione temporis,* and *ratione materiae*). It is not enough merely to be aware of this, nor yet even to be able, whenever necessary, to translate the old measures into their metric equivalents; what is addi-

---

* Throughout the book, the author posits an important distinction (see his note 5, chap. 17) between, on the one hand, the metric system and its units, which he sees as basically abstract and agreed upon by convention, and, on the other, a variety of pre-metric measures, mostly derived and transferred from man's limbs and labor, e.g. the foot. For the latter, he has coined the adjective *znaczeniowy* from the Polish *znaczenie*, "meaning" or "significance." After considering and trying out a number of translations, it appeared that the term *representational* was the most serviceable—TRANS.

tionally necessary is an understanding of their varied, but hidden, social content. The core of this understanding is to be found in their derivation from concrete phenomena in daily life, in contrast to our meter, which has been agreed by convention.

Today's standard measures signify nothing more than a common denominator for all the dimensions measured, e.g., length, area, mass, time, and exchange value. The size of the unit is a matter of indifference; what matters is that the unit should be invariable. The fact that the kilogram stands for the weight of ten cubic centimeters of water at the temperature of 0° centigrade, or that the meter stands—or, strictly speaking, initially stood—for 1/40,000,000 part of the meridian, has no inherent social significance whatsoever. The overwhelming majority of people who employ measures are ignorant of these facts—and none are mindful of them when actually using such measures. By contrast, the measures of primitive societies, the early medieval European measures, and folk measures that ethnographers have uncovered for us do have a certain definite social significance that explains the size of units, their variety across space, and, in some cases, their mutability over time. To appreciate this is more valuable for historians generally, and for economic historians in particular, than is the otherwise highly serviceable ability to translate traditional measures into the units of the metric system. After all, the metric system, whose acceptance meant that the unit of measurement was based on an astronomical phenomenon independent of man, has been with us only for a century and a half.

Old measures, when we stop to consider them, may appear to us to be very inexact and to offer much scope for misunderstanding. But let us not look at them just through late twentieth-century spectacles, for under different circumstances, different degrees of exactitude are socially requisite. The very great exactitude of the metric system was more than sufficient for the construction of ferro-concrete buildings and of airplanes, but it proved far from adequate in the planning of lunar landings by interplanetary rockets.[2] However, the representational nature of old measures—especially anthropometric ones [see the title of chap. 5—TRANS.]— always ensures that different measures are used in the measuring of different kinds of objects. For instance, in discussing old Slavonic measures, K. Moszyński writes: "Every measure served a different purpose. The foot marked the distances when potatoes were sown, the pace, distances in general,[3] and the elbow, abbreviated to 'ell,' was applied to cloth. The peasant fisherman would refer to his net as being 30 fathoms long and 20 ells wide."[4] Further examples of this kind could readily be provided. The matter is of the greatest importance. The system was derived from nothing other than daily tasks: it was easier to measure the length of the net by the fathom and the width by the ell.

To simplify, and to present the matter in strictly evolutionist terms, we may say that the earliest stage in the development of man's metrological concepts is the anthropomorphic, in which the most important measures correspond to parts of the human body. It is in a later stage that reference is made to units of measure derived from the conditions, objectives, and outcomes of human labor. Naturally, the conditions of life and work dictated the lines along which the metrological system, or its component parts, would develop. In societies where land was relatively abundant, the system of area measures tended to be poorly developed. The Ashanti of Ghana, in whose economy the extraction of gold dust played a major part, had a very advanced system of weights.[5] On the other hand, the Saharan nomads, for whom the exact distance from one water hole to another may be a matter of life or death, have a rich vocabulary of measures for long distances. Thus, they reckon in terms of a stick's throw or a bowshot; or the carrying distance of the voice; or the distance seen with the naked eye from ground level, or from a camel's back; or walking distances from sunrise to sunset, or from early morning, mid-morning, or late morning; or a man's walking distance with no load to carry, or with a load to carry, or with a laden ass, or with an ox; or a walk across an easy or a difficult terrain. Such measures are still in use today, and we find reference to them in historical sources going back a thousand years or more.[6]

The social content inherent in the prevailing system of measures served to ensure their longevity. The Gauls, although they received from the Romans their art of surveying and the institution of the cadaster, retained their own traditional unit of measurement (the *arepennis*, the unit of land for one plow to till; *penn-os*, "the head"; hence, today's *arpent*).[7] Similarly, the Roman mile (equal to 1,000 paces or double steps) found no use in Gaul, since that country— renowned for horse breeding, the manufacture of wagons, and for its cavalry—preferred to retain the traditional *leugae* (the later French *lieue*) of approximately four kilometers. "Even the Roman administration of highways had to accept as official the country's unit of measurement, and took to marking on the milestones distances in *leugae* as well as in miles, or sometimes in *leugae* alone, for the drivers of the imperial postal services, the road overseers, and, in general, all those associated with transport in Gaul were, after all, local people."[8]

The association of measures with the techniques of production and the productivity of labor is, of course, also apparent outside agriculture. This is quite striking as far as textiles are concerned, for the width of the cloth is determined by the width of the loom, and the length of the piece also partly determined by technical considerations and partly by those relating to the social organization of production. The length of the piece commonly coming off the loom becomes, in turn, the customary unit of length for textiles. When the determinants change, so does the length of the unit in question—albeit without a change of name; and naturally, that length will not be uniform for different textiles, such as linen and woolen cloth, since the looms employed in their making differ.[9] If the *pane* is the measure of glass, then the size of the pane is determined by that of the milling equipment in the glassworks. The size of the *pig* with which raw iron is measured depends upon the technique of releasing molten metal from the primitive smelting furnace (or, in later time, from the blast furnace). The *bar*, however, which is the measure of wrought iron, has dimensions derived from a particular technique of smelting. The same applies to the *kiln* as the measure of lime, or again, to the *load* in measuring charcoal.

Transport arrangements may also determine the units employed to measure certain products. Units of measurement derived from this source are found in the commodity market. In the wholesale trade of commodities whose production is dispersed, it is the large units that are so determined, but some commodities produced within a small area and sold by retailers are measured in relatively small units that are also derived from transport. Corn is an example of the former category: here, transport determines the very large unit of the *last*. On the other side, charcoal exhibits the latter characteristic, being measured by the *basket*, and so do the various products measured by the *wagon-load*. A *boatload*, by which sand is often sold, is, again, a transport-determined measure. And if salt from the Ruthenian saltmines is sold by the *cartful*, then, to be sure, the cartful will before long be standardized.[10]

An interesting case was that of the use of the *strip* in the Cracow market by the sellers of cabbage and turnips.[11] A measure derived from production is here used in commercial transactions.

The practical functional character of such measures is often striking. For instance, in Geneva in the fifteenth and sixteenth centuries, the measure employed in the assessment of taxes on traveling traders arriving in the town was the *sack*, as carried in double pairs by the pack-ass.[12] Here, we have a system imposing upon the taxpayer himself, namely the trader, the necessity to observe the standard. After all, it is primarily in his interest to see that the ass should be neither too lightly loaded nor overburdened, for he might well end up on an Alpine pass with his wares strewn on the rocky ground beside his fallen beast of burden.[13] To take another example of a measure that is derived from the forces of production in transport, in Brazil (Nordeste) today, the *cartful* serves as the unit of measurement. It is a small unit, but the interested party dare not increase it, for the cart's construction is flimsy and the draft animals weak.[14]

More generally, the sack (or bag), being the instrument of both preservation and transport, would be used as a measure and, with the passage of time, would come to be accepted as a more-or-less standardized conventional measure. One example of this process is found in medieval Pisa,[15] and another comes from Silesia.[16] Sometimes the functional character of a measure is linked with the nature of the product it measures: wine is measured by the cask, which also serves as the container for it; accordingly, in areas such as Languedoc, where wine does not keep well and goes sour quickly, the cask tends to be small.[17] Salt affords an interesting example of a commodity where the determinant of measure is, on the one hand, production-related, and, on the other, institutional: the dimensions of a block of salt in the Wieliczka mines in Cracow were dependent upon natural factors and the techniques of hacking the salt out of the deposit, but large blocks were prized partly because they would be proportionately less damaged in transport. The limiting factor here, however, was the strength of the carts as well as reluctance to incur a higher rate of customs tariff.[18]

Finally, we come across measures determined by the needs of consumption. In the sixteenth century, French agriculturalists advised landowners to bake individual loaves for their laborers to insure that nobody would cut an unfairly large portion, as happened when a large loaf had to be divided.[19]

The diversity of representational measures in different periods and countries is astonishing. In old Ethiopian recipes, we find the following description of the measure of salt: "enough to cook a chicken."[20] The *bowshot* has been found almost everywhere as a measure of distance,[21] but the measure disappears with the demise of the bow. A hatchet's throw as a measure will not surprise us; but a hatchet's throw backwards from a sitting posture is a more unusual unit.[22] A *stone's throw* in the measuring of distance is still used today in Slovakia.[23] Ethnographers confirm that both the *stone's throw* and the *bowshot* were still in use as measures in Latvia in the fifteenth century, although the bow had long since fallen into disuse.[24] Indeed, an ethnographic expedition in Latvia as recent as 1947 found distance measured by the neighing of the horse or the bellowing of the bull ("two bull's bellows away").

We shall be discussing the most important groups of representational and functional measures, namely anthropometric and agrarian ones, in more specialized chapters. Certainly, here too, the real-life context was a source, though not the only one (as chap. 2 will show), of the diversity and variability of measures. The area measured by the time taken to work it may depend on the quality of soil, or the quality of tools and the main produce. Dry measures of capacity related to sowing will vary for different grains, and measures of cloth will vary depending on the model of the loom commonly used in the locality, and so forth. For instance, in 1790, within the single *département* of Basses-Pyrénées, nine different sizes of the *arpent* were found, the ratio of the smallest to the largest being 1:5;[25] and in the district that was to form the *département* of Calvados there were as many as sixteen of them![26] The "length" of foreign cloth in Poland in the sixteenth century varied between 32 and 60 ells.[27] Thus the range and diversity of measures simultaneously in use, within even a small district, is often truly amazing.[28] This is not a matter to deplore, but one whose social, indeed human, significance demands our attention.

· 2 ·

REALISTIC AND SYMBOLIC
CONCEPTIONS OF MEASURES
AND MEASURING

In stratified societies, even in very early stages of development, honesty in the employment of weights and measures is highly regarded and given all manner of guarantees. Therefore, in addition to a guarantee from the secular authority, one of a sacred nature emerges as well. Very early, too, we find that "the just measure" becomes symbolic of justice in general. Practices bound up with man's attitude to measurement assume the character of a symbolic expression of many elements of popular "social philosophy."

We can easily follow this evolution in the Bible. In the Books of Moses, which constitute a code of social conduct hedged with sacred sanctions, the norms relating to measures are still put literally. Thus we read: "Ye shall do no unrighteousness in judgment, in meteyard, in weight, or in measure. Just balances, just weights, a just ephah, and a just hin, shall ye have."[1] Or again: "Thou shalt not have in thy bag divers weights, a great and a small. Thou shalt not have in thine house divers measures, a great and a small. But thou shalt have a perfect and just weight, a perfect and just measure shalt thou have: that thy days may be lengthened in the land which the Lord thy God giveth thee."[2] The religious sanction against metrological transgressions is therefore death. However, by the time of Solomon and the Prophets, the phraseology becomes symbolic. For example, Solomon writes: "A just weight and balance are the Lord's; all the weights of the bag are His work."[3] And in threatening the people of Israel with divine punishment for their sins, the Prophets give this example of a sinful thought: "When will the new moon be gone, that we may sell corn and the Sabbath, that we may set forth wheat, making the ephah small, and the shekel great, and falsifying the balances by deceit?"[4] And they make the Almighty say: "Shall I count them pure with the wicked balances, and with the bag of deceitful weights?"[5] Clearly, the weights and measures serve here as symbols: "the weights in the bag" stand for men's deeds, just or unjust. This evolution reaches its apogee with the metaphorical words of Christ in the New Testament: "With what measure ye mete, it shall be measured to you: and unto you that

9

hear shall more be given";[6] or more elegantly: "Give, and it shall be given unto you; good measure, pressed down, and shaken together, and running over, shall men give into your bosom. For with the same measure that ye mete withal, it shall be measured to you again."[7]

It is also interesting that the Koran, although it came even later than the New Testament, condemns metrological offenses in a totally realistic fashion (as was the case in the oldest books of the Old Testament). The comparatively late eighty-third Sura, dating from the early Medina period and devoted to this theme, reads as follows:

> Those who stint
> In the name of Allah, the Compassionate, the Merciful.
> Woe to those who stint the measure.
> Who when they take by measure from others, exact the full,
> But when they mete to others [i.e. to their own advantage], or
>   weigh to them, diminish.
> What! Have they no thought that they shall be raised again
> For the great day,
> The day when mankind shall stand before the Lord of the
>   worlds?[8]

The acme of man's symbolic conception of measurement is found in those widespread references, in various civilizations, to the "great day"—as it is called here by Mohammed, or the "awful day," the day of the Last Judgment. Amon weighed the deeds of the ancient Egyptians in his scales, as Archangel Michael did for Christians. Countless tympanums of Romanesque and Gothic cathedrals show the latter on the day of the Last Judgment, scales in hand, at the side of Christ the Judge.[9] As they approach the throne—kings and princes, bishops, abbots, knights and ladies, craftsmen and peasants, many of whom had doubtless cheated or been cheated in their life on earth—their deeds will be weighed by Archangel Michael in his just scales, and their just deserts will accordingly be meted out to them.

All three monotheistic religions of Near Eastern origin—Judaism, Christianity, and Islam—share this belief. It is, therefore, not at all surprising that in those civilizations, the knowledge of measures symbolizes—is even synonymous with—civilization itself. A companion of the sixteenth-century explorer Magellan, Antonio Pigafetta, in describing an Indian tribe in South America and praising its civilized way of life, writes: "Those kinds of people live with justice, weights and measures."[10] And his namesake, Filippo Pigafetta, describing jointly with Duarte Lopez in 1591 the fabulous

Congo, refers to a tribe distinguished by its high degree of civilization, and notes that its members "employ numbers, weights, and measures—things known only in those parts of the Congo."[11] Montaigne, however, in writing of the recently discovered New World, asserts that "as recently as fifty years ago the written word, weights and measures, clothes, corn or wine were unknown there."[12] Although he differs from Pigafetta in his appraisal of the peoples of the newly discovered lands, Montaigne nonetheless subscribes to the same hierarchy of values, acknowledging the employment of weights and measures as a mark of civilization, equal to the knowledge of writing, the cultivation of corn or the making of wine, or the weaving of cloth.

Using weights and measures as a criterion by which to distinguish civilization from barbarism is not unknown even today. It is a matter of common knowledge that the majority of newly independent African countries are presently engaged in an "Africanization," as it were, of their history. Among the investigations seeking to revise the view imposed by colonizing powers of a "barbarous" pre-colonial Africa, into which European colonists first brought the torchlight of civilization, we encounter studies of pre-colonial, indigenous weights and measures. The artistically made weights used by the Ashanti in the weighing of gold, known and admired by Europeans since the seventeenth century, are a favorite revisionist example. Asserting the traditional values of their culture, African students hold that the little ornamental weights form a coherent system, and since the Negroes were able to create an indigenous coherent system in no way inferior to the old European systems, there must have been Negro scientists, contrary to the myth of a lack of any conception of science among the blacks. European students of an anti-African disposition had tried, by hook or by crook, to establish that that system of weights was "too complex to have been created by the Negro brain" and had been borrowed from the Arabs or even from Greece, Rome, or Portugal. Yet all archaeological evidence refutes this. An African scholar writes: "These weights open a window into our past: they are a fragment of the Negro contribution to the world's treasure-house of art and culture. Africa was not just a cultural borrower but constantly created and invented." And so, as in the writings of Pigafetta or Montaigne, the presence of a system of weights and measures—with emphasis on the concept of "system"—constitutes a criterion of civilization.[13]

It is only relatively recently that to the medley of symbolic conceptions of measures yet another has been added, that of measure as a symbol of prosaic pedantry. In his memoirs, Ehrenburg describes a visit to Duhamel shortly after the First World War. "For some unknown reason, he [i.e. Duhamel] recollected the *verst* of the old Russian novels; he then remarked that France had given the world the metric system, and was it not splendid that the Russians, too, had now adopted it . . . When Duhamel left, we [i.e. Ehrenburg and Olga Forsz] burst out laughing. . . . What made us laugh was his naiveté: he truly believed that his centimeter was capable of measuring our roads."[14] And yet the reasons for the association in Duhamel's mind were not all that unfathomable. Duhamel acted like a true French bourgeois, brought up in the tradition of the Great French Revolution. He quite properly saw in the metric system a symbol of victory over feudal survivals, a symbol of the modernization of his country. And surely what was seen in it by the leaders of the October Revolution, in their hurry to promulgate the decree introducing the metric system in socialist Russia, was no different. But then Ehrenburg did not appreciate the fact that the French *centimetre* was quite equal to the task of measuring even the mighty Russian roads at least as efficiently, if not more so, than did the *verst*. Let us not laugh, however, at his contempt for the "prosaic" measures, and for the pedants who used them. After all, it was our own national bard [the Polish poet, Adam Mickiewicz, 1798–1855—Trans.] who wrote: "The compass, the scales, the yardstick—apply but to lifeless bodies."

Finally, it seems worthwhile to ponder certain anti-egalitarian connotations, not so much of measures as of the process of measuring, especially the practice of levelling or "striking." Idiomatic phrases for "employing one (striked) standard for all" are found universally, and Kolberg has recorded the analogous and more pithily formulated Polish proverb: "Stick your head out of the bushel and you'll get it levelled by the strickle."[15]

The metric system has produced a cult of its own as an achievement of *homo sapiens*, rational and complete in its intellectual purity, free from all superstitions and traditionalism, good for all—in short, perfect. All that is needed for people to appreciate its perfection is education. Like any other cult, this secular one has its own martyr, one Pierre F. A. Mechain. When nearing the completion of his labors to ascertain the length of the meridian along the line running from Dunkirk to Barcelona, he contracted malaria in Catalonia and died. Here was a victim not of superstition, not of obscurantism and prejudice, but one sacrificed upon the altar of science.

· 3 ·

BELIEFS ASSOCIATED WITH
MEASURES AND MEASURING

It was Cain, then, who devised measures. To count and to measure is sinful. Since it is a well-known fact that the devil himself gave David the idea of counting God's people,[1] it is clear that to count, and especially to count people, is sinful. Similiarly, it is sinful to measure a human being. "Among the Czechs, at the end of the eighteenth century, a belief was prevalent that a child under six years of age would cease growing, become stunted, a 'measure-ling,'* if the cloth intended for his shirt or outer garment was measured."[2] Taking a man's measure, or the measure of some part of his body, invests him with symbolic and ambivalent significance. A thread or a ribbon the length of the circumference of the head may be offered as a votive offering for the recovery of the person so measured. It may be used as well, in the practice of black magic, to harm him, and if the subject is already deceased, then it may be hung on the altar to prevent his unwished-for return.[3]

Superstitions about measurement were not limited to measuring parts of the body. In Macedonia, at the end of the nineteenth century, peasants would not eat what had been measured for fear of developing a goitre. In the Vladimir *gubernya*, at the beginning of the second half of the nineteenth century, peasants were inimical to the practice of calculating what they had harvested: "What the Lord has provided, even without counting, will find its way into our barns; it is not for us to assess the verdicts of Providence. There is no way in which you can summon God to a court of law. He who calculates the yield of the harvest from our fields, sins. We gain nothing by counting." At about the same time, ethnographers work-ing in Krzywicze, in northern Poland, recorded the popular belief that harvests had been more abundant before people began to measure land and calculate its produce. And from the same period, from the Cuiavia district [northwest of Warsaw—TRANS.] comes the information that local people, when purchasing drugs, would ask that they be poured into the bottle as the eye saw it rather than

---

* I have made up this word in order to render more clearly the sense of the passage, though the word in the Polish original (i.e. translated from the Czech) is more properly translated as "runt"—TRANS.

measured, "for the sick man will get well only if the medicine for him has been given with a free hand and honest heart—unmeasured."[4]

We may venture the hypothesis that the distrust of counting and measuring in the examples above is typical of a great many agricultural societies, whose basic economic unit is the family farmstead, geared primarily to its own subsistence needs. But to substantiate this hypothesis we need far more evidence; it is not enough to offer examples from parts of central, eastern, and southern Europe, for even if very numerous, they have come to us piecemeal, from widely scattered localities, and often from fieldwork notes left by chance by ethnographers of the past. The matter is different when we deal with pastoral societies or with commercial and mercantile communities. For as we have already remarked among the pastoral-commercial Semites, the arts of measuring and weighing in those societies were taken as a matter of course, provided that the measuring and weighing were "just."

Yet ethnological findings bring home to us the ambivalence found in beliefs relating to associations and interdependences. If it should happen that the act of measuring is detrimental to health, then the inference is that there is a direct relation between the two; so if measuring can be detrimental, then why might it not also prove beneficial? The arsenals of black and white magic are, in principle, one and the same. We have just referred to the votive, curative ribbon equal to the circumference of the patient's head, and in medieval Poland there was an ailment known as *myera* [cf. *miara*, "measure," in Polish—TRANS.], the remedy for which was to measure the stricken person.[5] The ambivalence applies to the very origin of measures, identified by some as the invention of Cain but linked by others with the most positive of legends. According to Greek tradition, it was the wise Phidon Arginus who invented measures and he was venerated for it.[6] The Romans traced the principles of land measurement to the nymph Vegoa, who revealed the practice to an Etruscan named Aruns Veltimnus;[7] in this story, the origin of measures is all but divine.

Equivocal associations cling even to the agent of supreme, ultimate, and—surely—fair justice, who, to repeat, is the Archangel Michael. Having discharged his function, the Archangel on the tympanum hands over the condemned to the devil who, be it noted, is competent to deal with the matter in his own right, to visit justice upon the unjust. As it has recently been demonstrated,[8] Polish folklore abounds in stories of diabolical punishments for dishonest measures, where the devil is shown to have a better sense of justice than the law courts! And if we remember that 90 percent of peasants in the pre-partition Poland (i.e. all, other than those on the royal domains) enjoyed no access at all to any law court, then it is scarcely surprising that both assessment and execution of punishment for injustices suffered should have been entrusted to the devil's relentless hands. Suffering the torments of hell, the well-known ironmaster, Jacob Gibboni, tells blacksmith Martin Mularczyk to go to his new master and implore him "to recover the false weights I left behind . . . and restore them as true hundredweights, just as they had been, for it was I who tampered with them—and for those weights I now suffer the severest torments amid unspeakable fires, and shall go on suffering until the weights have been put right."[9]

Innkeepers guilty of selling short measure of vodka were very frequently consigned to hell. The wall painting in the church at Słopanów (Szamotuły district), dating from 1699, depicts the devil seizing a woman innkeeper who had cheated; just to remove any doubts as to the propriety of the devil's action, a laconic inscription affirms: "She never poured full measure." Let us not laugh at the peasants' widespread idea that some innkeepers deserved to end up in the devil's talons: we must bear in mind the part played by the innkeeper in the organization of seigneurial exploitation, and also the importance of vodka as a source of calories needed by the undernourished peasant. Those drops our lady innkeeper held back contained the very calories the peasants failed to get in other ways.

The Greeks of Crete saw in the dung beetle making dung balls, the soul of the baker whose bread had been weighed short and whose penalty was to have to turn dung into bread forever.[10] Millers were very often commended to the devil's attention by the peasants. There were three stages in the milling process that afforded the miller opportunities to cheat the peasant: first, when he measured the grain brought to the mill; second, when he weighed the ground flour he was returning; and third and most important, when he took his—and indirectly the lord's—tolls of corn in "measures" as payment in kind for his service of grinding. Since in Poland it was the rule virtually everywhere that the corn had to be brought for grinding to the lord's mill, the peasant had no choice of miller nor any say in the matter of such "measures." The lord would endeavor to exact from the miller for the mill he "held" the maximum rent possible, which the miller would then have to recover from the "measures" he charged. In addition, the miller would superimpose something for his own profit. Now since the number of "measures" due to him for the grinding would be set by tradition, written down in old inventories, and since no miller, however well entrenched, would readily dare increase it, there was only one other way: to increase the size or capacity of the "measure." Thus everything conspired to produce conflict about measures; historical sources relating to Polish villages in recent centuries abound in corroborative information. However, since the miller's position was often strong, and the manor, mindful of its own interest, would support him, the peasants tended to lose most of the disputes and were left only the consolation that the sure and certain torments of hell awaited the miller. That this belief was very firm indeed is attested by the fact, noted by B. Baranowski, that even after World War II, peasants would tell of long-since deceased millers haunting villages by night: the whole village would hear the millstones grinding and know full well that a dishonest miller of long ago was condemned for using false measures to turn the millstones forever in vain.

Finally—and here we come to a more complex matter—the peasants would often invoke the devil to lay his talons on the land surveyors. It is known that in the eighteenth century and the first half of the nineteenth century, surveyors many times sided with the peasants—even helped them, or led their attempts at resistance.[11] However, this did not alter the fact that the peasants always feared land surveys. They felt that nothing good would come of them, only a great measure (*sic!*) of trouble. The survey might actually show that some peasant family enjoyed more land than it had right to, having "accidentally" tilled over the boundary into an adjacent fallow or deserted tract of farmland. In such a case, the family would either suffer a diminution of its holding or an increase in the feudal dues it owed. Worse still, the surveyor might falsely demonstrate an encroachment of this kind by measuring land with smaller than customary "rods" or "yardsticks," with a similar eventual outcome; or, he might prepare the way for enclosure by the lord, an action invariably resulting in the peasants being given some inferior land in exchange for their present holding, and never the reverse. Consequently, although at least a century separates us today from feudal practices, countless "injuries of enclosure" live freshly in rural memories. Like the miller, therefore, the wicked surveyor was to enjoy no rest in afterlife. Yet the peasants' imagination—if we judge the examples collected by B. Baranowski as typical—demands relatively lighter penalties for the surveyor: his

16

posthumous fate was to be a harmless "candleman," a will-o'-the-wisp meandering over bogs and swamps night after night. In the Łódź *voivodeship** such will-o'-the-wisps were themselves until quite recently called "measurers" or "metermen" as well as "candlemen" or "candlesticks."

Measure is intimately connected with man and the things he values above all others: land, food, and drink. It metes out to him what his destiny has failed to afford him in abundance. Sometimes fate will give him a full measure, but often, it will be a short one. Measure is not a convention but a value. It is never neutral: it is good or bad; or rather, there are countless bad measures, and only one, the one "of old," that is just, and "true," and good.

* Roughly, the equivalent of an English county—TRANS.

· 4 ·

MEASURES AS AN ATTRIBUTE
OF AUTHORITY

The right to determine measures is an attribute of authority in all advanced societies. It is a prerogative of the ruler to make measures mandatory and to retain the custody of the standards, which are here and there invested with sacral character. The controlling authority, moreover, seeks to unify all measures within its territory and claims the right to punish metrological transgressions. It is not by chance that in the Old Testament we find references to "measures of the sanctuary"[1] in periods of ecclesiastical domination, and to "the King's weight"[2] in periods when the rule of the King prevailed.

The assumption by the ruling power of this attribute and the determination of the scope of its metrological jurisdiction had come about by a lengthy historical process, spreading economically "downward" from the early regulation of the measurement of precious metals. It is a process, moreover, that to date has not been much researched. In classical Greek antiquity, no one harbored any doubt that measures, like coinage, were an attribute of the sovereign power.[3] In Athens, the standards of weights and measures were in safekeeping on the Acropolis, additionally secure in their dedication to the gods (in Rome they were kept in custody on the Capitoline Hill), and specialist officials were employed to authenticate them. Nevertheless, the political particularism of ancient Greece was faithfully reflected in the particularity of weights and measures. Newly emergent city-states created their own standards as symbols of their sovereignty, while those that had the misfortune to be conquered had the measures of the conqueror imposed upon them as symbols of the new domination. Here we should also mention a practice of some relevance to the conceptualization of measures in more primitive societies: the king, in certain instances—as among the Ashanti until fairly recently[4]—used weights larger than the normal ones; as a fiscal device, this practice was as good as any.

In historic times, the attribute of authority we are discussing here is generally already established beyond question. The frequent struggles centered about metrological competence of the consti-

18

tuted power are but a manifestation of the rivalry between various organs of authority aspiring to control measures in order to bolster their standing. The competing parties represent different sections of the ruling class; in the feudal system, they include the royal power as well as that of regional princes and larger seigneurs. The strife in question may also be an expression of the rivalry between groups representing different and differentially privileged social classes—for example, the clash between state and municipal authorities.[5]

Elsewhere we shall discuss attempts to standardize measures; for the time being, it is enough to point out that attempts to control measures have been an ever-present element in the struggle for power between interested representatives of the privileged class. Charlemagne's standardization of measures was integral to his general unifying policy, and he did not, of course, create any new measures, but bestowed his sanction upon those already in existence and extended the area of their use. In later times, unification of measures would similarly be an essential element in the subordination of regional potentates and in the general unifying activity of the absolute rulers of the Renaissance. This was also the way things were managed by the revolutionary authority at the end of the eighteenth century in France. Again, the rivalry between representatives of different privileged feudal estates expressed itself in the struggle for the right to determine and control measures, which inevitably entailed jurisdiction over metrological offenses. This thesis is well substantiated in the history of rivalries between municipalities and their lay or ecclesiastical overlords, as well as conflicts between the last two.

What was unknown in the feudal system, however, was the inalienable and exclusive right of private property, which does obtain under Roman law or Napoleonic civil code. The rights over land, houses, and people were variously claimed and exercised by the parish, the seigneur, the Church, and the sovereign ruler—and significantly for our theme, sovereignty included the right to determine measures. This configuration of authorities led to situations that must have appeared chaotic, but by no means were so. In fact, they were governed by strictly observed rules, recognized by all and maintained over centuries; if the rules were complex, this but reflected the complexity of the feudal social structure. Thus, we know of situations where, within a single village, one measure was used in the market, another in the payment of church tithes, and yet a third in rendering dues to the manor. Such arrangements were quite usual in the context of feudalism and similar social structures and, in principle, need entail no abuses or protests. If in practice the coexistence of several different measures enabled the stronger party to perpetrate many abuses, then that was a different matter.

Even in the leading cities of the Netherlands, England, and Switzerland, such practices continued, although in large urban centers the modernization of metrological systems tended to occur relatively early, and commercial activity aided metrological standardization. Thus, Lier and Malines in the Netherlands had two corn measures, municipal and ducal.[6] In Winchester, the College, being an ecclesiastical foundation, had one measure and the town, another.[7] In Geneva, there was a prolonged dispute over the control of measures between the municipality and the local bishop, and as far as we can judge, the final result was inconclusive.[8] Again, the contest between the municipal masters of Lyon and the archbishop took almost a century from 1389 before the burghers won the day,[9] and there were many other similar cases.

Authority in metrological matters, that is, the right of independent determination and oversight of the units of measure, has been so much the symbol of sovereignty in general, or even of "freedom" in general, that it was at the very center of the learned debate between Schmoller and von Belov on the rise of German towns and their municipal liberties.[10] It is not for us to revive their debate, but there are hundreds of instances where measures and metrological rights of seigneurial origin had to be won or, more often, bought by the municipalities, just as were many other immunities from feudal impositions. The matter is seen in the same way by H. van der Wee with respect to the Netherlands,[11] and by J. W. Thompson as far as England is concerned.[12] Thereafter, the municipalities jealously guarded their right of metrological self-determination, just as they staunchly defended their other privileges and "liberties"—perhaps even more so, since a right to their own measures, like the right to mint coinage, was an external symbol of freedom that was readily seen by the whole world. If one city was overrun by another, then the latter would duly impose its own measures. Florence forced its system on Pisa, and for Pisa the changeover not only entailed the loss of a symbol of sovereignty but constituted a major commercial handicap, because in the thirteenth century, when she was an important port, Pisa's measures enjoyed wider international recognition than did those of Florence.[13]

Great cities, once they had won their sovereignty, were clever at defending it, and they erected effective systems of control to guard the integrity of their measures.[14] In due course, however, municipal metrological sovereignty, as a facet of municipal sovereignty in general, was threatened by a new foe—the state.[15] As usual in the feudal period, sovereignty "in general" meant the monopoly of jurisdiction in particular areas. As far as punishing metrological offenses was concerned, the commonest penalties for "bad" measures were aimed at the offender's pocket, and for repeated offenses expulsion from the town was sometimes the ultimate sanction. The use of the pillory was also quite common,[16] and corporal punishments were in use, too—for example, in Gdańsk, the lopping off of two fingers for employing dishonest measure.[17] An extreme and curious penalty was applied in Latvia in the thirteenth century, the law implying a certain ratio of tolerance: "bad" measure was punishable by death, but only if the false ell understated the correct length by more than a finger's breadth.[18]

The great diversity of measures and methods of measuring in the feudal period did indeed open a vast area for disagreement. Hence, the authority that enjoyed the right to determine and control measures was able to gain further prestige by arbitrating such conflicts. Pope Gregory the Great wrote to his officials on this matter: "Above all, it is my wish that you should watch carefully that no one, in collecting charges, should use false scales. Should you encounter any such scales, destroy them and bring into use new true ones."[19] In supervising fairs, municipal authorities had to institute their own scales, which they often looked upon as a source of revenue, charging fees for the use of standards or for renting out certified scales.[20] Moreover, in some places the use of officially approved scales was compulsory[21] (some of these standards survived long enough to be remeasured with more precision in the early nineteenth century). In some towns, the authorities permitted direct access to the standards, as in Chełmno, where they have remained to this day riveted to the outer wall of the town hall, or in the episcopal town of Kielce, where we can still inspect the standards of the measures of length graven in a table set into the wall of the Cathedral near the main entrance. The municipal law laid down penalties (confiscation of the objects in question or a fine) for craftsmen found in possession of ells or weights that did not conform to the standards.[22]

An interesting situation prevailed in Silesia, which was notorious for the fragmentation of landed property under feudalism. There was a widespread differentiation of measures into seigneurial (actually ducal, and thus, in a sense, "state" measures), ecclesiastical, and municipal. Newly enfranchised towns would determine their own bushels as a symbol of liberty and sovereignty.[23] The Church, in particular, was ever ready to assert its sovereignty and independence from the state and had measures of its own.[24] The political fragmentation was marked by a proliferation of weights and measures peculiar to the numerous sovereign duchies. When the Raciborz region was split into a number of independent duchies, only the duchy of Raciborz itself retained the former units,[25] while the others devised new measures. Wielkie Strzelce, which in the fourteenth century was the principal town of an independent duchy for some fifty years, enjoyed during that period the use of a measure of its own, which fell into disuse later on.[26] Altogether, a state of shocking confusion reigned: in the single village of Jastrzębie, Upper Jastrzębie used the Pszczyna measure while Lower Jastrzębie used the measure of Wodzisław, and the Vicar kept both measures available until the 1830s;[27] in the district of Lubliniec, peasants paid their church dues by the Lubliniec measure, but their manorial dues by the Wrocław measure.[28] Yet the apparent chaos obeyed strict, long-established, "organic" rules, which left no room for arbitrary conduct. Historical understanding of these rules, however, requires a thorough knowledge of the succession of political, ecclesiastical, and seigneurial divisions and subdivisions. Some students see in this process an evolution from uniformity to diversity,[29] but this view may be illusory. Admittedly, Polish nobles, deliberating in their local assemblies, did believe that "in the beginning" measures had been uniform, and that belief was shared by the third estate in France at the time of the preparation of the *cahiers de doléances* and by many others. Nevertheless, it is doubtful that this belief was well founded. The uniform measure binding throughout the vast territories of the Carolingian Empire had not, after all, been any "primeval" standard, but had resulted from a very early attempt to standardize a mass of diverse local measures, many of which survived the reform and remained in use for centuries to come.

In Poland, the contest between the towns and their lords was settled fairly early in favor of the latter, and oversight and control of municipal measures was then handed over to the representatives of the feudal order—the *voivodes* and their deputies. The struggle was far from over, however, and it continued—in different and less overt ways—right until the end of the eighteenth century, when, as a result of the partitions, the lands of Poland came under the rule of absolute monarchs. For the contender ultimately destined to triumph was the state, with all its absolutism and its quest for centralization. This is a matter to which we shall devote an entire chapter, but, for the present, it is enough to point out that when the state grew strong again, it sought—as it had under Charlemagne—to gain a monopoly of metrological jurisdiction as one of the fundamental attributes of sovereign power. It is, finally, worth noting that modern constitutions of federal states assign the competence in metrological matters to the central authority; witness the Soviet Union's constitution of 1924 (the question was no longer on the agenda when the constitution of 1936 was promulgated) and also the constitution of Yugoslavia.

· 5 ·

MAN AS THE MEASURE OF ALL THINGS
(ANTHROPOMETRIC* MEASURES)

"Man is the measure of all things." This sentence of Protagoras had, of course, a dual significance. On the one hand, it was the synthesis of the anthropocentric philosophical stance, perhaps also a declaration of cognitive faith. At the same time, it was a simple statement of the existing state of affairs, a generalized view of a system in which man used himself, the parts of his body, to measure all other objects.

The author of a recent Greek novel in the realistic genre writes of his heroine, a Cretan peasant woman: "The world had been cut in man's measure. . . . She herself stood at the heart of creation, and let it filter through her inner being, as the threads of warp filter through the comb of the weaver's loom. I might even venture to say that she measured the world with her body: with the palm of her hand, digit, forearm, outstretched arms, foot and pace, by her embrace, her cast of the stone, her heartbeat, the warmth of her body and its weight, the reach of her eye and her voice: those were her weights and her measures."[1]

Had Protagoras been able, twenty-five centuries later, to watch that countrywoman of his, he would have felt no need to rephrase his immortal statement, for people still use themselves and their limbs to measure the world. Measures so derived we term *anthropometric* (the derivation being, of course, not from *meter* but from *metrum*).

The emergence of man's metrological concepts and habits is a very important aspect of his apprehension of the world, and of the formulation of his taxonomic systems and abstract concepts. Primitive man measured the world by himself; to measure objects independent of him, he employed his own parts: his foot, arm, finger, palm of his hand, outstretched arms, pace, etc.[2] There were a great many potential units, since there are a great number of measurable parts of the human body.[3] However, the intellectual turning point came with the transition from concrete to abstract concepts, from

---

* The author's word *antropometryczne* for measures derived from the human body has been translated as "anthropometric," although the meaning of this term is somewhat different in English—TRANS.

the particular *my finger, your finger,* to the general *the finger.* Measures such as the *ell* (elbow), the *span,* and the *foot* enjoyed currency in our civilization until quite recently—pending the complete dominance of the metric system—but they had become abstractions. The unit was *the-foot-in-general* or *the-foot-as-such;* at any one time, its length was fixed (though it was variable over time), and it would be somewhat longer than *my foot* or somewhat shorter than *your foot.* On the other hand, Ethiopian medical prescriptions that refer to *your finger* as a measure as late as the sixteenth or seventeenth century are singularly significant.[4] Naturally, the author of the prescription would take into consideration the fact that one patient's finger might be longer than another's. We may safely assume that the range of differences would not materially affect the dose— would not reduce it to ineffectiveness or increase it to the point of danger. What is important to us is the phrasing itself as evidence that the stage of employing *the finger* as an abstract measure at large had yet to be reached.

Nevertheless, man is but a grain of sand in the vast world he inhabits, and the multiples of his limbs were too small to enable him to grasp nature's dimensions—hence the use of measures like the carrying distance of a man's voice, or the bowshot.[5] But the world man inhabits is also full of objects too small to be measured reasonably accurately by a man's limbs. Here, probably the most widely encountered solution is the use of a grain of cereal in local cultivation as a unit of the length, or width, or weight. This unit of measure, known in France until the Revolution, is still to be found in many societies. The differences in the length and weight of this unit are very considerable, depending on the principal cereal—and indeed its particular variety—in question.

The system of anthropometric measures was very convenient. Not only were they generally understood, they were always available. Minor discrepancies stemming from individual differences in length of leg and hence the pace, or of finger and hence the span, were surely of little consequence, for it was not often that a very great degree of accuracy was required, and a compromise would be readily reached when the rare dispute arose.[6] The major inconvenience of the anthropometric measures was the lack of simple multiples. The pace could not be divided into the whole number of ells (elbows, or forearms), nor the ell into so many spans, etc. In Ethiopia, the ell was equal to two spans plus two digits,[7] and in Latvia, as late as the seventeenth century, 16 feet equaled 7½ ells.[8]

Arguably, in the earlier stages of economic development, this difficulty was of little account because different kinds of things were measured by paces, by ells, by spans, etc.[9] But, as economic development proceeded, and as the process of the transformation of anthropometric measures from individual-concrete to abstract progressed, the crystallization of a system of anthropometric measures with simple multiples and simple, fractionless divisors was bound to take place, too. That this change should have been accomplished may indeed appear wonderful. How, seemingly in contradiction of the facts, could the pace, or the "reach," or the fathom, be made equivalent to a whole number of ells, and the ell to a whole number of spans, without using fractions in either case? And yet, in many civilizations, because of a variety of methods for calculating anthropometric measures with the possibility of selecting those which satisfied the demands of real-life situations, the problem was resolved. The process was investigated, using Russian evidence, in a splendid work by B. A. Rybakov.[10]

Rybakov's method was, on the one hand, to analyze various sources that referred to anthropometric measures and, on the other, to carry out anthropometric measurements of live males of 1.70 meters in height; he then combined the two to show a variety of methods for determining each anthropometric measure of length. For instance, the fathom would be measured either with the arms outstretched, from one tip of the middle finger to the other, or, in other places, with the arms outstretched, but from one wrist to the other, or, here and there, with one arm stretched upwards, from the floor to the tip of the middle finger. His measurements of live subjects established that the first method yielded a fathom of 1.76 meters, the second method, 1.52 meters, and the third, 2.16 meters. The ell would sometimes be measured from the elbow to the tip of the middle finger of an outstretched palm of the hand, at 44 centimeters; sometimes, from the elbow to the knuckles, with the fingers closed, at 38 centimeters; and sometimes, from the arm to the tip of the thumb, at 54 centimeters. Measured accordingly in the different ways, the span could be 19 centimeters, or 22 centimeters, or 27 centimeters. Even the measure defined as the digit can be used in different ways and therefore have different values.[11]

The picture that results is apparently one of chaos, with the same appellations for very different values; but the sources, as a rule, tell us nothing of the system of measuring that was applied, for people at the time, in a given region, took the matter for granted. In fact, they suffered from no confusion at all, but rather were served by a system that had evolved empirically out of the experience of generations and represented a great achievement of the mathematical culture of the common people. To clarify: where the fathom was measured in a way that produced a length of *ca.* 1.52 centimeters, the ell could not be measured so as to result in a length of 44 centimeters, but would be 38 centimeters long; whereas in the same circumstances the span had to measure 19 centimeters, not 22, nor 27. *Mutatis mutandis*, lengths for the measures would be adopted to fit in with a fathom of 1.76 meters or of 2.16 meters. The manipulation of the methods of measuring anthropometric magnitudes offered considerable scope for choice of units and is known to have produced a few (three, at least) internally consistent systems. In each of them the ell measured a quarter, and the span an eighth, of the fathom. Rare indeed were cases where the integration of two anthropometric measures into a single system proved beyond human wit and "piecing up" was needed. Presumably the Russian concept of the *piad' s kuvyrkom*,[12] or the Polish term "the ell-and-the-knuckle"[13] are survivals of such failures.

Thus homespun wisdom, working by trial and error over many generations and many centuries, devised a general system of diverse anthropometric measures, a system wherein all measures, being multiples or fractionless divisors of one another, were commensurable. At the same time, they all remained—indispensably so for the primitive mind—multiples of, or divisible by, two. Once those measures changed from individual-concrete to abstract in such a system, they were perfectly suited to interpersonal dealings within small communities.[14]

Once firmly established, the system of anthropometric metrology was an all-embracing one. It reduced to common measures nature and culture the world around, and man's artifacts as well. It not only enabled man to measure fields, trees, and roads, but also imposed its proportions on the dimensions of the weaver's loom, and on bricks and church belfries, the dimensions for the bricks being part of the same system as the proportions of the church architecture.[15] Reconstructions of medieval buildings in France by Viollet-le-Duc were criticized, *inter alia*, for using the meter in the rebuilding of an architecture that had been governed by the span and the foot, and thus distorting fundamental proportions.[16] We gather that much trouble was taken over the recovery of old-style bricks for reuse in the rebuilding of Warsaw's Old City quarter [after World War II—Trans.], although we cannot vouch for the truth of this report. It would seem, *a priori*, that the metric system should enable us to express all proportions, even between incommensurable quantities. One thing, at any rate, is certain: anthropometric measures, which had begun to evolve in prehistoric times and which had been endlessly improved over thousands of years, when once they evolved into a coherent system, served man well in his work. They enabled him to satisfy his daily wants and to create immortal works of art: the noble proportions of the Romanesque, Gothic, and baroque cathedrals still astound us today.

· 6 ·

HOW WAS LAND MEASURED?
(AGRARIAN MEASURES)

Relatively underdeveloped systems of surface measures for cultivable land have generally been associated with low density of population. In the Spanish New World colonies of the sixteenth century, pastures were marked out with the aid of a circular measure. There were but two sizes, large (*hato*) and small (*corral*). Such measures would be unthinkable in long-settled and densely populated regions. In Europe, from the early Middle Ages until the introduction of the metric system, there were two types of measures for cultivable areas: those derived from the labor-time for plowing and those derived from the amount of seed required.[1]

Measurement based on labor is found all over Christian Europe. The Lubiń inspection records of the early seventeenth century furnish the following rigorous, if scarcely original, definition: *iuger autem est, quod duobus bobus uno die arari potest*[2] [the *iuger* is thus the amount of land that a pair of oxen can tackle in one day]. A similar but looser definition comes from a French source of the same period: "The measure which the French call *arpent*, and the inhabitants of Burgundy, Champagne, and the other provinces call *journeau*, is derived from the Latin word *iugerum* [*sic!* the etymology is absurd—W. K.], which signifies an area of land that can be plowed with the aid of either two oxen or two horses [the definition is lax here—W. K.] yoked together, within a single day."[3] In Brittany, there were units of *journal à charru* (for plowland), *journal à foucher* (for meadows), and *journal à bêcheur* (for gardens and vineyards).[4] In Germany, the unit of area capable of being tilled in a single day was named *Morgenland*.[5] In Catalonia, until the eighteenth century, two *journals* were distinguished: ordinary, applied to cornfields, and the *journal de cavadura*, applying to the area of vineyard one man was able to tend in a single day.[6] Similarly, in Burgundy from the second half of the eleventh century, the practice of measuring cornfields by the *journal*, vineyards by the *ouvrée*, and meadows by the *soiture*[7] (all three measures relating to labor) was widespread, while in Italy there was a measure of land called *giornata*.[8] Again, in Latvia land was measured in terms of days of work using one horse,[9] and in the Ukraine there was a measure of land defined as a "day" of field.[10] In Russia, *obzha* stood for a day of plowing with a single horse[11] (as confirmed by a document dating from 1478), and among the Slavs at large, the area of cultivated soil measured in terms of human labor was, apparently, divided not into halves but into three parts separated by mealtimes during the working day.[12]

Thus, from Spain to Russia,[13] a system of measuring agricultural land by the amount of man's labor prevailed.[14] The differences in time and space were of secondary importance; what mattered was the general emphasis on the relation of man to land. The fact that the principle of mensuration was based on this relation tells us that the input of labor needed to yield a harvest was of paramount importance. Correspondingly, when we come shortly to consider the other primary system of mensuration—that by the amount of seed required—the characteristic most heavily emphasized is fertility.

Huge variations in harvests caused by the vagaries of climate might invalidate the adoption, as standard, of the quantities of produce yielded, whereas the amount of seed used for sowing would be the same in good years and in bad. It seems, however, that measuring land by the quantity of seed was of later origin than measurement by the amount of labor, the latter method being of ancient Roman lineage. The Lubiń inspectors' definition, or that of the "Maison Rustique," was but a slightly modified version of definitions given by Pliny and by Varro. The former stated: "What a pair of oxen can plow up in one day is called a *iuger*."[15] The longevity of this system in some parts of Europe was most impressive, for on the eve of the French Revolution, in a *cahier de doléances* from the vicinity of Bourges, we find the following: "The *arpent* is not divisible by rods or feet but by *journées*, which means by fields that one man is able to plow in one day; according to the local custom, one *arpent* of land is equal to sixteen *journées*."[16]

The system based on the amount of seed coexisted with the older one over many centuries throughout Europe,[17] and there is evidence that it was used on other continents as well. From the vicinity of Bourges, on the eve of the French Revolution, comes the following definition of the measure of arable land:

> A *seterée* of land is the only measure known in this canton. It is larger or smaller depending on the quality of the soil; it thus signifies the area of land to be sown by one *setier* of seed. A *seterée* of land in a fertile district counts as approximately one

arpent of one hundred *perches*, each of them consisting of 22 feet; in sandy stretches, or on other poor soils, one *arpent* consists of no less than six *boisselées*. Thus, one *seterée* usually counts as 1⅓ *arpent* in Sologne.[18]

In fact, the sower's measure was more accurate than may appear to us now. In one sense, it constituted a linear measure based on the number of a man's steps, for in sowing by hand the seed is cast at every other step.[19] The important practical consideration, however, was that, depending on the quality of the soil, an experienced sower would take larger or smaller steps and cast tightly packed or loose handfuls of seed. A sixteenth-century manual, in referring to a given area, states that more or less seed should be sown depending on the quality of the soil and the topography of the terrain.[20] An eighteenth-century French manual on the compilation of estate inventories clarifies the matter well, if schematically: for soils of middling quality, the proper measure is one-fifth as large again as for good soils, whereas for poor soils, it is larger by a sixth; the reason for the difference, says the author, is that the seed is sown relatively thickly on good soils and sparsely on poor soils.[21] Thus, if we appraise the matter in terms of the economic value of a piece of arable land, the measuring by the amount of seed had considerable merit because the value of one hectare may be far from equal to that of another, though the two are identical in area. The seed measure would offset the differences, and two plots of unequal area might thereby be "equated," that is, shown to have virtually the same productive potential. It seems, however, that the usefulness of this system of measuring would be related to the density of the farming population and the fragmentation of land holdings. In his investigation of the agrarian measures of the Pisa region in Italy, M. Luzzati found that the seed units survived longest in mountainous, sparsely populated districts, where extensive cultivation methods were employed; they were more speedily ousted by geometric units in the plains, nearer the center of the region, where the population density was greater, as was also the fragmentation of landed property, and where more intensive husbandry prevailed.[22]

There were also instances of coexistence, without any direct interdependence, of seed and labor systems of measuring land. For example, in Gironde *le journal à boeufs* (that is, a day's plowing with a pair of oxen) was no simple multiple of the *seterée* (the area it took one *setier* of seed to sow).[23] It is, nevertheless, to be born in mind that units of all pre-metric metrological systems did contain some element of convention. In France, for instance, there were cases of measuring the border of fields *en grande pas* or *en pas raisonable.*[24] We should not therefore be surprised that the foot which we find in some contracts had been fixed *ad hoc,* and each party would receive one attested wooden copy as an "enclosure" to the document.[25] The widespread employment of the amount-of-seed system in measuring farm land is astonishing: we encounter it throughout Europe, and also in India in the period corresponding to our late antiquity, where a very different crop, namely rice, was in question. The underlying purpose, however, was no different: the practical advantages (notably the reduction of soils of varying quality to a common denominator of the greatest importance for the farmer) were the same.[26]

Country	Measures of Dry Capacity	Measures of Area
France (Bourges)	boisseau	boisselée
France (Bourges)	setier	seterée
Rome ,	rubbio	rubbio[27]
France (Noyon)	setier	setier[28]
France (Burgundy)	bichot	bichetée
France (Burgundy)	boisseau	boisselée[29]
India[30]	kula	kula
Capo Verde[31]	quarta	quarta
Capo Verde	alqueire	alqueres
Columbia	fanega	fanegada
Hong Kong	dau	dau chung
Ifni	fanega	fanega
Japan	bu	bu
Japan	sun	sun
Japan	shaku	shaku
Libya	sáa	sáa
Malta	tomna	tomna
Malta	modd	modd
Nepal	mana	matomana
Nepal	pathi	matopathi
Holland	mud	mud
Ryukyu	shaku	shaku
Ryukyu	go	go
Vietnam	than	than
Vietnam	miéng	miéng
Vietnam	sáo	sáo
Vietnam	máu	máu

Circumstantial but reliable evidence of the use in numberless countries of the seed system of measuring fields is found in the terminology—widespread until the adoption of the metric system and persisting even later—in which the basic unit of measurement for agricultural land and the basic unit of dry measure often bear the same or very similar names. According to Hauser, the labor measures would be employed only in situations where the type of cultivation precluded the use of seed measures—as in the case of meadows or of viniculture. Hauser's thesis, however, is patently incorrect; for one thing, it was not just in the cultivation of meadows and vineyards that labor measures were applied (witness our evidence above from Catalonia and Burgundy), and secondly, it seems that in fact seed measures appeared chronologically later, and in the early Middle Ages work measures alone were used.[32] In modern times, the former have been used throughout Europe, coexisting with the latter.[33] In modern Poland, nobles generally calculated the area of their estates in terms of sowing,[34] and the system was employed as late as the first quarter of the nineteenth century as the basis of a truly modern statistical description of the agriculture of the Grand Duchy of Warsaw and the Congress Kingdom of Poland.[35]

We have a great many examples of seed measures being used as measures of land in a manner that made them correspond to the unit of dry capacity; e.g. Rutkowski has shown their use in Brittany. In Burgundy, the *bichetée* was the term for an area that was sown by one *bichot* of seed grain and the *boisselée* was the area that absorbed one *boisseau*.[36] In Russia, the area unit of *diesyatina* stood for a plot sown with one *tchetvyert*.[37] The practice of measuring the area of land by the amount of seed needed is, as it were, reversible: the system allows us to measure an amount of seed grain by the area of land that is to be sown. Measures of agricultural land and (dry) measures of corn, therefore, simply melt into one system grounded in farming practice and representing most cleverly the level of development—at that particular stage, in a given country—of the productive forces in agriculture. To be sure, an individual's apprehension of a metrological system of this kind is quite different from his view of a system involving a *tertium comparationis* unifying disparate systems that is derived not from daily work on the land but from the astronomical unit of the meridian—for the meridian is beyond his conceptual powers and has no associations with his daily life.

S. Strumilin has put forward an interesting hypothesis to explain

33

old agrarian measures. He suggests that the area unit measured by one day's plowing was increased as the productivity of agricultural labor increased: specifically, that Pliny's *iuger* in the first century A.D. was equal to a quarter hectare; that the medieval *morg* in Germany and the Rhineland equaled 0.31, and in the Moselle basin 0.34, hectare; that the thirteenth-century English acre was 0.4 hectare, and the corresponding unit in Georgia was 0.5 hectare.[38] On the assumption that the basic unit of farm land corresponds to the amount of labor expended in plowing, the enlargement of that unit in the course of history may constitute an objective measure of the growth rate of the productivity of labor. A comparison of such units from different countries would afford a much-needed objective measure to compare their respective improvements in productivity (e.g. as between England and Georgia). Strumilin's thesis, despite some similarity to the notion of the philosopher's stone, is not implausible. His argument is, quite patently, cut-and-dried: he does not allow for differences in the work rates in agriculture caused by differences in soils, climates, and—above all—in crops. Otherwise, he would be unable to maintain that the size of the agrarian unit of area provides automatically an index for the productivity of labor. Moreover, Strumilin's line of thought fails to recognize the "dual" nature of the progress of productive resources in agriculture—that sometimes its essence lays in saving labor, but at other times in using more labor-intensive cultivation. Labor saving per unit of area, or, to put it the other way round, increasing the area cultivated within a given period (usually one day), constitutes a measure of increasing labor productivity, but only if "other things are equal"—that is, soil, climatic conditions, and, in particular, the crop. Failing that, we would be making mistaken inferences, since progress in agriculture is quite often marked by a changeover to more labor-intensive crops, for example, market gardening. Strumilin's argument may be overly schematic, but it holds out some promise to the student of labor productivity in precapitalist societies, and investigations of basic customary units of cultivated land may prove useful, provided the conditions required for comparability are duly observed.

As the example we have cited shows, the fact that traditional measures were firmly rooted in the realities of life and labor is of great importance for social and economic historians. If measures represent (or "express") a particular economy, then their analysis will throw light upon certain aspects of the economy under investigation. Naturally, these measures "express" such quantities as labor-time or the amount of seed sown, and it is scarcely surprising that today's standard conventional abstract units are not matched by any corresponding customary land measures. Land measures vary, depending on the quality and type of soil, the location (level or slope), and the chief crop grown on it. The name of a particular geometric area unit may long have remained the same—*iuger* or *journal*—but the actual area specified may have changed over time and certainly varied at any particular time between provinces and villages, or even among farmsteads within particular villages. It is not for the historian to bemoan the variety when he seeks to translate these measures into metric units, and to eschew their use as evidence just because the conversion is so difficult, sometimes impossible. He must rather exert himself to grasp the social meaning of representational measures, and he will then be rewarded by the information embedded in them. Indeed, he will no doubt learn more from them than from our arbitrary measures set by convention, for the latter have nothing to say of social realities.

Historical studies abound in references to the "primitivism" and "crudity" of former metrological systems. But, as far as measures of agricultural land areas are concerned, it is difficult to accept the view that traditional systems are inferior to the metric system, despite the "perfection" its advocates have claimed for it. On the contrary, the principal measure applied today to arable land is the hectare; but we have established that, in the social and economic sense, one hectare may not be equal to another. Consequently, the hectare is ill-suited to fiscal assessments or commercial transactions, given the unequal quality of soils and inevitable boundary disputes. Technically then, as well as economically, adding up hectares whose value and profitability vary does not mean adding like to like, and there is nothing here that makes valid the assumption that the owner of ten hectares of land is twice as rich as the owner of five hectares. We must, therefore, conclude that the old measures based on labor-time and on the amounts of seed are definitely more commensurable and "addable." Given their arithmetical inequality—indeed, because of their arithmetical inequality—such measures are more homogeneous with respect to their social and economic significance. In some cases, they may even—to put it plainly—serve better than do our hectares as a basis for statistical censuses of landownership. They are more serviceable to tax assessors, more helpful in dividing inherited lands and in other transactions—in a word, they are more "functional."

The amount-of-work and the amount-of-seed measures remained in use side by side throughout Europe over many centuries. The surveyor's square measures underwent changes, though the principle governing such systems remained the same. The system was expressive of "the main factors in the balance between man, climate and land"[39]—in other words, expressive of the knowledge, acquired by empirical experience over generations, of the relationship of man to nature, mediated through work. Yet this intimate relationship is not entered into by a person as an isolated being but as a member of society, and a differentiated society at that. The differentiation would result both from the social division of labor and the existence of an ever-changing social hierarchy. In the feudal system, man established his contact with nature, through his agricultural work, as a member of three social structures: first, as a member of his village parish and community, acting interdependently with his fellow parishioners; second, as a serf of a given lord, within the framework of the latter's estates; and third, as a subject of the state. The interdependence among the parishioners would engender levelling tendencies within the village community itself ("levelling" in the economic sense of comparative affluence). The feudal lord's actions would be determined by the structure of the dues he wished to levy on his serfs; if his preference was for rent, then he would favor much larger farmsteads for the peasants than would be the case if he preferred the *corvée* system. Finally, the state would seek a unification of measures for the sake of a "just" apportionment of taxes.

Now, in Poland, of the three parties [Church, lord, and state] the most powerful was the lord of the manor. In the period of settlement upon land, a settlement organized "from above" within the framework of a rent economy, the lord had to allocate to the settlers large holdings, each capable of feeding the peasant family, of reproducing the traction power [that is, the dray animals], of which a large holding required a great deal, and of producing a surplus from the sale of which in the market the peasant would obtain enough money to pay at least his rent.[40] Provided that land was abundant relative to the size of the population, and given the natural variety of soils and the concomitant varied types of husbandry, the holdings allocated to settlers would tend to represent approximately equal economic potential, but they inevitably differed considerably in their area.[41] In the dispute on this issue between the Polish historians Bujak and Piekosiński, in which the latter emphasized the socio-agrarian character of the primeval *łan* of Lesser Poland whereas Bujak maintained that, even before colonization, the *łans* were geometric measures used in surveying, the present author is inclined to agree with Piekosiński.[42] Be that as it may, and regardless of whether the peasant's holding was rent-based or *corvée*-based, what mattered to the manor was that it should be neither too large nor too small to ensure subsistence for the peasant and dues for the lord. This was the avowed objective of the entire process of settling and colonizing land, and this, too, was what in later times the manor strove for in its interventions related to the peasants' possessions. Viewed thus, the peasant holding was of a *nadział* type (i.e., owing the full complement of dues),[43] both during the period of being rent-based and even more during the period of the *corvée*-based manorial demesne farming (the *folwark*). This was particularly in evidence on the not infrequent occasions when the manor, taking advantage or making pretext of the need to survey the estate, would attempt economically to reduce the peasants to the common level that had formerly obtained but obtained no longer.[44]

The factors we have discussed were not, however, the only determinants of the size of the basic unit of cultivable area; they were overlaid by two others. First, there were the metrological abuses perpetrated by the stronger interested party, i.e. the manor, in its endeavor to diminish the acreage of peasant holdings. In extreme cases, this would result in a reduction of the size of the basic unit of cultivable land, and thereby of the peasant holdings, to a level where the peasants could no longer discharge their crucial "subsistence and reproduction" functions. Second, there was the tendency to underscore the "sovereignty" of the great landed estate, both in general and in metrological matters, including the imposition of the same measure upon all its villages.

These tendencies pulled this way and that, with varying outcomes. However, traces of the initial agronomic character of agrarian measures are often discernible in modern times. An enquiry by Śreniowski led him to conclude that the size of the *łan* varied inversely with the quality of the soil, but he also found that where the *łans* were smaller, a larger crop per *łan* was harvested.[45] Here and there, the basic agrarian measure was uniform throughout the village, despite the obvious fact that the holdings did not possess soil of uniform quality; in such cases the measure in question differed considerably from its namesakes in other villages of the same district.[46] Again, here and there the measure would be standardized within estates belonging to a single landowner, while remaining different in size from the corresponding measure in the village owned by another. Finally, measures might be standardized within a certain group of estates, while remaining at variance with measures in other such groups, even those belonging to the same lord.[47] All these tendencies emerged slowly, even imperceptibly, and preserving the *status quo* at any given time was the conflict of interests between the village and the manor as well as within the village. Overt and widely heralded attempts at reform of agrarian measures encountered firm resistance from the peasantry, a good example being their rejection of the so-called "*vloka* survey."[48]

Finally, we have to consider the activity of the state, which aimed primarily at uniformity of fiscal burdens. This policy was pursued in numerous countries, including Poland, where, admittedly, it was scarcely pursued very determinedly because of the long tradition of weak central administration and of inviolability of the sovereign rights of the nobleman upon his estates. From the time of the granting of the Košice privileges in 1374, the *łan* of land was the basic Polish unit of tax assessment, and as early as the mid-sixteenth century, Modrzewski demanded that *łans* be compared in order to avoid unjust assessments. Social determinants of this customary basic unit of area of arable land were many and varied—the nutritional needs of the peasant family, the importance of having traction power reproduced, the overt and covert aspirations of the great feudal landowners, and last, state policy. All these, although acting with uneven effect and variously balanced, were in evidence in many countries. The activity of the state tended to be more pronounced and effective where the central authority was relatively strong, while the activity of the manor was more marked where its sovereignty was relatively unrestricted by the prerogatives of the Crown. The basic unit of land measure was of one size where the rent system of landholding prevailed, and of a different size where labor dues were the norm. Ultimately, the social determinants of agrarian measures were effective less uniformly, and for a longer period, in Eastern Europe than in the West.[49] However, both the work-based and the seed-based measures (*agrotechnical* is perhaps the right term) were universally in use in Europe and were known outside Europe, too. In the present writer's view, the impact of the various social forces that were the determinants of the size of the basic agrarian unit of measurement has yet to receive due consideration from historians.

Polish peasants were familar with both types of agrotechnical measures. Village records of judicial proceedings (an excellent source for students of rural society) frequently define pieces of land in terms of the amount of seed that can be sown upon them. Thus, one document states: "I assign to and bestow upon my son Caspar, in the middle of my holding, . . . a piece of land to be sown with one bushel of seed"; and another: "They agreed . . . to allocate to Wojciech Stec from their part enough ground to be covered by one bushel"; etc. Occasionally, the wording is more precise, for example: "Casimir, the elder brother of Nicholas, gives over to him one bushel and 24 gallons (*garniec*) of land."[50] Moszyński maintains that in cases of dispute the amount of land would be decided "at sight" and thereafter with the aid of the most honest and experienced sower, who could be trusted to be right to within a gallon.[51] Moszyński further holds that the seed measures were more accurate than units like "the strip" or "the patch," and that accounted for their persistence. Until recently, for instance, the Tatra mountaineers would refer to "a patch to take half a bushel of oats" or "patches to be covered by two bushels of spuds."[52] Judicial records thus leave no doubt that seed measures have been used in dealings between Polish peasants in modern times, and that they were accurate not just to the bushel but to the gallon.[53] Measures based on labor, however, were in use along with them, most often at the time of plowing and in relation to meadows according to the time it took to mow them. The Croats, apparently, used two measures based on a day's work; the summer *ral* and the winter *ral* (*ral*, cf. Polish *radło*, a simple wooden hoe)—the latter being naturally smaller as the day was then shorter. It is of interest that the representational character of the measure would sometimes lead to certain representational portions of land, such as the Ukrainian "day's measure" (i.e. the area that could be plowed in one day) divided into three *uprukhs* or shifts—morning, afternoon, and evening—partitioned by two mealtimes for the plowman, when the oxen would naturally rest, too.[54]

It is the considered view of this writer, however, that along with time-of-work and seed measures, there has been in operation yet another representational measure, namely the area necessary for the subsistence of a landholder and his family at a socially acceptable level. This may be termed a subsistence-and-reproduction plot of land; the "reproduction" refers to the implied sufficiency of seed for sowing, of feed for an appropriate number of cattle, fowl, etc., and a right to draw upon woodland for cooking, heating, and maintenance of buildings. The extent of the given area would thus be a function of the development of the productive forces: the less advanced the forces of production, the larger the plot (or farmstead) would have to be. In addition, a holding would have to be able to yield enough for the usual dues and taxes owing to the state, to the Church, and, above all, to the feudal lord. Labor dues would favor relatively small holdings, but dues converted into money would dictate a larger size. Whether the Polish *łan* of the period of colonization functioned in that manner has yet to be ascertained. The present author has no doubt that this was the case.

All three systems of land measures that we have described have undergone various modifications dictated by historical realities. The topographic factor is particularly important: as the soil of the Podhale region [in the Tatra mountains, southern Poland—TRANS.] is poor, we would expect the Podhale *łans* to be comparatively large, but, in fact, as S. Grzepski has found,[55] they tend to be small because, given the topography of the district, large units are not feasible.[56] Nevertheless, units of agricultural land area, like other traditional measures, exist within social contexts and change with their societies. They may, at first, bear some representational significance derived from certain facets of real life, but in the course of time, they may lose that significance or, indeed, assume a different significance. Since they operate in societies, they are circumscribed by institutional frameworks, and the institutions in question involve such measures in complex clashes of interests among various social groups.

The basic unit of area in the Polish countryside was the *łan*. It was sometimes defined in terms of man-hours of plowing, sometimes in terms of the seed sown, and sometimes appraised as the area "sufficing" for the needs of one settler. The *łan*, however, would also become the basic unit of assessment of feudal dues, and hence, if a peasant managed to increase his *łan* by plowing up an adjacent piece of fallow land or the odd scraps of some deserted holding, then the proportion of his (enlarged) income absorbed by his (unaltered) dues would be the less. Consequently, it was ever the direction of the peasants' endeavors to "expand" their *łans*. Contrariwise, if the manor succeeded in reducing the *łan* and, say, carved ten *łans* out of an eight-*łan* village, then the amount of dues accruing to the lord would, of course, increase. In 1599, the Crown Referendary suspected that the villagers of Kazom (Kazuń) in the Zakroczym district enjoyed outsize *łans*.[57] There is no way for us today to decide whether the charge was true or false, that is, whether the Kazom peasants had somehow "collared" some extra land on the sly, or whether the manor had brought up the charge of "collaring" as a pretext to reduce the peasants' *łans*. Given the

Polish manorial demesne farming with its labor dues, either might equally well have been the case; but there is less doubt as to how the judgment would go: given the existing imbalance of power, the stronger party would readily have gained the day. We know that in a similar case, Queen Bona [Italian second wife of Sigismund the Old from 1518—TRANS.] ordered that the *łans* be remeasured, and, sure enough, they were adjudged to be "overlarge."[58]

The eighteenth-century writer Haur, who was no denigrator of the nobility, is worth quoting on this issue: "*Vlokas* and *łans* are of two kinds: some defined by Crown statute or other law, and others defined by no law, but by custom or whim."[59] And he goes on to indulge himself in historical reflections:

> It is a question of consequence, why the divisions of land determined by the laws of our Commonwealth long ago should nowadays (admittedly, not everywhere) be departed from, and a just *vloka* is but rarely found. The answer is: first, in some places, overzealous stewards and their like, seeking to curry favor with their masters, diminished the large, true *vlokas* so as to increase their numbers and thereby the labor dues, too. . . . Fourth, there were places where some, who coveted more land, readily made out a case that large holdings were of no use to serfs who could not cope with them, but were of use to the manor, or indeed acquired more land by some other means that I would not recommend to any God-fearing and justice-loving man, for they are wrong in law and in conscience. Even he that has a million *łans* will have but three ells of earth after his funeral, which others will till and sow.[60]

We can trust Haur, for he well knew the economic practices of landownership and, by and large, approved of them. The development he describes in its two aspects took place over three centuries, although its intensity varied. The nobles, admittedly, knew very little of land surveying or map making,[61] but could avail themselves of the services of obliging surveyors or, at least, stewards. The excerpt from Haur's book suggests, however, that although the practices he elucidates were very common, they were perceived as less than morally correct.

However, even the most learned members of the Cracow Academy were unable to determine just what that "primeval," "antique," "defined by Crown statutes," and "just" *łan* was.[62] We cannot, consequently, pretend to any surprise at the distrust the peasants showed towards surveyors and all attempts to measure land; their

41

fears that, for them, no good would result from such attempts were by no means unfounded. Nor, indeed, should we consider it strange that the abortive attempt to assassinate King Sigismund Augustus had been engendered by injustices suffered by the peasants when lands were surveyed.[63]

To sum up: all pre-metric systems of agrarian measures, despite their many differences, were systems of representational measures that "signified." Of the various characteristics of any piece of land, that of its square dimensions was the least taken into account, and qualitative aspects were of major importance. When the numerous relevant quantitative aspects were due for assessment, the real task was therefore to agree on measurable criteria for nonmeasurable values. Two qualitative aspects of any cultivable field are of crucial importance: the time it takes to cultivate it, and the harvest it is capable of yielding. Since in those times harvests greatly fluctuated from season to season, the average amount of seed taken to sow the field stood for the average harvest. The dominance of qualitative over quantitative considerations in the social thinking of preindustrial societies, so strikingly present in relation to agrarian measures, applies to other spheres of traditional metrology too. It is, in fact, a truly fundamental characteristic of traditional mentality, which would not disappear until the onset of industrialization and the coming of the metric system took hold—and even then slowly and reluctantly. This is nothing to wonder at, for upon this characteristic the agrarian system of measures rested. Considered in the context of man's relation to land as his "workshop," the measure that speaks in terms of the man-hours of work or amounts of seed surely says more than does a certain number of hectares. In the chapters that follow we shall, accordingly, deal with what amounted to nothing less than a revolution in the traditional rural way of thinking. This revolution had to be effected in order truly and universally to bring into use the modern, strictly quantitative measures with their utter disregard of all those qualitative aspects that do interest man.

· 7 ·

HOW WAS GRAIN MEASURED?

Scales are an instrument of ancient lineage; they were wielded by the archangel on the day of the Last Judgment—as depicted for the admiration of the faithful on the pediments of many Gothic cathedrals—and earlier still, they were an attribute of Amon, the god of Justice personified. However, before the meter came to reign supreme, they were used by virtually no one but merchants, and even by them only for a limited range of articles. Many goods that we are now accustomed to buy by weight used to be purchased by customary measures of capacity or by the piece; thus, cheese was sold by the chunk or slice, butter by the "round," wool by the fleece, yarn by the skein, honey by the "hand," nails by the dozen or *kopa* [sixty, in Poland—Trans.], eggs by the *mendel* [fifteen, in Poland—Trans.], etc.

As already suggested, the creation of a "measure" is a complex mental act; it demands that we abstract from a great many qualitatively different objects a single property common to them all, such as length or weight, and compare them with one another in that respect. In modern Europe, a wide variety of goods are reduced to a common denominator and sold by weight. The qualitative diversity of cheese, butter, honey, wool, and nails is so great, and the traditional techniques of production so varied (e.g. the curious shape of curd cheeses among the Tatra mountaineers), that they blinded contemporaries to the fact that these products do share the property of weight.[1] The difficulty is compounded by the fact that scales are, after all, a clever instrument, relatively costly at that, and not everyone can afford them; furthermore, they can readily be tampered with and therefore tend to arouse suspicion.[2]

Throughout Europe, in medieval as well as in modern pre-metric times (and even later, here and there), both liquids and dry goods were measured by volume or measures of capacity, such as the bushel, the gallon, etc. For us, a measure of capacity presents no great difficulty in use; it is easy to control and there is nothing contentious about it. Yet daily use over centuries, and particularly in societies subject to sharp conflicts, has complicated the matter. On the one hand, numerous abuses and ways of cheating and, on the other, appropriate forms of defense against them came into

43

being. Of the many abuses associated with altering the size of the bushel, we shall treat elsewhere; here, let us consider practices associated with the methods of measuring.

The elements to be considered were the substance of which the measure was made; its shape, that is, its three dimensions; the thickness of the rim; the ways of safeguarding it against distortion through age, damp, drying, or indeed against deliberate distortion by devices such as ferrules; the manner of fitting it; last, but certainly not least, the problem of the "heap" and the manner of applying the strickle. Standard bushels, both state and municipal, were, to be sure, made of metal. Thus, when the 1565 reform came to Cracow, the *voivode* decreed that no one might possess a bushel "other than the copper kind that had long been at the Guildhall."[3] A few bushels of that kind have indeed survived and found their way into museums. One can scarcely believe, however, that metal bushels were used by every petty trader, or that they even graced the guildhall of every country town. We have it on the authority of K. Kluk that the measures used in measuring grain were either turned or hollowed out.[4] Now, there are major problems here. First, the problem of production: however honest and willing he may have been, was the village turner or cooper able to calculate the size of a measure with complete accuracy? Bearing in mind that the radius and the circumference of the circle are not related to each other by a common measure, what was wanted was such a measure relating the radius (or the diameter) to the height of the container. Secondly, it was important to make bushels of the right material and technically sound, and to safeguard them as well as possible against accidental or deliberate distortion. What was of crucial importance was that the prescribed shape should be observed in respect of all three dimensions. In his regulation of 1565, the Cracow *voivode* continued his order to copy the copper standard at the Guildhall: ". . . of which bushel . . . the magnitude, height, width, and depth must be reproduced the same in the ordinary bushel, so that neither its height, nor depth, nor width shall be either larger or smaller by aught."[5]

It is difficult to say how, in fact, the technical difficulties were coped with. One tradition has it that prism-shaped bushels were commonly in use in Poland,[6] but we have no firm evidence of this. The story seems unlikely and, if true, would have amounted to a practice unique in Europe. Ethnographers tell us of some relevant customs practiced by coopers that have survived to this day. For example, a calibrated measure, which was most likely passed on from father to son, "defines the length of the staves in relation to the number of gallons that will make up the container";[7] if so, then the assumption would be that a fixed-size bottom and variable height would go with measures of different size. But this would be practicable only within a limited compass, for, when dealing with large amounts, the measure as container would have had to be shaped like a narrow cylinder, whereas the actual practice was quite the opposite: large dry-measure containers used to be relatively flat. Since it was unrealistic to demand that all bushels be made of metal, the Cracow legislators of 1565 decreed metal reinforcements: "The quarter and the bushel, too, shall be reinforced with iron, crosswise, both at the top and at the bottom."[8] Again, the statute of 1764 decreed that dry measures had to be "iron-bound at the top and bottom, and with a bar fastened across the middle."[9] In practice, the townsmen (e.g. in Poznań) would generally keep in their homes iron or copper measures, or wooden ones with iron reinforcement.[10]

The Prussian edict of 1796, which was the next pertinent enactment, laid down the following practice: "The measure for dry goods must be perfectly circular. . . . All wooden measures for grain must have iron binding going crosswise across the entire bottom, and there should also be iron hoops round the top and round the bottom. Moreover, to prevent any deformation of the circular shape with the passage of time and frequent handling and to prevent any spillage of grain with the strickle, the circumference must be stretched out with an iron bar fixed to an iron fastening in the middle of the bottom, the bar being soldered to the appropriate iron bindings and being laid level at the height of the circumference at the top." This passage is repeated, virtually word for word, by Nakwaski in his scheme of 1810, except for his additional concluding sentence, "To make sure the intended purpose is achieved, the bar always has to be quite straight," with the additional demand, "The bottom has to be quite level and tight-fitting to the sides, and made of one piece of wood, no more, as the coopers' wont is."[11] Doubt was being cast here upon the very competence of the coopers or turners. The head of the Warsaw police, whose familiarity with dishonest market practices can scarcely be doubted, questioned it as late as 1847 (this being the date of our source, but surely the practice had been known earlier, too): "The measure ought to be made of the inner tree-bark, well and truly reinforced with metal and never just made of staves, for these—being made of unseasoned wood—in drying out diminish the measure a good deal."[12]

To conclude: the apparently simple measuring implement appears now to have been highly complex, and its production was far from an easy task, nor was securing its invariability.

Let us now turn to questions of the shape and dimensions of the bushel and above all, to the all-important ratio of its diameter to its height. In geometrical terms, the matter is one of indifference: the number of permutations of ratios between diameter and height that are capable of yielding the same capacity of the cylinder is infinite. Yet, as we have already noted, it was by no means a matter of indifference to the users, with their appreciation of the practical implications, for it was the actual ratio that determined the magnitude of the "heap." The bushel could never be a very tall vessel: its height could not exceed the distance between the hand of a man standing upright with his arm hanging by his side and the floor. For, as we shall see, the height from which a man poured grain into the container was of crucial import and it was considered ethically proper to pour it from dropped-arm height. Furthermore, to pour from shoulder height would add to the considerable physical effort involved in the task; it was hard work, and the young man to whom it was entrusted was admired for his strength.[13] In practice, judging by the few surviving artefacts, the bushel was a good deal less in height than the "maximum" we have suggested. Over the centuries, in fact, the shape of the bushel grew flatter. It would appear that this tendency applied less, if at all, to urban measures than to rural measures, which in the feudal system were controlled by the lord, who also used them in the levying of taxes *in natura*, flattened down almost to something like a washtub.

It is quite astonishing that the same process occurred in a variety of countries without any direct reciprocal influences. Given the same social system and the same general situation, the privileged responded with the same simple rules of conduct. In the volumes of customary law of the French provinces—the so-called *coutumes*—one frequently encounters prescriptions for the ratio of the diameter to the height of the *boisseau*, the commonest ratio being 3:1. In Anjou the ruling was that the depth of the *boisseau* should be equal to one-third its diameter;[14] in Touraine, likewise;[15] while in the great city of Paris, in the eighteenth century the diameter was more than the height, but only slightly.[16] The care taken in the *coutumes* over specifying the proportions suggests that the objective was to check the seigneur's attempt to flatten the *boisseau*.

In Poland, recent historical researches arrived at the following ratios between the dimensions of the bushel: the Cracow bushel had a diameter of 51.28 centimeters and a height of 18.31 centimeters; the Wrocław bushel, 60.68 and 24.68 centimeters, respectively; and the Gdańsk bushel, 64.54 and 16.73, respectively.[17] The last had undergone the most flattening, for it was altered on three occasions early in the eighteenth century—in 1701, 1720, and 1729.[18] Since, however, Gdańsk merchants measured the grain that they bought with a "bit above,"[19] then, given their whiphand over the seller, the development is easy to explain.

On the other hand, it was not unknown in town markets for the grain retailers themselves to reduce the diameter and increase the height of retail measures in order to decrease the "heap" with which, by custom, they were obliged to measure. This appears to be implied in the following words of the royal procurator of Nantes:

> As regards a variety of foodstuffs that are sold by the heaped measure, grave abuses are perpetrated daily, and steps should be taken to counter them. The Nantes *boisseau*, which ought to contain 446 *pouces cubiques*, should indeed do so invariably and independently of the ratio between its diameter and height. However, the heap that surmounts it does vary in size depending on the diameter. It is, therefore, of consequence to the general public and in particular to those members of it whose circumstances necessitate that they depend for their sustenance on low-grade grains, such as barley, peas, broad beans, oats, etc. . . . , which are sold by the heaped measure, that the actual dimensions of the Nantes *boisseau* be determined once and for all—its width and its depth.[20]

Thus did an official of the Enlightenment bring into focus commonplace everyday problems with which the masses had long been familiar. The problem is one to which we shall return in discussing the resistance of the urban populace to the metric system itself because its adoption entailed the practice of striking the measures.

Let us pass on to consider the manner of pouring grain into the measure, and the way this affected the amount contained. The more effort, or "impetus," that was employed in the process, the more firmly pressed down were the contents and the more grain that would enter the measure; and for the force of a man's arm, the force of gravity could be substituted, with the grain being poured from a greater height. Dozens of peasant complaints have survived in our sources (see chap. 18, below) relating to the demands by the collectors of taxes in kind that grain be poured from shoulder height, and not, as was proper, from dropped-arm height, that is, from the height of the hand, with the arm hanging down. Only the latter manner was deemed ethically right, both by the peasants and the nobles; that this was the view of the former is borne out in their complaints and petitions, while for the latter we have analogous testimony in the verdicts of the Referendary's Court. We can also argue *ex absentia*, for in no royal or seigneurial demesne inventory has the author in his researches come across the view that, in rendering their grain taxes, the peasants should pour it from shoulder height, while the very presence of statements by the peasants relating to the practice of pouring in that way constitutes an accusation. Where a dispute did arise, the statement would be made explicitly that the grain was to be poured "from dropped-arm height."[21] Pressing it with the hand was another way of cramming much more grain into the bushel.

In the main, it seems legitimate to put forward the generalization that in keeping with the ethics of the time it was fair to measure grain "from dropped-arm height," "striked," and "unpressed." At the time of the survey of 1789, the villages of Kosmalów, Zimnodola, and Osiek (in the Rabsztyn district) were complaining to the inspectors that although they belonged to the same lord as the villages of Racławice and Zederman, yet they enjoyed no "equal happiness," for only the latter rendered their grain striked, as the law says, whereas "the way ours is collected 'heaped,' pressed, and poured from shoulder height, adds up to an extra quarter."[22] It often fell to the inspectors to insist that the pouring be done from dropped-arm height,[23] or to prohibit "compressing."[24] The latter prohibition may also be found in many village judicial records[25] and even in some manuals.[26]

What exercised the nobles in cases like that, however, was a conflict of moral principles. For, on the one hand, the proper procedure was to use striked measures, pouring from dropped-arm height, and without "compressing," but, on the other, the fair practice was whatever was sanctioned by long-established usage. If, therefore, there was some circumstantial evidence, readily available, of the *osep* [basic grain tax—TRANS.] being formerly rendered heaped, then the heap would be deemed proper and lawful, and this indeed was the usual practice. Instances of rulings that the grain should be poured from shoulder height or "compressed" were few and far between, and suggestions of "over-head" pouring are quite unknown. Such practices, in the main, were offensive to the morality of the age. Furthermore, it was ethically correct to stand the measure on level ground and not to touch it during the pouring of the grain, for to place it on a slope, to knock it, or to shake it down might substantially affect the actual amount of grain it would hold. It is stated in St. Luke's gospel that God will repay a man's deeds with a "good measure, pressed down, and shaken together, and running over";[27] this means that God's reward will be somewhat over and above what is strictly due.[28]

The whole question was not, apparently, settled by law until the Prussian edict of 1796, which dealt with it as follows: "No one, moreover, is entitled to demand that the grain be poured into the measure from a height, with 'impetus,' " and then, "At the time of measuring, the measure must be in no way knocked, nor must the grain in it be shaken by any violent movement of the floor." The matter, however, continued to be troublesome for a long time. From a piece of evidence dating from 1846, we learn of an interesting procedure used by the millers:

> Almost every miller who buys grain to be delivered to his mill will have a store for it inside, and when the grain is being remeasured, it is usual for the mill to be busily grinding. The movement of the wheels causes continuous shaking of the building, which ensures that the grain being poured into the measure will be most tightly compressed, the amount at the end of the remeasuring being considerably short, despite the fact that the supplier may have measured it at his home most honestly before delivery.[29]

That the practice was widespread is borne out by the fact that it was mentioned in a widely used manual of husbandry: "In selling grain, when it is due for remeasurement, it is advisable to avoid places where the grain will be shaken down through some knocking, e.g. mills; even in other places, sometimes the buyer will have arranged for someone to cause similar clattering, using this pretext or that to move about or to stamp on the floor."[30] It does not seem unduly speculative to suppose that this practice, which we have found referred to in a mid-nineteenth-century source, was one with a long history behind it.

At this stage, we have now to consider the crucial question of "the heap." The genesis of the custom of measuring either "heaped" or "striked" does not concern us here and may properly be left to medievalists. The situation that obtained in the middle of the sixteenth century makes it plain, however, that the uncertain and frequently questioned compromise arrangements of the day had resulted from an agelong conflict between, on the one hand, pressure from the manor in favor of the largest possible "heap," and, on the other, resistance to this from the village. Thus, the *lustratio** of 1565 always made careful note of whether the peasants were supposed to render their dues by heaped or by striked measure. There was no fixed principle relative to the issue, and even the same year's statute did not stipulate one, leaving it all to "ancient measure and propriety." However, long-established custom had, in fact, resulted in a set usage in each village, and the very record of it made by the inspectors thereafter gave it the force of law in that the record in question would later be referred to in cases of dispute. The contentions, haggling, and halfway-house agreements eventually arrived at, would, over generations, complicate the matters so greatly that we know of instances where individuals were obliged to render part of their grain tax by heaped measure and part by striked.[31] All those concerned treated the matter as one of major importance.

Let us now consider the conduct of the inspectors engaged in the very first *lustratio* in 1565. In the Cracow and Sandomierz *voivodeships*, wherever the inspectors converted heaped bushels into striked but retained the same measure—thereby enabling us today to carry out the necessary calculations—the "heap" is quite consistently 33 percent of the bushel for wheat and rye, and 50 percent for oats.[32] In one case, the inspector went so far as to proclaim it as a general rule: "*Item* wheat and rye: 3 heaped bushels will make exactly 4 striked ones; oats: 2 heaped bushels will make exactly 3 striked."[33] Clearly, the quantities involved were very considerable; and it is not out of the question that occasionally they were exaggerated by the inspectors. For, in truth, just as the insistence on collecting ever larger "heaps" from the serfs might well be a way of increasing their dues, so in later times, when the heap in question had become customarily accepted, it would be lucrative, in converting to level measures, to increase it yet again. Thereafter the procedure would be repeated all over again when the new measure, in its turn, came no longer to be questioned—that is, a "heap" on it would be demanded.[34] Let us also bear in mind that the heap was proportionately larger for the cheaper grains such as oats, as compared with wheat and rye.[35] This is a matter to which we shall return.

* The *lustracja* (pl. *lustracje*, from Latin *lustratio*) of estates, by inspectors/surveyors appointed *ad hoc*, was a common Domesday-Book-like procedure in sixteenth- to eighteenth-century Poland, especially on the royal demesnes, following the relevant statute of 1562—TRANS.

However, the nobles did not use the bushel only to collect their grain tax—they also sold grain by the bushel. Since heaped bushels were associated with all manner of arbitrary decisions, they preferred to sell in the market by the striked bushel. We shall describe in due course (chap. 19) the inconsistent conduct of Ruthenian nobles relative to the striking of measures in the urban market in their struggle with the bourgeoisie. On one occasion, they were in favor of all grains being sold striked, on another, they preferred that an exception should be made for the spring crops, and, on yet another, they wanted oats alone to be excepted. Again, in legislative terms, it was apparently the statute of 1764 that eventually settled the issue by decreeing that all grains, both winter and spring crops, should be sold striked. Yet the evidence left by the *lustratio* of 1789 leaves no doubt that "heaps," "tops," and "heads" were still in use a generation later, and while the inspectors, for the sake of bookkeeping, converted the figures obtained into "level" or "striked" bushels, they nonetheless refrained from prohibiting the use of such "extras." The commissioners of the Crown Referendary, ever watchful of proper observance of all the 1764 statutes, in their efforts to remove a source of endless disputes, would on each occasion "reduce" the customary heaped measures to striked ones, but with no loss to the lord. And if this was the normal practice on the royal estates, then the situation was more likely worse rather than better in villages owned by hereditary lords.

The statute of 1764, moreover, made a point of stating that the striking should be carried out by means of "an ordinary, broad, wooden strickle, and not a cylindrical one. . . . The iron bar must be visible." This was important, too. The Prussian edict of 1796 put the matter more clearly, describing in great detail the shape of "the strickle or iron," the manner of holding it and applying it to different varieties of grain, and insisting that all measures of grain be used in conjunction with a duly stamped strickle, "for badly made strickles are a cause of serious injustice and fraud." Despite this legislative care and precision, the matter was still less than settled half a century later. In 1846, the Department of Industry and the Arts stated that the process of measuring offered scope for all manner of swindling, and explained the diverse ways of holding and applying the strickle while also building the "heap."[36]

About the same time, two leading corn wholesalers, Ludwik Biernacki and Konstanty Wolicki, put it to the authorities of the Con-

gress Kingdom [the Russian part of Poland, centered on Warsaw—
TRANS.] that the selling and buying of corn should be by weight
and not by measures of capacity,[37] supporting their proposal by
reference to practice abroad, especially in Holland. They also discussed at length the abuses and frauds that occurred in commercial
transactions, to the detriment of the poorer classes of people, due
to the use of measures of capacity—although one finds it difficult
to think of the authors as philanthropists! The whole idea was,
indeed, scarcely novel: it had been aired in 1815, when a superintendent of army stores remarked of his supplies that "my grain
is received by measure, but there must be a certain fixed amount
of it; yet the weight is not the same for different qualities."[38] That
the scope for abuses was considerable goes without saying.

In reviving the issue in the 1840s, Biernacki and Wolicki worked
on three assumptions: first, that "measures throughout the country
are not often standardized—every manor, every town, has its own";
second, "the remeasurements of grain everywhere is carried out
by Jews, without the accountability that should be imposed upon
the measurers, whereas the manner of pouring grain into measures,
or of striking it, may cause a loss of several percent to the buyer
or the seller"; and third, "wooden measures undergo distortion
through frequent handling, or drying, and even those that the local
authorities look after as standards are borrowed by buyers or sellers
for the purposes of remeasurement of grain and then more often
than not are returned tampered with." The central government's
Commission for Internal Affairs sought advice on the matter from
the administrations of the *gubernyas*, and their answers are of some
interest.

The *gubernya* of Kielce, in stressing the unfamiliar nature and
novelty of the system to the people, did not see its adoption as
being out of the question, provided that use of municipal scales
would then be made obligatory and overseen by officials, rather
than leasing the scales. The Radom governor felt that the sale of
grain by weight would be far preferable for the buyer because, in
selling by volume, the seller "can charge as much for poor-quality
as for best-quality stuff." The *gubernya* of Augustów foresaw considerable difficulty in introducing the reform, advising that it be
done step by step: first, all towns should be provided with scales
(the implication being that some had none!), and the parishes likewise (only capacity measures being likely to be owned by peasants);
second, all landowners of standing should then be encouraged to
employ the new system; and, finally, the peasants must be per-

suaded of its advantages. From Warsaw came the suggestion that the people should be at liberty to sell either by measure or by weight. Finally, the authorities at Płock held that "the sale of grain by weight and sack would produce more harm than would sale by measure . . . for buyers unfamiliar with the scales, countryfolk in particular, dealing with smart tradesmen—nearly all of whom, in this *gubernya*, are Jews—would be duped, and, conversely, the sellers might tamper with the grain so as to increase its weight." They went on to explain how grain might be dry or damp, clean or impure, and how the best or "white" variety wheat was dearer yet lighter than inferior types, and, finally, how selling the grain in sacks, especially when large quantities are bought at the manors, may enable the seller to spread some best-quality grain only on top—whereas this cannot be done when grain is sold by measure, and the entire amount of grain has to be poured back and forth.

The views of the governmental Commission were divided, with the Department of General Administration opposed to the reform and the Department of Industry and the Arts in favor of it, the latter office providing a graphic description of the abuses associated with the various ways of measuring. We read in the description that "measures are attested with a forged official stamp, and while in use, continue to dry out, thus undergoing some distortion and slightly altering their capacity. But the speculators are clever at covering up the defects, and the fraud remains undisclosed." Selling by weight might, however, merely reduce rather than eliminate opportunity for such frauds, since forged weight pieces are not unknown, and their corrosion by rust may be over ten percent. Above all, the Department of Industry and the Arts stated that "the value of the grain sold by measure often works out unjustly, for although its quality is associated with its heaviness, the measure does not show this, but in itself constitutes the price of grain, regardless"—an assertion that, apparently, more often than not was correct. The Department of Industry and the Arts, however, in its support for the reform, advised against its compulsory introduction, for "it would encounter the difficulties of eradicating the widespread and deep-rooted custom of dealing in grain by measure, the simplicity of which enables the countryman readily to check the value of the grain he puts up for sale; contrariwise, the far more difficult process of weighing, with its complex and not always portable machinery, might prove impractical for the majority of our producers." Furthermore, the Department drew attention to the possibilities for the sellers of dishonestly moistening the grain or soiling it with sand or grit, although such practices might be easier to detect than might fraudulent measures; finally, experience and precedent were available to learn from: for example, how did the pioneering steam mill in Warsaw, which had already for some time bought grain and sold its finished products by weight, guard against forgeries?

Meanwhile, Wolicki was pressing his cause. Five years after his first submission in 1846, he wrote to point out that selling by weight was gaining ground in Russia. He also added that "those carrying out the measuring are so expert that out of a last of 30 bushels they are able to measure off—when their palm is greased—29, 28, or even 27½ bushels by means of employing different ways of pouring the grain into the measure and applying the strickle"—a wholesaler's worries, to be sure. Against him, the Warsaw superintendent of police argued that "there are so many permutations available in using the scales that the countrymen and the poorer buyers are unlikely to appreciate that the sellers have so many ways of doing them down." Sadly, we cannot but agree with the policeman! The weighing was undoubtedly a method that favored the seller, and particularly the wholesaler, and threatened the interest of the peasants. The latter would possess no scales of their own and were therefore unable to check the deal at all, even by weighing at home beforehand with their own implements the grain to be sold. The matter was duly filed away and did not surface again from the depths of the responsible officials' desks.[39] As far as this writer is aware, the custom of selling grain by weight was never introduced by any prescriptive piece of legislation; it gained ground gradually, however, at the behest of merchants rather than the police.

The landed noble's own feelings, meanwhile, were desperately divided on the question of measures: for one thing, as collector of dues in kind, his preference was for a waxing bushel—the more so, the merrier—and also for having it measured heaped. However, as seller of corn in the urban market, he preferred a stable bushel (since it was hardly possible there to cut its size) and demanded that it be bought striked.[40] Again, as exporter of corn through the Baltic port of Gdańsk, he had in practice no say in the matter, at most being able only to tell his scribe accompanying the transport to be "personally present at the measuring, and to seek to persuade those in charge [sic!] to perform their duty justly and conscientiously."[41] Doubtless, any outlay incurred in the "persuading" would not be omitted by the scribe in the expense account presented on return home.

Apart from his relations with peasants, local townfolk, and the merchants of Gdańsk, our noble had also to consider his position *vis-à-vis* the administrative personnel of his own estates, a relationship that grew more important and more contentious as time went on. The administrative element increased more than proportionately as the concentration of lands owned by the magnates increased and the passive resistance of the peasantry called for redoubled control. Given the prevailing ideology of the *szlachta*, with its idealization of landownership, it was plain to all that the déclassé noble, who formed the bulk of the personnel of estate administration would, overtly or covertly, seek to acquire his own land. It was no less obvious that his objective would remain unattainable if he relied merely on putting by a portion of his annual emoluments; it could be achieved only by services so faithful that the master's grace might—just might—reward the servant in old age, or by theft. The bulk of the prescriptive activity of the magnates within their own estates, as expressed in the many manuals of instruction and guidance for their stewards, aimed at creating conditions that would prevent, at least partly, the perpetration of abuses through tighter control over the administrative personnel. The process, however, formed an endless spiral: in order to cut the costs of administration, more control personnel had to be engaged; this was costly, and the additional "controllers" were all potential perpetrators of new abuses, since their social position, as non-landowning nobles, could only drive them towards the same socially determined target: the possession of their own manorial estates (*folwarks*).

This widespread and often paradoxical situation impinged on metrological issues, too. The nobles, who generally wanted their grain taxes rendered "heaped" or piled up by the peasants, would, from the moment of collection, apply striked measures. The lord's own grain, yielded by the manor's demesne, would also be measured applying the strickle. Those preferences were universal, and their motivation is plainly revealed on occasions. For instance, instructions issued in the estate of Stanków in 1760 specified that "both the receipt of grains at the granary and their release for any purpose should be administered by the currently adopted [?] Lvov measure, and striked with no top whatsoever, because it is easier for the auditor to estimate the short measure properly as due to ravages by rats and mice rather than to account for the unusual surplus brought about by heaping."[42] Again, the steward of the estate of Tworyczów was instructed in 1763 that "for storage in the granary, do not accept heaped measures, for they will result in surpluses of grain, which the watchman will count on being able to steal with impunity. Keep, therefore, careful watch against any heaping for, if you allow it, then a surplus (if marked) against you will show in the accounts. You will show your honesty better by making sure that the grain goes from the barn to the granary, and then leaves the granary, always striked."[43] The common threats of severe penalties for "surplus" which we find in the stewards' manuals need baffle us no longer.

Although explicit statements like those we have just cited, which plainly reveal the underlying concerns, are rare in the sources, instructions to use striked measures at all stages, in selling and for local consumption, are very numerous. The Bazalia estate manual of 1800 states: "The steward must, finally, see to it that level weights and measures are used in receiving and giving out all produce at the stores."[44] The manager of the manorial villages of Endrychowice, Bablowo, and Zubowszczyzna was told in 1759 that "grain is to be accepted for the granary and, equally, distributed from the granary, striked."[45] Again, we know of similar injunctions at the manors of Buczemla and Oleszkowice in 1758–1759,[46] on the estates of Nieborów in 1783,[47] at Czernawczyce in the late 1790s,[48] and at countless other places.

Admittedly, there were exceptions to this rule. The custom of measuring grain without the strickle was too deep-rooted in the countryside for at least some landowners not to perceive the benefits of compromise. Perhaps they themselves felt uneasy about the use of the strickle as the norm in rural dealings. Nevertheless, as the real issue was always how to enhance their control, they sought, even while rejecting striked measures, to limit and standardize "the heap." To exemplify: the instructions that Piotr Małachowski issued to his steward in 1774 insist that "from the barn, all grains are to be taken striked, ordinarily with a head, to the granary,"[49] while a corresponding document from Nieborów, dated 1787, lays down that "the barn scribe shall, in the presence of the farm overseer, measure all grains leaving the barn for the granary, applying the strickle and leaving a small head of one-eighth on top of the bushel. . . . as for the overseer . . . his job is only to give out the grain, whereas the remeasuring is to be carried out again by the said scribe . . . again using the strickle and leaving a small head."[50] Parenthetically, let us note here the tangible yet generally unnoticed distinction between the concepts of "top," "heap," and "crown," and that of the "head."* The "head" signifies here a much smaller overflow above the level of the rim. We hear thus of measuring "with the strickle, but allowing a small head" [*kołnierzyk*, the diminutive of *kołnierz*—TRANS.] of an eighth, or approximately 12 percent of the measure, whereas other evidence tells us that the true "heap" (*czub*) amounted to 33 percent for wheat and as much as 50 percent for oats. There is a nice linguistic differentiation here among the terms rendering metrological customs. To be sure, though, internal manorial accounts never mention the true large heap or top (*wierzch*); at most, they contain, just here and there, references to the smallish and standardized "head."

In endeavoring to safeguard themselves against abuses, of which the administrators of large estates were ever suspected—all too often correctly—the magnates sought, naturally, to ensure uniformity in the methods of measuring and, in particular, in the measures themselves. The stand that they all took here referred to their lordly rights, that is, emphasizing and externally manifesting them. Hence the continually repeated orders that all measures used within a given estate must be stamped with the owner's "mark." The estate clerk at Pratulin was enjoined to see to it that "all measures of corn as well as dairy produce are to be checked, throughout the estates of Pratulin and Trościany, to ensure that they are uniform and stamped with the lord's mark";[51] the steward at Tworyczów was given by his lord a standard measure "sealed with my own seal";[52] and, in the important Zamość honor** estates, the 1798 administrative manual announced: "Desirous of ensuring justice in respect of corn measures, I recommend that the head steward, as soon as possible and no later than St. John's day, should have appropriate measures made of seasoned wood, then dispatch the same to my office at Zamość, where they shall be stamped with my seal and thereafter distributed around all the manors, accompanied by an injunction that no one should dare sell, buy, or borrow corn by any measure other than those established for the purpose";[53] finally, in the Nieborów estate: "The administrators and overseers must take every care that . . . pure grain be deposited in the granary after being measured by the stamped treasury measure."[54]

* I have translated *kołnierz*, literally "the collar," as "the head," mainly because the latter term is generally applied to beer in England as is *kołnierz* in Poland—TRANS.
** The term "honor" seems to render well the meaning of *ordynacja*, whereby—unusually in Poland—the principle of primogeniture was applied in law to ensure that the estates of the Zamoyskis would be inherited intact—TRANS.

The "lord's mark" was to be stamped even in situations where the manorial accounts relied on the measure of the main town of the *voivodeship*, for example, of Lvov[55] or of Sandomierz.[56] In one instance—unlikely to be unique even though this writer knows of no other—the reliance upon the known urban measure meant, simultaneously, a reliance upon the measure as standardized in 1764. This occurred at Boćki in 1767, where it was enjoined that "all measures in the town be defined in terms of the existing royal metal-reinforced gallon" (*garniec*), although what seems to have been intended was a uniformity of measures in a privately owned township, especially in "all the inns and hostelries where people stay,"[57] but not in the countryside. The conflict of interests between the landowner and his own officials would even, on occasion, affect the relationships between the manor house and the local peasants. After all, the insistence that the peasants should deliver their *osep* heaped afforded an excellent opportunity for collecting a large heap from them and then passing on a small heap to the lord's granary. Hence, at Boćki it was specified in 1800 for the clerk in charge "to collect fairly the *osep* due from the countryfolk and townsfolk, requiring no extras nor measuring heaped, and if ever he is found to have done so, he will have to compensate those injured."[58] We have, however, come across no other cases like that.

To complete the tale derived from Polish sources, let us briefly consider the course of events in Silesia under Austrian and Prussian rule, and also in the city of Gdańsk. In Silesia, at first glance there appears to have been an extraordinary proliferation of corn measures, but a closer examination reveals that "there was method in this madness"; the apparent confusion is governed by strictly observed rules. At Bytom, there coexisted three units: the old market measure, equal to 207 liters; the old castle measure, equal to 184 liters; and the Church measure, equal to 224 liters;[59] at Nysa, there were as many as five units.[60] The nomenclature sometimes hints at the causes of such variety: for example, there was the *Decemscheffel* or tithe measure, and the *Dominialscheffel* for measuring the lord's due.[61] The "heap" amounted usually to one-third,[62] but could be less—a quarter[63] or two-sevenths.[64] We know of disputes occasioned by the peasants' reluctance to render their dues heaped; in one case, they yielded only on the promise of some vodka *pro consolatione*,[65] and generally taxes in kind were measured with a heap until the end of the eighteenth century or even longer.[66]

All the practices that we have found in Poland are met with in Silesia, too. To consider first the permanence of measures: in the town of Ścinawa, the Niemodlin measure remained in use for two hundred years after the town had been transferred from the Niemodlin district to the Prudnice district;[67] the free mining settlement of Tarnowskie Góry, established in 1526, retained its distinctive outsize measure originally brought in by forebears from Bavaria.[68] Again, Silesia had its own measures and its peculiar methods of measuring in regard to oats: generally, they would be measured heaped, but there was also in use a special "oats bushel," whose relation to the ordinary bushel was indeterminate, sometimes exceeding it by merely one-tenth and sometimes being double its size![69] In Silesia, too, measures depended on "economic" distances: in country towns they tended to be larger than in sizeable urban centers, which depended on food "imports."[70] Furthermore, the influence of commercial links is well evidenced in Silesia; for example, the town of Uszyce, in the Olesko administrative district, used the Byczyna measure—for was not Byczyna within 10 kilometers whereas Olesko was 25 kilometers away?[71] The town of Głogówek had its own measure, but sold corn in 1685 by the Opole measure, wishing to accommodate a large purchaser.[72]

Last but not least, we must note the occasional contest between forces seeking to change measures and those defending the *status quo*. The Church and the manor, the town and particular segments of its population—all these, whether overtly or covertly, brought pressure to bear. In Bieruń, the entail which the town received from the then lord of Pszczyna, Jan Torzon, states in 1547 that "the aldermen, along with the council of elders, shall establish measure, ensuring that neither the rich nor the poor will be harmed; and if it appears to the council that the existing measures are not just, then it is for them, and not for the aldermen, to put things right."[73] What precisely was at stake eludes us; but clearly there was a dispute here between the better-off and the poorer townsmen, with the lord siding with the former. Yet in Lubliniec, in 1639, when the council—the mayor, to be precise—changed measures, the townsmen rioted.[74] Of the contestants, the Church was probably the strongest; its measures were most often the largest;[75] but the Church, too, had at times to yield to tradition. For instance the parish priest of Radzionków collected his dues from some of his parishioners by the Bytom measure, and from others by the Tarnowskie Góry measure—keeping, of course, copies of both at home.[76] A record, of sorts, was surely established by the parish of Jodłownik, where, even as recently as 1881 (!), the tax in respect of the celebration of Mass was collected partly by the measure of

Raciborz and partly by that of Wodzisław, again still keeping the two standards to hand.[77] At Żywiec, some agreement must have been arrived at between the town and the lord of the manor to throw up the practice of doubly stamping the local bushel with the municipal mark and the lord's mark.[78] At Rogoźno, in 1688, two systems of measures apparently coexisted: one used in market dealings and the other in transacting business with the manor.[79] And there were also priests who journeyed from village to village with their own bushel, collecting their tithes.[80] Like the manors and the parishes, peasant families also piously kept their measures of bygone days well into the nineteenth century.[81] A Silesian saying sneered at the peasants: "Those peasant measures lack any faith,"[82] but, truth to tell, their "faith" was of the same type as that represented by the lord's or the Church's measures, only that their measures stood for something different.

Silesia had belonged to Austria, and thereafter to Prussia; it had obeyed first Austrian and then Prussian laws, and it had been under Austrian and then Prussian administration. And yet, as far as matters metrological were concerned in the silent agelong struggle among the manor, parish, and village, there was nothing found in Silesia that was not also found in the lands under the scepter of the Polish kings. Although no two villages are the same in this respect, nonetheless their diverse situations are governed by certain basically simple rules and exhibit certain typical features. These common features in the struggle are patterned by the manor's might, the sacred authority of the parish, and the solidarity bonding the peasants in their acts of passive resistance to authority. Negotiations and compromise agreements by the score would eventually produce in every place a state of affairs peculiar to it, which would before long come to be underpinned by the power of tradition. The principal guardian of that tradition would be the weaker party to the struggle, namely the peasants, acting on the assumption that any change could not but be a change for the worse for them. The traditionalism of the peasantry is thus a rational social attitude.

In the city of Gdańsk, however, quite different problems claimed the most attention. The cases we have cited earlier exemplified relations inside villages, or between villages and towns—usually small country towns. Now, in Gdańsk we are primarily dealing with relations between the town and the nobles—the municipality acting from strength *vis-à-vis* the individual noble, or his "lieutenant" supervising the transport of corn from the hinterland to the great seaport.[83] Even if the noble should be quite dissatisfied with the terms of trade, he finds on arrival in Gdańsk that he can scarcely refuse to sell his grain, since he would then have to take his still-laden rafts back home upstream. The noble, moreover, enjoyed no support from a strong state authority that might have gained something by aiding him; he and his fellows, at local dietines from Prussia to Volhynia, demanded state intervention to check metrological abuses at Gdańsk, but the state was too weak to help them effectively.[84] The merchants of Gdańsk would only heed action by other merchants, be they customers or competitors. There was an attempt in the sixteenth century to render the wholesale corn measures uniform in Gdańsk, Elbląg, and Toruń, while leaving alone the measures in nearby smaller towns, where as a rule they were larger than in those three major centers of wholesale grain trade.[85] Uniformity of their measures with those of Amsterdam also mattered to the wholesale merchants of Gdańsk,[86] but the measures in various regions of the kingdom of Poland did not concern them at all.

In view of the huge volume of wholesale grain trade passing through Gdańsk, the organization of measuring facilities was of some complexity—particularly as it had sometimes to accommodate clashing interests of several merchants involved in a single transaction—and municipal authorities had to keep a tight rein on the arrangements. The measurement of corn was a time-consuming business involving teams of from five to seven men. In 1695, a total of 199 men were employed as members of such teams in Gdańsk; by 1764, the number increased to 288.[87] While the nobles wanted their grain to be measured in Gdańsk by their own skippers rather than by the local specialists,[88] the Gdańsk merchants naturally insisted on the latter. The manner of measuring was subject to a variety of regulations;[89] for example, the measurers were not allowed to carry and to measure the grain simultaneously; or, the grain that had been measured had to be placed aside for twenty-four hours to enable any interested party to check the quantity.[90] Nothing, however, could wholly prevent dishonest practices; just as the nobles told their skippers and functionaries to "win over" the measurers,[91] so the merchants would offer them bribes,[92] and the merchants' opportunities for effective action of this kind were by far the greater. This advantage was the product of economic and geographical factors—of the fact that the noble had no option but to sell his corn once it had been brought to Gdańsk, and of the strong, indeed monopolistic, position of the business community as buyers *vis-à-vis* the unorganized landowner-suppliers.

Given the high degree of commercialization of the trade [i.e.

production for the market—Trans.], it is no surprise to learn that striked measures were obligatory in Gdańsk. Yet the different methods of applying the strickle still offered much scope for abuse.[93] Thus, a heavy cylindrical strickle could cram into the measure a goodly amount of extra grain, and even more could be gained by the "shaking-down" of the container. Many disputes eventually established in Gdańsk the lawfulness of the custom of measuring by taking two blows at the bushel with the strickle. But even with the high degree of commercialization and the careful bookkeeping in Gdańsk, in disputes between nobles and merchants there recurred a metrological argument typical of a less advanced economy: the merchants demanded a "surplus" and justified this demand by contending that the grain the nobles sent them had not been well rid of impurities.[94] One would have expected in a well-developed market situation that the quality of the commodity in question would have been settled at the same time as the prices, and thereafter taken for granted!

Polish nobles were adept at taking advantage of the variety of bushels and methods of measuring. They also took advantage of their stronger position in their dealings with the peasants by foisting "their" measures upon them, and did likewise in dealing with the bourgeoisie by influencing, if not determining, municipal measures. The same advantage was pursued in relations with the employees on their own estates by controlling—on balance, effectively—the measures the latter used. But, come the climax of the year's economic activity and effort, come the time when the noble's maximum disposable amount of grain arrived in Gdańsk—then the position of privileged strength obtained no longer. It was the Gdańsk merchant who now enjoyed superiority, taking from the noble—by fair means or foul—part of what the latter had, by similar means, painstakingly accumulated.

We have thus far concentrated on the situation and practices prevailing in the grain trade in Poland, but our findings and analysis, in the main, apply equally to other countries insofar as metrological phenomena in them were the product of similar values and ideologies. Let us, for example, consider France. Here, our analysis may very fruitfully take as a point of departure the answers received from the parishes to the questionnaire that the Academy of Sciences distributed throughout the country via the *départements* in 1791. The questions related to the names of measures, their legal "titles" or foundations, their standards, and the methods of measuring. If the answers from the *départements* tend to generalize, and thus to simplify, those from the parishes always present a full and detailed picture; however, to find them all would have required visiting some ninety departmental archives. The present author sampled two, namely those of the *département* of Doubs at Besançon[95] and of Meurthe-et-Moselle at Nancy;[96] however modest his harvest, there is yet enough in it to throw light on the metrological issues discussed in this chapter.

The metrological confusion revealed by the answers is shocking, as is the peasants' helplessness in coping with it. Almost no parish possessed a standard (*étalon*) of its own and for checking those in use had to rely on the models kept in the nearest township or by the landlord; more often still, no testing would be carried out at all. The problem whose importance we have just emphasized, namely that of the "heap," had no uniform solution; there were variations even between neighboring parishes, although the variety was always clearly codified. Let us cite a handful of rather typical examples among a great many culled from the two archives named above. In the *département* of Doubs, the parish of Droitfontaine states: "Grains are generally measured striked, and oats heaped"; the parish of Chamescy notes that cereals are striked except for barley and oats, which "it is normal to measure heaped"; the parish of Rosureux says: "The strickle is applied to wheat, barley, oats and vegetables"; from the parishes of La Grange and Les Belivoir we learn: "According to old custom, here we always strike wheat, peas, lentils and other 'round' cereals but we measure barley and oats sometimes striked and sometimes 'topping the rim' (*grain sur bord*)"[97] the parish of Cheint states: "The accepted custom is to use striked measure, except as regards oats, where the measure is three parts for two"; the parish of Vaucluse informs us that wheat is measured striked but barley "semi-heaped" (*demi-comble*, a rare term); the parish of Herimoncourt applies the strickle to all cereals "without any grain above the rim" (*sans grain sur bord*)—except that the quarter of wheat here weighs 42 pounds and the quarter of oats is larger by a third"; next, we have the parish of Dambelier where "wheat and all other cereals are measured striked, save oats, which are measured heaped but not compressed" (*combler sans presser*, another rare term); in the parish of Gleze wheat is striked, barley and oats have the same striked measure applied to them, but it had been the custom of the former seigneur of Vaufrey to remeasure the oats supplied to him with a measure larger by a quarter. In the same parish, potatoes (*les pomedetaire!*) were measured just as wheat—sometimes heaped and sometimes "grain topping the rim"; in the parish of Blamont, the measure employed for oats was larger by a third than that for wheat, with both coming under the strickle; the parish of Provenchere customarily measured wheat striked, peas *grain sur bord*, oats semi-heaped, and barley "with a full heap"; next, in Dampierre-les-Monbeliar everything was measured striked except for "heaping the oats"; in the parish of Solemont, "wheat is measured using the strickle, barley and peas too, but oats heaped"; the parish of Froidevaux measured wheat striked, peas *sur bord*, while barley and oats were measured heaped; the parish of Bellefort striked wheat but heaped oats; the parish of Curcey generally applied the strickle, and gave three measures for two in oats; in Estonnans, wheat, rye, and barley would be striked, while oats were heaped; the parish of St. Mauris-sur-le-Doubs customarily striked its measures, the oats alone being measured heaped or else three bushels for two; the parish of Escot striked wheat, and also oats, whose measure, however, was larger by a third; finally, in the parish of Pont-de-Roide "all varieties of grain are measured striked, save oats which are heaped, as are also potatoes and other spices" [*sic!*].

In the *département* of Meurthe-et-Moselle almost all the parishes went on record as measuring wheat with a striked measure, and other cereals heaped. The few exceptions included the parish of Vezelere, where rye as well as wheat had the strickle applied to it, but barley and oats did not, and Lunéville where "wheat, *meteil* (a mixture of wheat and rye), rye, dried peas, beans, lentils and gypsum [!] are measured striked; whereas barley, oats, small broad beans [?], millet, rape seed, and potatoes are measured heaped."

There are also, admittedly, cases where all cereals were measured in the same manner. For example, the parish of Goux stated that its measure for wheat also served for oats and each was measured heaped. More often, though, the strickle was applied throughout—equally to wheat, barley, oats, and vegetables—as in the parish of Rosureux. Again, in Mondeval, all cereals were measured striked, *sans grain sur bord*, and the same practice was specified by the parish of Russey, while the parish of Laval used striked measures for all produce. Finally, there were instances of striking all cereals, but treating potatoes and fruit otherwise. For example, the parish of Noirfontainne had a single measure for all cereals and it would always be used striked, but fruit—for example, apples or pears—was measured with a heap. In the parish of Montandon "all cereals and vegetables are measured striked, except for potatoes"; in Belfay "all types of corn are sold striked, but potatoes and hard fruit such as apples or pears are sold by the heaped measure"; lastly, the parish of Frambouhon applied the striked measure to wheat and other cereals, but excepted "apples, potatoes, hard fruit, nuts, and hazelnuts, to all of which the same measure is applied but heaped."

Although the examples we have cited far from exhaust the list of cases where different methods of measuring were used in connection with different cereals, the cases where there was no such differentiation (i.e. either the striking or the heaping was the rule throughout), or those where all cereals were measured striked but other products were not, have all been given. Clearly, they were few and far between. Furthermore, although our cases come from but two *départements*, the situation they exemplify must have been nationwide: witness the request by Condorcet in his letter of 11 November 1790 to the Chairman of the National Assembly that the question of whether different types of cereals were differently measured be investigated throughout France.[98]

To conclude; the geographical restriction of our primary source material to just two areas does not mean that similar practices were unknown in other parts of France. Thus, from the district that was to become the *département* of L'Indre-et-Loire we hear that whereas there were, in fact, some parishes that customarily measured grain by the striked measure, others did not; while in the region's principal town of Tours oats were always measured *comble*. Market usages resulted in some bizarre measurement methods. For example, in Amboise, when large amounts were concerned, then every twelfth *boisseau* would be measured heaped but the rest striked; similar, but stranger still, was the variation obtaining in Chinon where every twelfth *boisseau* out of thirteen would alone be measured heaped. The proliferation of units and measurement methods here was so confusing for its members that the chapter had no option but to issue conversion tables.[99] And later in the same region, during the Revolution, the Commission for Weights and Measures converted the customary measures of *boisseaux*, with and without the heap, and found that the range extended from 2582 milliliters at Chinon to 3739 at St. Christophe.[100]

Within our civilization, the problem of "the heap" is met with everywhere. Even where metric reform has formally eliminated it in the eyes of the law, in reality the process of getting rid of it takes centuries: this process commences earlier than metric reform and continues beyond it, to be completed only within the context of advancing urbanization and industrialization. Poland and France apart, in Antwerp the heap was considerably smaller, amounting to between 6.25 percent and 13 percent of the measure in question. The conflict relating to the practice was typical and was extremely long in disappearing: the burghers wanted striked measures, whereas the large manorial landowners favored "the heap," and oats were still measured heaped at the time of the French Revolution.[101] In Florence, the heap was apparently reduced to a mere 3.5 percent already in the Middle Ages.[102] And it is worth mentioning that the Byzantine "heap" overtopping the *medius* had been reckoned at between one-sixth and half (!) of the measure itself, and manipulating its width was not unknown.[103] Since, to repeat yet again, the many customary ways of measuring had emerged out of conflicts between discordant social forces, their diversity should not surprise us. France experienced the coexistence of striked and heaped measures and the *demi-comble* variety, too;[104] similarly, in Silesia men applied their measure "and a bit" or with a "semi-heap,"[105] and behind such unstraightforward practices long-drawn disputes and compromise agreements can readily be guessed at. This goes, too, for the very widespread practice of applying one measure—or manner of measuring—to high-quality grain, and another to inferior varieties: in the Netherlands, except for equal treatment in Amsterdam, a larger measure would be applied to oats than to wheat or barley.[106]

Several generalizations may be ventured here, all based upon the foregoing evidence:

1. The metrological differentiation was tremendous, even within the relatively small geographical areas that we have investigated and despite the fact that we have not considered any data relating to the sizes and nomenclature of particular measures, which is a problem outside our scope. Admittedly, in the areas that we have discussed, the French monarchy did not seek standardization as early as Francis I, but only from Louis XIV in Franche-Comté and from Louis XV in Lorraine; nevertheless, it is unlikely that the relationships in question were radically different in the other *départements*, though this is a matter that has yet to be fully investigated.

2. Our evidence points to seigneurial dominance in full swing. In time, however, the disappearance of the seigneur would deprive the local system of measures of its frame of reference; this development, although hardly ever remarked, was another factor making metric reform essential. In no case that we are familiar with did the peasants possess any written titles setting out the relations between the measures binding in their village.

3. While there was an endless variety, among the parishes, in the dimensions and names of measures, in the rights to test them, and in the ways of measuring, nonetheless the metrological usages were—in their very complexity—clearly defined in each parish. One thing was measured *ras*, another, *comble*; there were any number of ways: *ras, comble, demi-comble, grain-sur-bord*, and even *ni comble ni ras*.[107] And it was not just by chance that, on occasion, the point would be made that corn that was being measured was not to be compressed; doubtless, there had been a history of controversy relating to the practice.

4. Finally, we note a certain general rule in the practice of measuring various products: the cheaper they were (among the grains, commonly oats and occasionally barley or rape, too), the more generous the measure, that is, either the measure itself was actually larger, or, more often, measuring with a heap was the normal method. In most of Franche-Comté, for instance, the oats bushel was comparatively large, but it was measured striked; in some parts, however, it was measured *comble, sans se presser*.[108] At the other end of France, in Angoumois the oats measures were relatively large, too.[109] The Great Encyclopaedia tells us indeed that the oats measures were double the size of those used for wheat.[110] We have evidence of large measures being used for oats also from the vicinity of Paris,[111] from Brittany,[112] and from the Loire basin.[113] In the region of Beauvais, the *mina* of oats was, by custom, half as large again, and the *mina* of barley a quarter as large again, as the *mina* of wheat.[114]

Countless disputes resulted in many places from the adoption of diverse yet strictly observed methods of pouring grain into the measure. Indeed, occasionally, a way of measuring *sui generis* came into being; here is one from the Loire region: "There is a special ritual for measuring grains in each market. At Baugé, the *boisseau* of corn is measured striked: when the corn is in the measure, the *boisseau* is twirled round and round; when it is measured poured from a sack, three blows of the hand are made on the side of the *boisseau* in filling and a fourth to shake it down."[115] In the eighteenth century the French had also to cope with the changeover from measuring grain by volume to weighing it. The problem is mentioned in the *cahiers des doléances* from the *bailliage* of Angers,[116] from Angoulême,[117] from the Paris district,[118] etc. In Alsace, the answers to the questionnaire probing attitudes to the change were about equally divided.[119] Ten years later, Geneva (now under

French rule) declared against the reform.[120] The matter remained unresolved prior to and during the Revolution.

Some other innovations proved no easier to carry through. The adoption of the metric system, with its novel unit and new systems of division and nomenclature, entailed also uniformity in methods of measuring—a matter of some significance as far as corn was concerned.[121] Our acquaintance with the process of the first metrication is derived from the questionnaire distributed by the Agence Temporaire des Poids et des Mesures in year III of the Republic,[122] and in analyzing the answers to it, we must bear in mind the circumstances of the time. For it is legitimate to assume that in year II the Notables quite often endeavored to guess the desires of those in power in Paris and to comply as promptly as possible with their suggestions. It is, therefore, scarcely surprising that we do find some positive answers, declaring support for the change-over from measures of capacity to dealing by weight. The dominant tone in the answers, however, is that of distrust mingled with anxiety. Furthermore, we should be mindful of the pressure upon the Notables (implied in the very phrasing of the questions) to be selective in their processing of answers, and of the fact that our evidence covers large districts and not individual parishes.

A few examples will not come amiss. In Lorraine, the district of Blamont answers: "In our view, it will be very difficult in the market to substitute weights of grain for the measures of capacity because of the difficulty of making enough weights and scales available, and also because of the tradition of following the local differences in measures of capacity when selling."[123] Another Lorraine district, Vézelise, states: "In our opinion it is far easier to measure corn than to weigh it. Difficulties and troubles associated with equipping ourselves with weights and scales will prevent the adoption of this method." From Franche-Comté, it is enough to quote the pregnant answer by the district of Baume: "All the parish councils from which we have inquired were convinced—and we share this view—that the replacement of measures by weighing will not fail to cause inconvenience. For one thing, the change will be expensive. We shall have to erect special premises (*des Engards*) in every market, provide ourselves with scales and purchase weights, and appoint and pay a number of persons entitled to act as weighmen; whereas at present everyone is in a position to have a proper measure, provided it be stamped as having been checked in conformity with the standard kept in the marketplace. This is a much more efficient method than going by weight, which may well waste a whole day."

There were many answers like that. Parenthetically moreover, we might note that as late as year III of the Republic even the No-tables—and not just the "common people"—were minded to point out "the prevalence of the custom of selling by local dry measures of capacity," and insisted on stamping measures in accordance with the marketplace standard.

What we have been seeking to demonstrate is that many problems were identical in France and in Poland; in both countries, ulti-mately, it was not so much the law as the needs of commerce that brought in the practice of selling grain by weight. All four general tendencies we have observed in France were also discernible in Poland. First, the differentiation of measures was enormous in Poland, too. Secondly, in the Polish countryside, it was the feudal landowners who exercised full control over measure. Third, in Poland, too, centuries-long strife and contention eventually re-sulted in different, but always firmly defined, metrological practices in each village, and these practices were protected by fears of clashes of interests and by the memories of old feuds; they were, accord-ingly, inflexible and durable, although from time to time overt and covert attempts would be made to alter them. Finally, in Poland too, the practice of measuring with a freer hand the cheaper ar-ticles, notably oats, was not uncommon.

That last issue calls for some further consideration. In every metrological system, the measure abstracts just one of the prop-erties of the objects measured—be it length, weight, or volume. This enables us to compare various objects in one particular respect, while ignoring all others. The "invention" of measures marks a significant step forward for civilization, testifying to a significant advance in social thinking. The advance, however, has to be con-tinued beyond the stage of the emergence of measures; historically, its continuation went on over centuries and even millennia, until it was completed with the adoption, for everyday purposes, of the metric system. Thus, although wheat, rye, and oats all share the property of occupying space and can therefore be measured by measures of volume or capacity, that "common denominator" does not wholly satisfy us. People find it conventional enough to close their eyes to the qualitative differences between, say, wheat and barley, and to treat them metrologically in the same way, but oats are in a different class qualitatively and functionally, and the way out is to apply either a different (larger) measure or a different method of measuring (heaped).

This important distinction merits consideration against a broader canvas, for we have pointed to it in relation to just one commodity, namely, grain, but it applies to many. In the pre-metric era, it was the norm that the apothecary's pound was minuscule, the spice merchant's pound somewhat larger, the butcher's pound larger still, and so on. The *łan* of barren soil in Poland was larger than the *łan* of fertile soil; and the bushel for measuring oats was larger than that used in selling wheat (or else the former would be heaped). The unit of measure, albeit bearing the same name, varies immensely in size with the value of the substance measured. Indeed, in the case of cereal grains, not only would the varieties be differentiated by value, but from one season to another as well: in Burgundy, the measures would be larger or smaller depending on the size of the harvest.[124] We are, therefore, far as yet from the concept of measure being perfectly abstracted as but a single property of the objects measured; the process of attaining that stage is fraught with difficulties and requires great intellectual effort from man in society, in the face of an unequitable social structure. For traditional measures and ways of measuring have always been associated with sectional interests of particular social groups, and among the weak the fear of any metrological change is deeply rooted in agelong experience and transmitted from generation to generation.

The final resolution of the problem of uniformity can only be effected by the adoption of the metric system coming in the wake of wider social reform, as, indeed, the system was first born of the events of the night of 4 August 1789 in France. Hence, in Poland, the constitution passed by the *Seym* in 1764 came to nought in its attempt to "reduce" all feudal dues in kind to the terms of the new, standardized measures. The reform was sabotaged by the lords— and probably by the peasants, too. The conflicting interests were too closely interwoven and overlaid by centuries of distrust, as well as by the age-old and far from groundless fear that every change was necessarily a change for the worse. The fate of this Polish reform affords yet another proof of the proposition that uniformity of measures and of ways of measuring cannot be achieved without a prior Declaration of the Rights of Man and Citizen, without first abolishing feudal rights, and without a well-developed market economy. In today's India the apparent chaos of diverse local grain measures prevails still.[125]

· 8 ·

HOW WAS BREAD MEASURED?

"Give us this day our daily bread. . . ." In medieval and modern Europe bread was not quite so ineluctably the basis of human existence as it may appear. A variety of biscuits, cakes of oats or of oats mixed with barley, and especially gruels and soups seasoned with fats—or soups, anyway—supplied the masses with their basic diet of carbohydrates if, or when, they could not afford bread. Yet those foodstuffs were always looked upon as substitutes, surrogates, *ersatz* bread, in the pejorative sense of these terms. It was bread that was considered the norm, the desirable staple to which man had a right. A good authority made sure that the people had bread. Whether or not the peasant enjoyed his bread depended on God's harvests and on the lords' dues. In towns, however, the plebs were isolated from nature, and bread was associated not with the corn in the fields but with the baker's shop, whose owner, in his turn, would obtain his grain from the wholesaler's granary. When supplies of bread failed, the blame was laid on those bakers and merchants, and ultimately on the municipal authorities who permitted bakers and merchants to raise their prices. The peasant threatened with a lack of bread sought to assuage the Almighty, but urban plebs faced with excessive prices for bread would riot against the magistrates or their overlord, loot the grain stores, and kill bakers.

For centuries, municipal authorities were obsessed with the charge laid upon them to even out fluctuations in the price of bread, and later state authorities followed suit. However, it was no easy task to eliminate or even reduce in the urban market the oscillations in the price of bread, given the large short-run swings, at times season by season, in the general market prices of grains in a precapitalist economy. The most effective policy in this sphere was, arguably, that pursued by the Vatican in relation to the bread supplies of Rome. The administration of the "annona" continuously supplied Roman bakers with grain at a stable price and it demanded that the loaves they baked should continue to be sold at the same weight and the same price of one *baiocco*. For its part, the administration did its best to buy as much corn as possible in years of good harvests and low prices.[1] To apply such a policy, however, very considerable reserves had to be held. In practice, a new way of tackling the problem—originating independently in several areas of Europe, and thus attesting to the social situation itself being responsible for it—provided a partial solution, which in the course of a few centuries became the normal system. The solution lay in neutralizing the fluctuations in the price of corn by adjusting the weight of the loaf.

Here was a system that, arguably, in M. Bogucka's words, "made it possible to conceal most effectively the rising cost of living. Its index, that is, the price, remained the same, and the alterations in weight, being less obvious and less immediately perceptible, gave rise to less bitterness among the townsfolk."[2] So far, so good, but it would be anachronistic to accept this explanation as adequate. There is no "proper" and no "ought to" about a rising cost of living. It reflects increases in the social value of a given product which result from a temporary decrease in the productivity of labor occasioned by a poor harvest. This decrease results in higher prices, and the system we have alluded to [of adjusting the loaf's weight] was but a "trick," perfidiously played on the consumer by the authorities in an attempt to conceal the rise in the cost of living. To express the changes in value by changing prices while the unit of measure stays fixed is the normal practice in industrial societies. It is, indeed, so deeply rooted in our way of thinking that any other resolution of the problem would strike us as an unworthy "trick" or fraud. Yet it is not the only solution; to resolve our problem by altering the amount of product while keeping the price constant is, after all, no less logical. More than that, in societies where the common mode of thinking is permeated by the Thomistic theory of the just price, even in its superficial form, the latter resolution would appear more "natural." It would be a compromise between the theory of invariable price, based on the inherent utility of the substance for sale, and the requirements of the commodity market; nonetheless, it would be a compromise effectively preserving the principle of constant prices.

Our argument is corroborated by the similarity in the treatment of the price and weight of bread, which we have referred to, and of other goods also. Grain containing impurities would, here and there, fetch no less a price than refined grain—but it would be measured with some "top"; again, in corn-exporting areas it would be measured with a larger bushel, its price being the same as for a smaller bushel in corn-importing localities, while the cost of transport and the trader's profit would be covered by the difference in the size of the two (such practices will be discussed in detail later).

72

Price of one last of wheat	Weight of a wheaten roll	Price (pfennigs)	Weight of a wheaten loaf	Price (pfennigs)
18 g.	ca. 0.25 lb.		ca. 0.5 lb.	
14 g.	ca. 0.35 lb.	3.75	ca. 0.58 lb.	3.75
10 g.	ca. 0.43 lb.		ca. 0.66 lb.	
6 g.	ca. 0.54 lb.		ca. 0.75 lb.	

Price of one last of rye	Weight of a rye loaf (white)	Price (pfennigs)	Weight of a rye loaf (wholemeal)	Price (pfennigs)
15 g.	0.75 lb.		1 lb.	
9 g.	1 lb.	3.75	1.25 lb.	3.75
3 g.	1.25 lb.		1.5 lb.	

[g = *grzywna*, a monetary unit of the period—Trans.]

In the particular case that we have just mentioned, the price per bushel would be kept constant, and the invariability of the price of bread, which is here of special interest to us, was no mere trick seeking to deceive the masses, but a manifestation of the economic mentality of the epoch in question.

Let us examine more closely the procedures that were followed when the price and weight of bread were fixed.[3] An "ideal-typical" case is that of bread tariffs promulgated in Gdańsk in 1433.[4] What is ideal here is that the price is to be constant forever; and it is set at 3.75 pfennigs, equally for rolls and bread as well as for wholemeal bread and fine wholemeal bread, for wheat bread and rye bread, and indeed regardless of changes in the level of prices of corn. The municipal council of Gdańsk saw the fluctuations of corn prices, in effect, as an independent variable. Even though the councillors had not read their Labrousse, they well knew from experience that the range of changes in prices was enormous, and greater for the inferior grains (rye) than for the best (wheat): thus, they anticipated oscillations of the order of 1:3 for wheat, and 1:5 for rye.[5] However, on a closer inspection, the striking feature of the table is that the range of oscillations in the weight of loaves which it provides for is considerably less than the range of fluctuations in corn prices. Thus, the expected fluctuations of wheat prices are located within the range of 1:3, but the weight of the wheaten roll of bread is to vary within the range of 1:2.16, and of wheaten bread, only from 1 to 1.5. Correspondingly, the "permitted" range for the price of rye was 1:5, but of white rye bread (fine wholemeal?) only 1:1.66. and of wholemeal bread 1:1.5. The intention was to shift on to the baker's shoulders part of the incidence of rising prices, so that when grain prices did rise, he was allowed to bake and sell smaller loaves, but reduced in size proportionately less than the increasing grain prices would have warranted.

In real life the consequences of such a policy cannot but have varied. If, as seems reasonable, we assume that in principle the sliding scale was so calculated as to leave the baker some profit, however meager, even in the years of rising prices, then his earnings must surely have risen by a huge margin in years of low grain prices. In considering the Gdańsk tariff, let us ascribe to the maximum price of wheat the value of 100, and let us assume the minimum feasible profit for bakers to be 10 percent on the bread baked from it, and therefore arrive at the figure of 110; then, in a year of the lowest prices for grain, valued at 33, the preordained increase in the size of the loaf will require 116 percent more of the raw material, and the earnings from the turnover—as far as wheaten bread is concerned—will drop from 110 to 66. In the case of rye bread (fine wholemeal) the valuation of the raw material will fall to 20 and the turnover to 72, and, consequently, the profit will have risen from 10 to 160 percent of the price of the raw material used! In money units—that is, no longer considering the relation to the price of the raw material—the increase will, of course, be much less, namely from 1 to 3.3 for the wheaten bread. Nevertheless, it will still be a considerable increase, especially bearing in mind that it will occur at a time of falling prices and therefore of rising value for money. The potent incentive to speculate that would result would manifest itself in the hoarding of grain in years of low prices to be baked and sold as bread in years of high prices. The opportunities, however, would be restricted by the difficulties of prognostication, by the perishability of corn and its susceptibility to rats or mice, and by the availability and capacity of stores (which should not be located provocatively for all to see!). Still, the authorities were able to direct the anger of the starving masses against the bakers, who were readily seen as the party guilty of causing high prices by hoarding grain.[6]

Technically, the method of tackling the problem that we have outlined would be associated with a variety of advantages and drawbacks. One indubitable advantage was that it accommodated slight changes in the price; a change of the order of 10 percent could always be expressed by altering the weight of the loaf, whereas on occasion there was no way of accomplishing it monetarily because of insufficient divisibility of the coinage. Yet, on the other hand, the technology of baking has never been equal to the task of weighing as precisely as that. Moreover, the problem would be exacerbated in times of rising prices when the raw material rose in value and the baker's profit waned, it being then in his interest to see that his apprentices took less rather than more dough every time they reached for it. Furthermore, we should be mindful of the fact that the weight of a loaf of bread diminishes rapidly once it is out of the oven and begins to dry. Bread baked overnight will hold its proper weight in the morning, but will be well short of it by the following evening. Since the moral precepts of the time held that the baker was entitled to charge for his bread the amount equal to his costs plus a "just profit," regardless of whether the bread had been baked an hour or ten hours since, there were instances (presumably rare) of regulating the size of loaves by volume rather than by weight. At the beginning of the nineteenth century, Magier saw in the Warsaw town hall surviving wooden standards bearing the words "unus obolus," and "duo oboli."[7] It is, however, likely that keeping check on the correct volume was neither easy nor rigorous.

A tariff of bread prices as "ideal" as the Gdańsk specimen of 1433 must assuredly have been a rarity.[8] The usual practice was to accept differences between wheat and rye bread, between wholemeal and white bread, and between bread and rolls, and to secure stable prices by adjusting the weight for each category as and when necessary. To determine the correct adjustment, test bakings would be arranged, enabling the municipal controllers to ascertain true costs, and there is no shortage of written protocols of such sessions surviving in municipal archives. There is, indeed, much evidence relating to Poland, and S. Hoszowski published a 1591 set of regulations issued by the Lvov councillors,[9] who determined the weight of rye loaves costing one *grosz* and one *szeląg*, respectively, as the price of a *kłoda*** of rye fluctuated from one *złoty* and 10 *grosze* to 8 *złotys* and 50 *grosze*, as well as the weight of the wheaten roll priced at one *grosz* as the price of a *kłoda* of wheat fluctuated from 3 *złote* to 10 *złotys*. Subsequent Lvov tariffs continually regulated the weight of the one-*grosz* loaf, and later, as the value of money depreciated and the general price level rose, of the 3-*grosze* loaf and eventually the 6-*grosze* loaf.[10] Here, we witness in action the practice of maintaining, for as long as possible, an unchanging price despite the long-run diminution of the weight of the standard loaf (with some

* An obsolete dry measure, especially of corn, usually equal to 4 bushels—TRANS.

short-term oscillations), until finally the time came for a sudden huge increase in price. The painful impact would be moderated by the simultaneous increase—albeit, less than proportionate—in the weight of the loaf. The same process is traceable from the evidence of the Cracow[11] or the Warsaw[12] tariffs; in principle, the practice was obligatory in all towns of Poland until the final partition in 1795 and beyond. Eleven tariffs were successively issued in Lvov during the eighteenth century relating to the weight of the one-*grosz* rye loaf and culminating in the "permanent" tariff of 1766, with a sliding scale of weights against fluctutations in the price of a bushel of rye from 4 *złotys* and 12 *złotys*,[13] while in Lublin a similar tariff was published even later, in 1795.[14]

Were the tariffs in question realistic? S. Hoszowski has attempted to prove otherwise by comparing the price of grain with the tariff price of the bread baked from it and adjudging the latter as unrealistic.[15] But Hoszowski assumed—deriving his assumptions from today's practice—that 100 kilograms of grain would yield 60 kilograms of flour, and the baking would produce an additional weight of 40 percent and therefore "one kilogram of grain will not yield even a full kilogram of bread." Yesteryear's practice, however, was different: a feature in a 1787 issue of the "Trade Gazette" (*Gazeta Handlowa*)[16] calculates that, after grinding, the flour will weigh 83 percent of the weight of the grain, and since the additional weight [of the bread] amounts to as much as 50 percent, a hundred pounds of grain will therefore yield not Hoszowski's "almost" 100 pounds of bread, but as much as 124.5 pounds. Quite likely, the flour then was not quite as free from impurities as our flour today, and oven-fresh bread was pretty moist. We have no reason to suppose that the normal proportions as detailed in the "Trade Gazette" diverged disadvantageously from those that would have been found in other towns, and in earlier periods, too; if, in fact, they did so diverge, then this would afford yet another proof of the calculations in the tariffs being less unrealistic than they appear at first glance.[17]

The system of a constant price for bread coupled with a variable weight for the loaf must have accorded well with the preindustrial mentality as well as with the social situation that obtained in urban markets, or else it would hardly have been found throughout Europe.[18] In practice, this surely meant constant control over the weight of bread by market officials, and frequent disputes whenever short weight was found; the control thus cannot have been easy, both because bread would become lighter by the hour as it was growing stale, and because it was technically well-nigh impossible to bake large numbers of loaves of identical weight. Many documents have survived containing complaints of short weight. Thus, the Lvov tariff of 1726 tells us: "Our master bakers have so far baked bread under-sized"; the tariff of 1738 mentions that "the bakers do not sell bread as per weight stated, but please themselves in the matter of loaves which they bake and sell"; in 1750, the tariff notes that the bakers "dare sell rolls well below the proper weight"; and in the tariff of 1765 they are said to "fleece the people shame-lessly . . . through baking undersized loaves."[19] The bakers, for their part, occasionally complained about the tariff being unfair to them,[20] and to decide such cases, a test baking trial was specified to be carried out under supervision.[21] In Cracow, a unique method evolved to allow for unavoidable inequalities in the size of loaves: the baker was obliged to display in his stall a row of ten larger loaves and another row of twenty smaller, and if a check on weight was deemed necessary, then all ten or all twenty would be placed on the scales,[22] reckoning that, in this way, slight, permissible in-accuracies would be cancelled out. We should, finally, note that despite the centuries-long battle waged by all competent municipal organs against the rising price of bread, it nevertheless rose more steeply in the long run than did the price of any other article.[23]

In our civilization, price represents the relation between a vari-able amount of money and an invariable quantity of a commodity. In saying that the price of bread is rising or falling, what we mean is the amount of money we have to offer in exchange for a constant quantity of bread. But this way of thinking is socially neither nec-essary nor universal. History also tells of other ways of expressing the same social phenomenon, and the history of the price of bread affords an excellent example. A very widespread method (almost universal in the feudal epoch) for expressing the price of bread states a variable quantity of it in relation to a constant amount of money, and this is generally acknowledged as a fair practice. *Voi-vodeship* tariffs in Poland regulated the price of bread on that very principle—as indeed the authorities were expected to do by the people. There was a profound justification and an important func-tion underlying this practice. Its ideological basis was St. Thomas's theory of the just price—just in the sense of being invariable, its invariability being dictated above all by its usefulness to man. The practice thus constituted a tolerable compromise between the the-ory of invariable price and the requirements of the commodity market, while preserving as constant the quantity of money paid. Technically, it would seem this method was favored by the frequent lack of small change and the limited divisibility of coinage.

In our view, however, the paramount importance of this system lay in the political sphere. For it made it possible to alter the price of the most basic article of diet in a manner that was not obvious, and therefore less offensive, to the urban plebs, whose wrath was often feared by the bakers' guild as well as by the municipal authorities and their feudal overlords. The people were accustomed to the price of a loaf of bread being one *grosz*—and for one *grosz* they would always have it. Naturally though, the system combining invariable price with an "adjustable" loaf could act as a social safety-valve only within certain limits. When in Barcelona, in 1788–1789, in the wake of a poor harvest, the size of the loaf was too blatantly reduced, the plebs rioted.[24] There were also disturbances "against the small loaf" in Gdańsk in 1561–1562.[25] To be sure, then, the system worked only up to a point. Major increases or decreases in the market supply of corn, or major monetary changes (e.g. in Poland in the 1620s and 1660s) would compel the acquiescence in a new price for the loaf of bread—two *grosze*, then three, and finally as much as six—and the people had to have it inculcated all over again that the new price was there for good. The social techniques of persuasion employed had to be cautious, and the people would be appeased by a simultaneous—albeit less than proportionate—increase in the weight of the loaf. It is thus reasonable to look upon the whole process, within limits, as a safety-valve or a buffer against social reaction to market developments. Where the safety-valve proved ineffective, the anger of the people would be vented upon those they deemed—rightly or wrongly—responsible for the rising prices: upon the bakers and millers, corn merchants and peasants, upon witches, and upon town councillors, too.

The "system" which we have been discussing was, however, no perfidious trick, designed to dupe the common people. It was a logical offspring of the preindustrial mentality and of the current economic theory, if that is not too grand a term. There was no question of the system being imposed by the privileged classes against the exploited masses, for it accorded well with the mentality of both and with the values of the epoch, although to say this is not to deny that it all redounded to the advantage of the "haves." In a class society, however, this applied also to other, sometimes far more important, facets of popular thinking. Insofar as the system did hold sway very widely and over many centuries, this was because it buttressed the existing social relations; indeed, it remained in existence as long as they did.

· 9 ·

STANDARDS AND THE GUARANTEES OF
THEIR IMMUTABILITY

Traditional doctrines required that measures be immutable. Yet there is no immutability in life, for time wears away all things; this is the context of man's eternal struggle with the destructive power of time. Polish ethnographers, however, are familiar with measures made of perishable material. The nomenclature alone, with several terms derived from *kora* (tree bark),[1] bears witness to this. In Silesia, some measures were made of copper, brass, or iron, but others of outer or inner tree bark, wood, woven straw or osiers.[2] Writing in the eighteenth century on the measures used in grain transactions, K. Kluk lists measures made by the coopers of oak, pine, and beech, and measures hollowed out of a single block, made of poplar, willow, alder, or aspen wood.[3] The commissioners of the Crown Referendary time and again experienced difficulties either in seeking to reconstruct in various localities measures that had once been binding but had failed to survive because they were made of perishable material, or in securing—against deliberate deformation by interested parties or by the ravages of time—new measures prepared under their supervision. These measures would be reinforced with iron and would carry the initials of the commissioners, but they could not resist friction, mildew, or rodents, and even in the nineteenth century the police of the Congress Kingdom of Poland had much trouble with these matters. And just as contemporary users deplored the perishability of materials, so, of course, does today's historian.

There were three types of guarantees aimed at securing fixed measures: social control, supervision by the authorities, and religious sanctions.

Social control sought to ensure, primarily, the ready accessibility of standards. Located in public places, in front of the town hall, in the market, or near the stalls, they were to be on display daily to the entire urban populace. For security's sake, they had to be difficult to remove; they were cut in stone, or cast in a heavy metal,[4] or—as in Geneva in the sixteenth century—they were cut in stone and then weighted with copper;[5] and if a standard ell of weight could not be produced that was immovable, then it would be riveted to the wall of the town hall, as was the case at Chełmno. At times, it was the costliness of the material from which the standard was made that served as the guarantee of its unchangeability: according to the diploma for Robestens (Tarn), the standard of the pound of Albi (in 1288!) had to be made of Levantine or Cypriot bronze;[6] certainly no one could afford to tamper with it!

Measures on public display, visible to all and sundry outside the town hall, are known to have existed in Cracow, Kleparz, Kazimierz, Sandomierz, Przemyśl,[7] Pszczyna,[8] Wrocław, and Ołomuniec.[9] Only major towns, however, could afford standards made of durable material. An unusual custom is met with in Silesia, where it was ordained that the largest towns of Wrocław and Głogów were to have in their possession standards of the bushel and its parts, whereas medium sized towns had only to have standards for quarters, and the very small towns, merely for the "little measure" (*miarka*).[10] The manifest accessibility of the standards, and their public display in places where commercial deals were transacted, offered the additional advantage of readily settling disputes. However, with the increasingly bitter struggle about measures, neither the stone of the standard nor the chain of the riveter was of much avail; either or both might disappear to suit the interests of the powerful.[11]

In the feudal era, there were as many forms of guarantees as there were types of authority. Sometimes it was the function of the communal authority to act as the guarantor; at other times, the function of the rural, or—especially—the urban, or the seigneurial, ducal, episcopal, or royal authority. The costly material, the often artistic workmanship of the standard kept by the authority, acted as an additional guarantee of its reliability, for it was then no easy matter to forge it. This accounts for the magnificently ornate standards we can still admire at the Musée du Conservatoire National des Arts et Métiers in Paris—the Carolingian pint, the ell of Philip the Fair, the Nantes *boisseau*[12]—as well as the standard bushel of Bristol kept at the Museum of Science in London.[13] Similar to the practice we have mentioned at Kazimierz, the mayor of St. Emillon, on assuming his office, would receive the borough seal and keys as well as "the iron ell, . . . the tin *pots* and *pintes*, the scales and weights used in the weighing of bread."[14] The immutability of the standard was all the more important when measuring pieces were commonly made of perishable material, and the very efforts undertaken repeatedly to restore them to their true dimensions would often damage them.[15]

The highest form of guarantee—although not necessarily always effective—was of sacral nature: the Jews kept their standards in the Temple, the Romans in the Capitol, and Justinian ordained that they be kept inside the Hagia Sophia.[16] In some cases, the temple where the standard was kept invested the measure with its own name. Thus, at Riga, the measures of "St. John's" and "St. Peter's coexisted."[17] And in Poland, the peasants would sometimes deposit standards in the church, hoping thereby to secure them against the landowner's arbitrary actions.

The introduction of the metric system, within the process of the evolution of the modern functions of the state, meant that the state guarantee was now of prime importance. Accordingly, standards have been preserved by state institutions competent in matters of weights and measures. Indeed, in later tsarist Russia, to ensure their safety, the standards were actually kept in the fortress of Sts. Peter and Paul! As time went on, more scientific methods of safeguarding them appeared. In particular, the standard meter at Sèvres, being the prototype or archstandard of the worldwide metric system, was surrounded by an intricate arrangement of antitheft devices and fire alarms. Yet even this was not enough for the world to enjoy a sense of security. The thought that someday, through an earthquake or a calamitous fire, the world might be "without the meter" was indeed a nightmare. The new regulations, introduced in 1961, have done away with the very concept of "standard." Today, the true or invariable meter is defined as "a length equal to 1,650,763.73 wavelengths of the orange light emitted by the Krypton atom of mass 86 in vacuo" and it is reproducible the world over in any properly equipped scientific laboratory.

· 10 ·

SYSTEMS OF DIVISION AND GROUPING
(MNEMOTECHNICS)

All modern quantitative thinking relies upon the decimal system. It is a system that, to us, appears perfect in its simplicity and ready applicability. Nevertheless, mastering its principles proved extremely difficult for the masses—a question we shall deal with at length in discussing difficulties encountered by the related metric system. As early as the seventeenth century, Leibnitz demonstrated that perfection was not so much a property of the decimal system but rather inhered in the "invention" of zero, and, indeed, that systems no less perfect could be constructed with the numbers eight or twelve at their center. But, for the time being, let us leave this issue aside.

It is common in discussions of these matters to hear the decimal system referred to as that "preferred by human nature," since the uneducated man counts on his fingers, and of these he has ten. This is an ignorant assertion and its advocates are to be pitied. The present author, for one, has twenty fingers [in Polish, the toe is termed "the leg's finger"—TRANS.]. If the other ten—the toes—are so readily forgotten, this is because they are not readily accessible to the modern man going about his business, but they were not at all difficult of access for men walking about barefooted or wearing sandals, and sitting, more often than not, cross-legged. Indeed, for such men, the "other fingers," that is their toes, are more serviceable in counting, since it is possible to count up to ten using one hand. It should not, therefore, surprise us to learn that despite the alleged "naturalness" of the decimal division, it is but infrequently found among simple peoples, whereas the vigesimal division is found among them quite often.[1] After all, Charlemagne did not invent it, even if he did sanction it in his domain.[2]

The major fault of the decimal system lies in the fact that the number 10 divides only by two (and by five, but this is of little practical use). The vigesimal system's basic unit of twenty divides twice by two. In the duodecimal system, the basic number twelve is divisible twice by two, and by three as well. In the sexdecimal system, we cannot divide sixteen by three, but we can divide it by two up to four times. All these systems—namely, the vigesimal, sexdecimal, and duodecimal—enable us to quarter numbers without resorting to fractions. Small wonder, then, that we come across the decimal grouping less often than duodecimal, sexdecimal, vigesimal, or even quadragesimal[3] and sexagesimal;[4] the basic units are very large, hard to apprehend as entities, but marvelously divisible—especially in the sexagesimal system!

Primitive man simplifies his arithmetical tasks by counting on his fingers, up to ten or twenty,[5] or on his knuckles—the latter way, strictly speaking, adds up to fourteen for one hand and to twenty-eight for two, but customarily is reckoned at fifteen and thirty, hence such units as the Polish *mendel* for 15 and *kopa* for 60.[6] Furthermore, primitive man simplifies the work of counting by resorting to well-worn phrases and counting their syllables: thus, Moszyński tells us that in the Ukraine the invocation "hospody pomyłuy"[7] [Lord, have mercy—Trans.] was used for this purpose, yielding six syllables, and with five repetitions adding up to thirty. For, ultimately, what matters most to primitive societies—indeed, the requirement was known in Europe until the early years of the nineteenth century—is dichotomous divisibility. Decimal grouping savors of perfection only to those familiar with the principle of multiplication and division by ten by means of moving the decimal point, and at the beginning of the nineteenth century only a negligible proportion of the European population was so informed. The Commission for Weights and Measures of the Cisalpine Republic put the problem very well in the letter to the Minister of Finance dated 18 *pluviôse*, year IX, when the difficulties of establishing the metric system appeared to be insurmountable: "Every girl and every unlettered tailor know what half a quarter-ell stands for; but we would lay a hundred to one that many professional accountants would be unable to assure you that half a quarter-ell is equal to one hundred and twenty-five thousandths."[8] The system of dichotomous divisions and successive dichotomous multiples constitutes, arguably, a universal phenomenon of the primitive mentality. Certainly, however, it undergoes such modifications in use as are dictated, on the one hand, by the needs of daily business and, on the other, by intercourse between different civilizations.[9]

As far as transactions involving counting are concerned, it would appear that the duodecimal system prevails throughout Europe: the dozen rules, assisted by its divisions and multiples. The unit of twelve dozen, or 144, has its own names, for example, "the large dozen," while the multiple of five relates the dozen to the sexagesimal system.[10] However, for practical purposes, transactions based on counting by the dozen, the *mendel*, the *kopa*, or the Russian *sotnya* did not always have to refer literally to the numerical values of 12, 15, 60, or 100, respectively. In Riga and the entire commercial area of the Hanseatic League there was current, besides the ordinary "hundred" (*sotnya*), the "long hundred" equal to 120;[11] as late as the seventeenth century, the customary "thousand" in that region's trade meant 1200, and the "long thousand" was 2,880.[12] We can safely assume that in practice, in counting by the piece, lengthy negotiations and hagglings would result in differences in quality being compensated for quantitatively; for example, the buyer of a dozen fish, if his dissatisfaction with their quality was accepted, would actually be given by the seller a "dozen" of 13 or 14 fish [cf. the "baker's dozen" in English—Trans.].

Systems of grouping and division are a more basic and durable feature of a given metrological system than are the absolute values of its measures. It would be more feasible to alter the latter than to effect change in the ways of dividing and multiplying that have been used for generations in the mental arithmetic practiced by the common people. It is precisely because the system of divisions and multiples is the fundamental property of every metrological system that we observe the coexistence of different systems of divisions and multiples in areas where different civilizations meet and overlap. In Riga, there were nine different groupings used in the sale of dried and salted fish alone![13] In Riga, too, two different systems of division of the last coexisted alongside each other, namely, the duodecimal and vigesimal;[14] while the *pura* (equivalent of the bushel) would be divided, on some occasions, dichotomously and, on others, into thirds, sixths, twelfths, and so on.[15] The complex art of calculating simultaneously in several different groupings was one of the basic qualifications of "the compleat merchant."[16]

Measures, as we shall see, tend to vary over time: they wax or wane, lengthen or grow shorter, etc. Many historians hold that even if particular measures do so vary, nevertheless systems of measures, with the proper divisions and multiples, do not undergo change[17]— a view that is frequently, but not always, borne out by the facts. We shall refer later to the variability of the barrel or of the Polish bushel or *korzec* reckoned in gallons (*garniec*); the case was no different in France where the increase of the *boisseau* would be expressed in the increasing number of virtually fixed *pots*.[18] Again, in West Africa, the basic medieval unit of dry capacity, the *mudd* ("the *mudd* of the Prophet"), towards the end of the Middle Ages was multiplied by 2, 4, 8, 12(!), 20, and 80.[19]

However, the commonest dichotomous division was of the pure variety. For instance, in Piedmont, the unification of measures at the beginning of the seventeenth century established as the basic unit of dry measure the *emina*, its divisions being $\frac{1}{2}$, $\frac{1}{4}$, $\frac{1}{8}$, $\frac{1}{16}$, $\frac{1}{32}$, $\frac{1}{64}$, and $\frac{1}{128}$.[20] On occasions the "pure" dichotomous division would be slightly modified so as to admit division by three. For example, the questionnaire sent out by the French Academy of Sciences in 1791 elicited the following reply from the town of Lunéville in Lorraine: "To measure dry goods we use the *resal* and its divisions. The *resal* consists of eight units [they apparently skipped the half and the quarter—W. K.] called *bichot*, each of these dividing into six [*sic!*—W. K.] *pots*. The *pot* divides into two *pintes*, the *pinte* into two *chopines*, the *chopine* into two *setiers*, and the *setier* into three [*sic!*—W. K.] *verres*."[21] They thus, eventually, arrived at $\frac{1}{1152}$ of the *resal*! Emboldened by the new Napoleonic intermediate system, the authorities of Geneva in 1812 introduced as lawful measures of capacity the double *boisseau*, one *boisseau*, and then $\frac{1}{2}$, $\frac{1}{4}$, $\frac{1}{8}$ and $\frac{1}{16}$ *boisseau*.[22] Similarly, dichotomous divisions and double multiples prevailed almost everywhere among measures of fluid capacity, linear measures, and surface measures of cultivated land. And the very ease with which this metrology lent itself to mental arithmetic without any writing was to be, among illiterate and innumerate people, one of the chief barriers to the adoption of the metric system.

It is evident that ancient and medieval societies produced works for which a knowledge of mathematics far exceeding that of the common people was requisite, a knowledge that enabled them to ascertain proportions and to grasp spatial relations (witness the structures they erected, notably the temples). In the absence of manuals the relevant technical expertise was structured, remembered, and transmitted by a variety of mnemotechnic devices which we can but admire. Thus, there existed, on the one hand, the "module," or the largest common denominator of all the dimensions occurring in a given building, and, on the other, "the magic square" and "the labyrinth," containing within themselves all the interdependent relations and proportions.[23] Analogous mnemotechnic methods of spatial planning may be found by analyzing medieval town plans.[24] However, it was not enough to carry through the calculation, its outcome had to be remembered; hence, in societies where illiteracy was the norm, mnemotechnic devices proliferated.

In the Polish countryside, two such methods in particular were

employed: the "chalk" used by the innkeeper to mark on a blackboard every dram consumed and the "notch" used by the overseer
to keep a tally as a type of receipt for every day of the *corvée*
completed.* That "chalk" and, indeed, those "notches," are, each
of them, worthy of a monograph. The inns all too often resounded
with furious quarrels about the "chalks"! And the crafty ways of
duping the illiterate were many—for example, the use of a grooved
stick of chalk, which would leave two strokes on the blackboard
after a single stroke of the hand, the customer in his cups only
noticing the latter![25] Elsewhere an autobiographical novel of a son
of the South American *Lumpenproletariat* recollects the high social
standing of the chemist in a suburb of Valparaiso thus: "The dis-
pensing chemist's shop is something quite different from a grocer's
or greengrocer's. . . . The chemist does not, as a rule, weigh whatever he dispenses—not in public, anyhow—and therefore, at least
to all appearances, does not give short measure."[26]

* The Polish word for the overseer in question, *karbowy*, was indeed derived from
karb, meaning "notch" or "tally," and hence, literally, "the tally-man"—TRANS.

·11·

THE MAGNITUDE OF THE MEASURE
AND THE VALUE OF THE
SUBSTANCE MEASURED

The attitude of today's civilized man towards measures reveals a highly developed capacity for abstract quantitative thinking. Of the many features exhibited by every object in a variety of contexts, we abstract one, and consequently, objects qualitatively as diverse as, say, a man's pace, a suit of clothing, a stretch of road, or the height of a tree, acquire a commensurability in our eyes, for we view them from but a single perspective, that of their length. The perfect divisibility and cumulativeness of the metric system enables us to "compare" very great magnitudes, such as the length of the terrestrial meridian, with very small ones, such as the thickness of a sheet of paper.

It took a thousand years to form this abstract-quantitative relation of certain properties of objects. The introduction of the metric system may have presupposed to some extent the development of this way of thinking, and is perhaps more likely to have accelerated it. The metric system not only expressed the transformation of the workings of the social and individual mind but, in its turn, contributed to the process. It would appear that the primitive mind conceived of objects in a synthetic-qualitative manner. Indeed, here, the quality synthesizes all the properties of the object. Conceived of thus, since there is nothing in common between a piece of linen cloth and a stretch of road, or the height of a tree, or the boundary strip dividing two fields, different measures have to be applied to such different objects.

The matter is bound up, to some extent, with the anthropomorphic nature of primitive measures that has already been considered. The finger serves to measure the width of a board; the span, the corner of a house; the elbow, a length of cloth; and a stretch of road is paced out. But the finger, the span, the elbow, and the pace did not themselves have to be of uniform measure (or share a common unit) and were therefore, in practice, unsuited to being integrated into a single system. How much less then, could we envisage the integration into one system of measures of length, of dry and fluid capacity, of weight and area? And yet, it was this

87

very task that the advocates of the metric reform had set themselves. They were to achieve their purpose—measures of time excepted. Nonetheless, and detracting nothing from their achievement, we ought to note that in some respects the reform never spread beyond the walls of the laboratory, for, even with our modern habit of abstract-quantitative thinking, we do not for the purposes of every-day practice "feel it in our bones" that there is a true bond between the meter and the kilogram. Even as far as objects of like nature are concerned, ones to which apparently the same unit of meas-urement, such as the pound or the bushel, may readily be applied, a close interdependence between the measure and the quality of the substance measured is manifest. That the unit of measurement bears the same name (pound, bushel) is here of some importance. It bears witness to a certain association in the social consciousness among the various objects whose particular common feature is being measured. And yet the pressure of qualitative thinking is so forceful that despite this association there are different measures for objects that differ in quality. To put it in a nutshell, the more valuable the object, the finer the measure employed in its meas-urement.

Naturally, the determinants of this procedure are partly practical. Medicines are purchased in smaller quantities than, say, flour. And since the pound is less perfectly divisible than the meter—with its hundredths and its thousandths having names of their own—the pharmacist's smaller pound is simply more convenient in use. But practical considerations afford only a partial explanation of this phenomenon. What we also witness here is yet another manifes-tation of the synthetic-qualitative mode of thinking about objects and of the mental difficulty of abstracting a particular feature there-from, a feature not necessarily of special importance to us: weight, or capacity, or length. Our emotive, "feeling" attitude to the object centers upon its "value" for man. The range of such "values" is enormous; we decrease it when a more valuable object is measured with a smaller measure. This practice applies to almost all objects that we measure.

We have already remarked that "cheaper" grains are measured with a larger measure—or, at any rate, are measured "more gen-erously." In the province of Ossola in Piedmont in 1826, when many attempts at standardization of measures had already taken place, and shortly after the first unsuccessful attempt to introduce the metric system, the standard weight in use was the pound (*libra*), and the standard measure of length, the ell (*braccio*). However, the pound used in the weighing of sugar, coffee, and groceries was one of twelve Milanese ounces; the pound applied to candles had fourteen such ounces; good-quality meat and cheese for retail sale were weighed by a 32-ounce pound; wholesale cheese (*sic!*) by the 36-ounce pound; silk materials were measured by an ell of 10½ Milanese inches, and other materials by an ell of 13½.[1] In medieval Brussels, salt used to be weighed by the standard of weight used in the grain trade, but halved.[2] Geneva's market allowed the co-existence of pounds of 18, 16, and 15 ounces (the last one applied to the weighing of silk).[3] In Troyes in 1538 the fluid measure was the *queue* equal to 45 *septiers*, but when oil was being measured, it consisted of 41 *septiers*.[4] Zeeland produced better-quality flax than did Brabant; consequently, in the markets of the Netherlands the former was measured by a smaller stone (this diminished the difference in price between "the stone" of one and "the stone" of the other), and this, in turn, or so the Zeeland merchants envisaged, softened the purchasers' resistance to the superior but costlier product.[5]

We have evidence from Latvia,[6] Poland,[7] and Russia[8] of the application of different measures to different wares, while in India, too, relatively precious objects are assessed by finer measures.[9] In Portuguese Angola, even today there are four varieties of the *saco*: the 61-kilogram for coffee, the 71-kilogram for rice, the 90-kilogram in the weighing of beans, and the 90-to-95-kilogram one when maize is traded.[10] Again, the more valuable the commodity in question, the finer the measure. We could multiply such examples endlessly.

We are, then, confronting here an extremely widespread practice, and this virtual universality entitles us to see in it a deep-rooted trait of the primitive mentality. To measure is to isolate a certain quantitative feature of the object measured—whatever its quality. But to the primitive mind, measure is something derived from the quality of the object or, at least, intimately associated with that quality—hence the need for a different measure for virtually every separate object, with none of the measures reducible to the others. Clearly, with the development of the commodity market, merchants would take advantage of that particular trait of the primitive mind in order to mask—one is tempted to say, to conceal—the differential range of prices that might otherwise discourage buyers from buying superior and therefore costlier articles.

· 12 ·

THE HISTORY OF HISTORICAL
METROLOGY

The history of the studies of past weights and measures is in itself
an interesting chapter of the history of historiography—one that
is not without some practical as well as ideological features. The
earliest large group of such studies was part and parcel of the
Renaissance burgeoning of textual criticism of the writings of antiquity,[1] in particular of the Bible. Thus, endless toil was expended
in attempts to ascertain the true weight of the pillars of Solomon's
temple! And countless uncommonly hirsute young men were forbidden to cut their hair for a year in order to establish the weight
of maximum annual growth of hair, since, given the statement that
the once-a-year cuttings of Absalom's hair weighed 200 shekels "of
the royal weight," this seemed to be the surest way to ascertain the
weight of the Israelite shekel! As late as the 1760s the entries in
Diderot's Encyclopaedia relating to weights and measures devoted
the most space to considerations of this very type.[2]

Another very large group of learned historical metrological studies, traceable, it would seem, to the twelfth century, belongs to
medical literature.[3] Traditional medieval medicine (and modern
European medicine to a large extent too) was derived from ancient
sources: Roman and especially Greek. But because this stock of
medical knowledge was transmitted to Europe by way of Arab
intermediaries, myriad problems arose from the translation and
retranslation. Interested parties had to engage in critical analysis
of the text in order to recover its original version. Establishing the
nomenclature of the substances in question (plants, minerals, etc.)
was relatively easy; far greater difficulties had to be surmounted
in establishing the dosages, that is, weights and measures, of med-
icaments. The studies of historical metrology in aid of medicine
thus proceeded on the assumption that "the ancients had known,"
"the ancients had been able to cope," better than the students' own
contemporaries; all that was needed was proper understanding of
Galen and, in order to gain it, a knowledge of weights and measures
used in the days of the Emperor Trajan.

The same assumption underpinned other historical investigations in metrology, too. Agronomic studies are a case in point. A

certain otherwise little-known Paucton, active in France in the second half of the eighteenth century, having embarked upon the study of ancient agronomy, found metrological difficulties in his path. In order to clear the ground for his agronomic disquisitions, he accordingly determined first to come to terms with ancient weights and measures. By 1780, he published a large work devoted to them,[4] but, sadly, it would seem that his life proved too short for him to avail himself of the fruits of his metrological labors by carrying out the larger task he had first envisaged. The conviction that the knowledge of the ancients was superior hardly testified to the progressiveness of scholarship in the second half of the eighteenth century in France. Nevertheless, the studies motivated by that conviction did contribute to the growth of learning.

Yet another, third group of early metrological studies was concerned with the social struggles of the final period of moribund feudalism. In France, these studies formed an integral part of the process we refer to as "the seigneurial reaction." What was crucially involved was a "return to the sources," involving a search in old diplomas and other documents for rights that the still-extant system made it possible to restore. Ideologically, this was a convenient method, since everybody wanted a "return to the sources," both the peasants, who were convinced (see below) that the "olden" burdens had been lighter, and the landlords who reckoned that they would be able—by fair means or foul—to prove the opposite. The resolution of the question was technically difficult, but the means at the disposal of the landlords were superior. Their castles now saw a new type of employee at work—the *feudistes*, men skilled in paleography, able to decipher old documents and sufficiently cognizant of law for their investigations to yield, each and every time, a verdict favorable to their employers. Among them—horror of horrors—was Gracchus Babeuf! He learned a good deal about feudal laws in this manner,[5] and his knowledge would one day be put to a scarcely foreseeable use.

The *feudistes* had two courses open to them: either to ferret out in the old documents rights that had fallen into oblivion, for example, an absolute right to exact corn-grinding dues, or a disused toll; or to make out a case for increasing the dues that were still being enforced. If the latter way was tried, then there was clearly a temptation to demonstrate that the measures, which had formerly been used to assess the dues, had been larger than those currently in use. This, too, led to historical metrological studies, albeit tendentious in their very conception.

To sum up, all three types of early historical metrological studies related to matters of consequence. Those concerned with Biblical criticism were, ultimately, about the freedom of thought. Errors in those concerned with medical issues would cost many a human life. And mistakes—more often than not, deliberate—in the historical metrological studies undertaken by the *feudistes* would be paid for in human toil and tears.

In Poland, as we shall later discuss in detail, the beginnings of historical metrology may be credited to certain late eighteenth-century officials, whose labors were patently oriented towards the practical needs of the judiciary, although credit is due to them for working out some "research" procedures of sorts. They did not publish the results, since they had no scientific objectives nor any desire to contribute to the body of knowledge. The practical need for investigations of this type derived from the class struggle of the period, with the peasants, on the one hand, disputing the assessments of their feudal dues, and the gentry, on the other, seeking to increase them by enlarging the measures. Both sides, however, in keeping with the then prevalent doctrine, accepted that the assessment ought to be "as it had been in old times." The peasants would seldom hold any documents and, when in dispute, would rely upon the testimony of old men; the gentry, however, would invariably produce documents. But, since these were not always easy to understand, not only the officials but the landlords themselves had to study historical metrology—if one may use the term. That this was indeed the case, is supported by a passage from a 1790 calendar;[6] under the heading *The Description of Measures used to Calculate Łans and Vlokas in our Country*, we read as follows:

> To be knowledgeable about measures used in surveying in our country is both useful and entertaining for the citizen. It is useful, since even a little knowledge enables him more readily to track down past transactions involving measures, to comprehend them, to divide the land in question into correct units of area, to assess dues thereon accurately, to compare past measures with today's, and easily to cope with all manner of related problems in the absence of a trained surveyor. It is entertaining because, in conversations upon statistical matters, time and again the talk turns to the measurements and dimensions of the *łan, vloka, morg,* and suchlike.[7]

From this, it would seem that matters of historical metrology would come under consideration in drawing-room conversation. However, this was still some way short of scientific treatment.

The scientific pioneer in Poland in this field apparently was Łoyko.[8] His researches were carried out according to all the rules of the historical method as then understood. The materials he collected, although of outstanding value, have not so far been published, although numerous students have had recourse to them. Łoyko's labors were continued by Tadeusz Czacki; measures were but one of his numerous interests in the institutions of the Polish state in former ages, a past that in his view was truly dead and gone, and of which he saw himself the inheritor, with a mission to cherish its memory for transmission to succeeding generations.[9] At the same time as Czacki's work, a group of studies appeared whose motivation was far different. These were the attempts to translate Polish measures into the newly introduced French metric ones, in particular the studies undertaken at Staszic's instigation by the Society of the Friends of Learning, aimed at working out a system of so-called "new Polish" measures.[10] The authors of those studies were A. Sapieha,[11] A. Chodkiewicz,[12] and J. Kolberg.[13]

The next wave of studies surged up in the second half of the nineteenth century and included works by W. A. Maciejowski,[14] J. T. Lubomirski,[15] M. Baraniecki,[16] and F. Piekosiński.[17] The opening years of the twentieth century saw a major outpouring of studies in the history of land settlements (by F. Bujak, K. Potkański, K. Tymieniecki, R. Grodecki, and, later, S. Arnold, K. Dobrowolski, and others), which paid most attention to the problems of the very earliest measures of land. Other fields of metrology as well as later historical periods were left almost uncultivated and not until the 1930s were studies in these areas published in any numbers. Among these, three pioneering works by E. Stamm were outstanding.[18] His expertise in theoretical metrology and mathematics enabled him to evolve certain methods and to point to a number of associations. Stamm, however, in our opinion, employs a formalistic method, which seeks to establish the metric equivalents of old measures notwithstanding the inadequate evidence from the available data; regional differentiation of measures is far from satisfactorily accounted for; and he fails to perceive, in all their vast entirety, the social determination and the social meaning hidden in the problems of historical metrology. Finally, a large and valuable contribution to the history of ancient measures was made in that period by all the students of the history of prices belonging to Bujak's school.

·13·

HISTORICAL METROLOGY
AS A BRANCH OF THE
STUDY OF HISTORY

Historical metrology is concerned with past systems of measurement. This definition, in which the emphasis is on the term "system," postulates that in our investigations we take into account all the elements associated with measuring: systems of counting, instruments of counting, methods of using these instruments (we have already noted that the methods may often be more important than the dimensions of the measuring instruments), the different methods of measuring in different social situations, and finally, the entire associated complex of interlinked, varied, and often conflicting social interests. Our definition incorporates also the conviction that all those elements combine into an internally articulated structured whole, and thus into a system. It is the task of science to investigate this system, and to locate it within the social totality that has produced it and within whose framework it functions.

The sources available to the student of historical metrology are numerous indeed and, perhaps more significantly, greatly varied. The symbolic role of the "just measure" (particularly of weight, an attribute of the Egyptian god Amon) has been responsible for extremely interesting and not infrequently beautiful iconographic conceptions bequeathed to us by various societies and epochs. Such works of art furnish us with the richest documentation to help us in seeking an understanding of measuring instruments and, occasionally, of the ways in which they were used. For instance, in many depictions of the Last Judgment, on the portals of Romanesque and Gothic cathedrals, the scales are commonly wielded by an angel and held in a particular manner. There is less evidence to aid our cognition of the measures of area; yet here, too, the significance attached to the activity produced valuable iconographic sources.[1] The practical importance and the difficulty of measuring area (fields, notably) gave rise to manuals of geometry, which abound in descriptions of considerable relevance to historical metrology.[2]

Next to the iconographic sources, there are various material relics. Ethnographic museums throughout Europe hold an abundance of them,[3] although catalogues or photographs are hard to come by and thus the wealth of evidence of this type is generally left unutilized, and the dating of the objects in question is not always wholly convincing. Central national offices of weights and measures sometimes house small specialized museums.[4] But the preservation of large numbers of metrological relics has tended to be defeated by the simple fact that most of them were not made of durable materials. This also, as indeed we have noted, facilitated metrological deception, and laws were passed that certain standards be reinforced with iron. Standards and measures preserved in their "original state" in specialized museum collections,[5] as well as those made of stone and still surviving in old market places or on the walls of guildhalls, are obviously priceless for the historian. However, there are not many of them to be found in museums—and among such as have survived there are hardly any but those bequeathed by towns and, in particular, by larger cities. This is hardly surprising because repeated attempts over the centuries to bring about uniformity of weights and measures, including metrication, have invariably been accompanied by demands that the old standards— no longer binding—be destroyed; indeed, not only would keeping them be thereafter deemed a punishable offense, but, if found, they would be broken up, deformed, chopped up if wooden, or melted down for scrap[6] (at times, the scrap was accepted in payment for the new standards)—all as a matter of duty incumbent upon local militiamen or state police. We do not find many specimens of local *boisseaux* in provincial museums in France, not only because they lost their value once new measures were introduced, but also because of the super-efficiency of the police forces under the successive régimes from the Jacobins to the restored monarchy.

But our researches relating to former measures do not rely only on the original surviving specimens as sources. For some purposes, all manner of objects, all sorts of folk artefacts, count as well. The very dimensions of the preserved objects inform us sometimes of the popular standards of measure, assuming, of course—and the assumption does not seem too bold—that the common people and the craftsmen among them, relied upon memory in matters of reckoning and kept clear of fractions. In this sense, historical metrology can use as a "source," old buildings,[7] the width of cloth pieces, the dimensions of bricks,[8] and so forth. If the three dimensions of a brick are commensurate, its width and length presenting simple fractionless multiples of its height, and if, moreover, the width of cloth goods is found also to be a multiple of the dimension of our brick—we may feel sure that this is no coincidence but a real-life folk standard of measure, operative in the particular community.

Next, and naturally of much later vintage, come manuscript sources, followed by printed manuals. Traders' manuals are known from the early medieval period (before print), and later, school textbooks, too, made their appearance.[9] The former tend to be more important for us, since they sometimes emulate one another in listing the weights and measures of the greatest possible number of countries, but their reliability poses a difficult problem. The forerunners of this type of compilation were probably Epiphanius of Cyprus in the fourth century A.D., or the Armenian mathematician Ananias of Shiraz in the seventh century A.D.[10] Of foremost importance here in the High Middle Ages were, of course, the Italian cities, whose commercial interests extended so widely and involved countries of such diverse metrological systems that no trader could memorize them all, nor could his assistant—whether son or apprentice—learn enough of them from the tales of men with firsthand experience. So the need for manuals became clear, and they duly appeared—first handwritten, then printed. Francesco Pegolotti Balducci's *Libro di divisamento di pesi e di misure per il mercatanti*, written *ca.* 1330, Giovanni da Uzzano's *Libro di gabelle, pesi e misure di più e diversi luoghi*, written *ca.* 1440,[11] the four volumes of *Libro di mercanzie e misure* by Francesco Dina, published in Florence in 1481, Luca da Borgo's *Summa di aritmetica*, 1494, or Bartolomeo Pasi's *Tariffa dei pesi e misure*, published in Venice in 1540, may be cited as but a selection. With some delay, at the end of the sixteenth century, an analogous volume appeared to serve Russian commerce, its concern being with the commodities and, accordingly, the measures of the south, of China, and of the Baltic countries: this was the famous *Targovuyu knigu.*[12]

It was not only traders who needed such manuals. They were also in demand from geodesists and land-surveyors, whose tasks multiplied and grew in complexity beyond the accustomed "carry-over" of traditions, after the custom of the guilds. In 1566, there appeared in Cracow *Geometry, that is the Science of Surveying* by Stanisław Grzepski.[13] A century later, in 1665, a parallel work was published at Reval [now Tallin—TRANS.] by J. Schelenius.

Finally, there is the rich and diverse category of manuscript sources, although to consider them fully is outside our scope here.[14] It includes decrees of municipal authorities, business account books, files of municipal court cases relating to metrological transgressions, as well as account books of large feudal estates of both lay landowners and churchmen. Perhaps we may add here that an exceptionally valuable source in many countries, with local variations, are records associated with the introduction of the metric system. This reform generally demanded that inventories be made of existing traditional systems and that their units be translated into metric units with the aid of modern methods; most significantly the reform would more often than not face all manner of social opposition. All this could not but produce quantities of source materials for the modern historian.

In the preceding chapters we have attempted to demonstrate the significance of sources emanating from the courts of law. As far as rural metrological problems are concerned, judicial sources are meaningful to us only insofar as the peasants enjoyed access to state courts. We know, however, that in Poland from the beginning of the sixteenth century, regrettably they did not, which was as unfortunate for them as it is for today's historian; hence the importance for Poland of the archives of the Referendary's Court. Admittedly this court dealt only with a small proportion of the peasant population, but we may well assume that many of the practices its files reveal were not confined to the royal domains. Hence, also, as far as Polish history is concerned, the relatively greater importance in this area of the archives left by the large estates, especially the latifundia, whereon the patrimonial courts of law were relatively well organized. Hence, also, the source value of the petitions and of the records generated by the proceedings they entailed.

The historian has, of course, no direct evidence on which to assess the justness of the peasants' grievances, but there is little doubt that the great majority were justified. That this was so is borne out, first, by the sheer mass of complaints; second, by the frequent wealth of detail in the descriptions of the methods employed to take advantage of the peasants; and, third, by the fact that the verdicts of the royal courts or the decisions of the landowners relatively often pronounced in favor of the peasants, for all that their interpretations would not side with the peasants and would be very sympathetic towards the *szlachta* tenants or the stewards.

As ever, in historical enquiries, given that the provenance and the survival of the primary sources have been determined by a multitude of complex processes, it goes without saying that the student has to face the problem of how to translate what he is able to establish from the "sample under investigation" into what is of chief interest to us, namely, a generalization referring to the whole "population." In the present instance, the "population" consists of the royal domains and the magnates' latifundia, our frame of reference being the totality of the world of the Polish village. What went on on the estates of the middling and minor nobles who administered their estates by themselves, where peasants wrote no petitions but merely came along to lodge their complaints verbally and enjoyed no access to the state courts—we do not know from extant evidence. Yet, since those landowners belonged to the same social stratum as the tenants and stewards of the royal domains or the magnates' estates, and given that they administered their villages in the context of similar social conditions and employed the same methods, we risk little in assuming that on the smaller estates, too, metrological abuses were the order of the day. Quite likely, they occurred more frequently since there the peasants were in a weaker position and less able to defend their interests. Now, considering that selected documents of the Referendary's Court, gleaned from a relatively short period and referring, in truth, only to a single wave of grievances from the estates of the primate, have yielded hundreds of peasants' complaints about metrological abuses, how many more might there not have been continuously taking place all over Poland?

The importance of historical metrology as an auxiliary science of history is obvious and manifold. Its foremost function is to ascertain the origin of documents (date them, establish their geographical and social provenance, detect forgeries, etc.); this function is strongly emphasized by L. W. Čerepnin in his manual of Russian metrology.[15] However, it is in the analysis of the content of documents that historical metrology is of incomparably wider applicability, for endowments or descriptions of sites can be quite incomprehensible without a degree of understanding of the measures involved; similarly, if we do not know the measures in question, we are unable to estimate peasants' dues listed on the old inventories or *lustracje*. Again, if we lack relevant metrological expertise, then the records of customs offices or of estate archives, as well as testaments and contracts of bygone days, can tell us nothing. Indeed, the same can be said for all the basic categories of historical primary sources. It is our opinion, however, that in order to serve us well as an auxiliary science of history, historical metrology must be treated as a branch of historical science in its own right, as should every auxiliary science of history.

Pre-metric measures, precisely because they are representational rather than conventional, precisely because they are an attribute of power and an instrument of asserting class privilege, precisely because they are, time and again, at the center of bitter class struggle,[16] are replete with important, concrete social meaning, the uncovering of which should become the chief task of historical metrology. This will not be achieved if its aims are narrowly restricted in the traditional manner as being "to ascertain precisely the terminology of former measures, to reconstruct the system of measurement, and to calculate the values of the measures of yesteryear, as well as to translate them into the units in use today."[17] For this conception of the scope of historical metrology has, on the one hand, deprived it of the opportunities of tackling problems of the greatest scientific interest and, on the other, has led on occasion to skepticism and cognitive pessimism among its students and, still more, among historians wishing to utilize the data from historical metrology. To convert oldtime measures into the units of the metric system is often, in fact, not a feasible task, and results of such attempts, however painstaking, are often of little practical use, because even the most meticulous determination of the dimensions of, say, the *łan* could not be extensively utilized when even neighboring villages in the same year, more often than not, would have *łans* of different sizes. The skepticism and the cognitive pessimism were therefore quite often by no means groundless.

Yet, when the historian succeeds in uncovering the social import of a given measure, although this may not tell him much of what he wants to know then (such as the correct metric equivalent), it may offer him an opening leading to many other, possibly more important, matters. If we should establish that the *łan* in the sixteenth century, albeit showing enormous variations in its surface area, does stand for a unit of the manorial economy based on the unpaid labor of serfs and full allotment of feudal dues, then this realization offers us new analytical openings.[18] Let us take, as an example, the statistics of agrarian structure. If the *tertium comparationis*, upon which we base our classification of the population in question, is the geometrical area of the land owned, then, quite patently, the skeptical voices warning us against the statistical analysis of the stratification of the rural population of Poland in the sixteenth to eighteenth centuries, using the methodology of Rutkowski and his school, are right. However, if we appreciate the full social meaning of the term *łan*, then the classification into groups of *łan*, half-*łan*, and quarter-*łan* peasants assumes a new significance, and statistical treatment based upon it assumes an altogether deeper meaning, more adequate for the understanding of the problem in hand. Naturally, both the student of metrology and the student utilizing his findings must realize that any geometrical commensurability is out of the question. Nonetheless, an understanding of the true meaning of the figures we find in statistical descriptions is a complex problem; its lack has given rise to many misconceptions, and in the light of our discussion so far, a great many problems in this area remain still to be investigated.

Since in texts as early as the Bible, we can perceive the transformation of the realistic conception of measures into the symbolic, we would wish better to understand the process and social conditions in which the idea of "just measure" becomes a symbol of "just man," of justice as such, and of just human relations. We would wish to know fully of when and how the authority of the state and other authorities gained competence in the control of measures, as well as to understand the struggle typically fought over those competences. Next, we would wish better to apprehend the role of measurement in market relations, or in barter transactions where a commodity market, reliant upon money, is not yet fully established; the studies by social anthropologists, and by economists specializing in the problems of the so-called underdeveloped countries, may well be relevant and helpful here. Also, we would wish better to understand the conflicts among the social trends we have drawn attention to, for example, the situation when the tendency in measures to persist clashes with their representational character and—as the commodity market develops—with the tendency of prices to stay fixed, and now one, now the other, tendency wins through (as in the case of bread referred to above). Again, we would wish to understand better the conflict between the tendency of prices to remain unchanged and the same tendency in measures, especially in cases where fixed prices gain the day and are used to obscure the discrepancies between the buyer's price and the seller's price (or between the prices in the exporting and importing regions, or between the wholesale and the retail prices). Furthermore, we would wish to comprehend better the functions of measures and methods of measurement in credit transactions, particularly as used to conceal interest on loans in medieval Europe—where it was banned by canon law—as well as in other periods and regions, including many of today's underdeveloped countries (the canon-law ban was a particular form of a phenomenon typical of pre-market economies). And, finally, we would desire to raise the level of our comprehension of the social conditions that are indispensable for metrological standardization; nothing is more illuminating

under this heading than the analysis of the many attempts at standardization that ended in failure, including analysis of the social
resistance to them, and the study of instances where the old practices somehow survived, notwithstanding the formally completed
standardization.

Every measure as a social institution is an expression of a particular configuration of human relations and may well throw light
upon these relations. Associations between measures, their diffu-
sion, and the peregrinations of their names serve to acquaint us
with the cultural links between nations and civilizations.[19] The increasing standardization of measures through time is an excellent
indicator of one of the most powerful, if, indeed, not the most
powerful, historic processes—the process of the waxing unity of
mankind. And, "metrological studies, so unyielding at first sight,
when handled by an intelligent student, become an instrument of
historical enquiry capable of bringing to light the major currents
of civilization."[20]

· 14 ·

THE FUNCTIONS OF MEASURES IN
THE PRE-CAPITALIST COMMODITY
AND CREDIT MARKETS

To repeat: modern man conceives of price as the relation between an amount of money and a quantity of a commodity, the former being variable and the latter fixed. Thus, if the price of bread goes up, or down, then to us this means that the amount of money we have to pay for a constant quantity of bread increases or decreases. Yet the assumption that a change in the market situation will be reflected in a change in the amount of money to be paid for a given measure of some commodity is socially neither necessary nor universal. Changes in the market situation have been known to express themselves in other ways, too. An obvious example was the method, widely used in the feudal epoch, of regulating the price of bread by altering the weight of the loaf rather than its price when conditions in the market changed (see also chap. 8, above). This practice had a profound justification and performed an important social function. St. Thomas provided the ideological justification in his theory of the "just price." Technically, the practice was apparently relatively easier to effect in view of the frequent lack of sufficiently small denominations of coinage (it was relatively easy to alter the size of a one-*grosz* loaf, but to alter the price by ten percent in the common coin might often be virtually impossible). However, in my view, it was the political function of such a system that was of special importance. For it enabled men to alter the price of the basic food in a manner that was less obvious, and therefore less offensive, to the urban populace, whose reaction was often feared by the bakers' guilds and the municipal authorities, as well as by feudal overlords. The populace was accustomed to paying one *grosz* for a loaf of bread; so be it, let it go on having it for one *grosz*.

Certainly, however, this practice could not continue indefinitely. Major changes in the supplies of corn, or in monetary relations (as happened in Poland in the mid-seventeenth century), necessitated the acceptance of increases in the price of a loaf of bread to two, three, or even six *grosze*, the community having thereafter to become accustomed all over again to the new price as immutable. When sizable alterations were made in the weight, the purchasers

102

refused to have "the wool pulled over their eyes." Within certain limits, however, the method we have discussed would serve as a safety-valve for the social reaction to market events, which was precisely what was wanted. The same method would be employed—if not by decree, then spontaneously—by sellers of many articles, notably those sold in certain specific traditional units, such as the *glon* of cheese or the *osełka* of butter, etc.*

The fact that the value system of the feudal society by no means insists that measures be immutable is confirmed also by, for example, the widespread view that it is legitimate for the tradesman to employ one measure (or, at any rate, one method of measurement) when buying and another when selling his wares. This is bound up with the conviction that the price of a commodity, being its inherent characteristic, as it were, may not be altered by man other than sinfully. Thus, the merchant paid for the bushel of corn the same price he asked for it when selling, save that in buying, the bushel would be heaped or "topped up," and in selling, striked or levelled. His profit lay in the "top"[1]—and it was no small profit.

Some examples we have encountered are even more striking. Sometimes, in buying and selling respectively, not just two different methods of using the same measure but indeed two different measures could be used. A traditional institution that had been for centuries regarded as normal, functional, and "ethical" would, if it outlived its usefulness, come to be seen as an abuse. Thus it happened, for instance, that as late as 1815, in the St. Étienne district of Toulouse, it was found that there were still two ways of measuring—the "measure of entry" to, and the "measure of exit" from, the stores; the retailer involved would thereby lose four hectoliters for every hundred *setiers* of corn.[2] Similarly, in Russia cases have been recorded of applying different measures for "giving out" and for "taking in" grain or, at any rate, of collecting it heaped and distributing it raked down.[3] This was common practice both in transactions between gentry and peasants and in those between merchants and peasants. And A. Gilewicz notes that, among early medieval Slavonic measures, the "commercial pound" and the "dealer's pound" were different—the latter, of course, being smaller.[4] This distinction is of considerable importance, for its deeper, latent meaning is the distinction between wholesale and retail trade, albeit expressed in terms of different measures rather than different prices.

* Cf. traditional English units of the clove of cheese or the roll of butter—TRANS.

103

From the Netherlands comes evidence of yet another interesting way of guaranteeing the shares of the wholesaler on the one hand and of the retailer on the other: the standards of large measures were larger than the corresponding sums of smaller measures. The wholesaler's profit subsisted in the discrepancy.[5] Changing the measure or the method of measurement might also maintain an apparent adherence to the principle of interest-free loans. This practice was especially common in the dealings between the manor house and the village: the lord would lend the peasants grain measured "striked" and have it returned "heaped."[6] Again, sometimes the principle of invariable prices would be upheld by altering the measure to cope with seasonal oscillations. Thus, in West Africa, corn would be measured one way after the harvest and another during the period preceding the new harvest.[7] We know also of cases of applying a larger measure to articles that were expected to lose weight as a result of drying (spices),[8] or to corn if sold fresh. The immutability of prices would be upheld on occasion by using different measures in the place of production and the place of consumption. Not infrequently, the price would be the same in both places, but it was usual to employ in the former a larger unit of measure. The difference between the two measures covered the cost of transport and the trader's profit.[9]

The importance of the metric reform in Poland lay in the fact that it unequivocally introduced but one measure and one method of measuring (this was provided for already by the statute of 1764). The measure was by then a convention, binding equally upon the lord and the peasant, upon the buyer and the seller, the supplier and the consumer, the creditor and debtor, wholesaler and retailer. In order to appreciate the revolutionary significance of this innovation, we must bear in mind that over the centuries of earlier history man's views of these matters had been far different.

Commercial ties, however, at times bring about standardization of measures in areas traditionally connected by exchange of commodities. The producer has to satisfy his buyers and this may lead him to adopt their units of measure. At the Wieliczka salt mines in Poland, blocks of salt of different sizes were produced for different buyers.[10] At Riga, one measure was used in trade with Sweden and the Hanseatic towns, another with Finland, another with Kiev, and yet another with Muscovy.[11] Just whose measures would be adopted presumably depended on the relative bargaining strength of the partners to the exchange.[12] Besides the commercial links, which affirmed the diversity of measures, there were also ties that encouraged their standardization.[13] But this side of the dialectical process of the transformation of measures is known so well that it need not occupy us here. The system of commerce in grain availing itself of differential measures has evolved over many centuries, and it has functioned effectively. Those historians who discern in the metrological "chaos" a hindrance to the development of internal trade are mistaken. There were indeed numerous obstacles—for example, poor condition of the roads, weakly constructed carriages, technical problems of storage, etc.—but the diversity of measures was not one of them. The contrary was the case,[14] and people at the time were fully aware of this and could not imagine things to be otherwise.

The Assemblé d'Élection at Sens in 1788, on being asked its attitude towards the standardization of measures, answered: "The differences noted everywhere between measures have proved to be a factor of prime importance in stimulating grain trade between towns and between provinces. Many merchants (*négociants*), and particularly those referred to as *blatiers*, choose to ply their trade solely on account of the differences in measures. . . . The differences are such that they defray transport costs and even the charges for use of scales. Hence, to establish a single system of uniform measures would wipe out this type of trade, ruining in the process countless small markets, which owe their existence to those differences, and which—while being of no great importance—satisfy the needs of the local consumers."[15] That the differences in measures, in case of standardization, might equally well find an outlet in differential prices, simply did not occur at all to the good citizens of Sens as late as 1788.

We can perceive here the contours of a consistent system, deeprooted from time immemorial in the psyche of preindustrial societies. In this system, given the great variety of economic situations, the differences between the place of purchase and the place of sale, or between regions with a surplus of foodstuffs and others that were not self-sufficient, or between a lender and a borrower were all of them expressed in a diversity of weights and measures. To put it another way, the quantity was variable, but the price was constant. However, in the industrial society, all these matters are reflected in the differentiation of prices, while the measures stay fixed. On this interpretation, the attitude in preindustrial societies to the measure-price relationship, or, strictly speaking, their formulation of prices not in terms of a changing amount of money *vis-à-vis* a fixed quantity of a commodity, but the other way round, namely the conception of price as a variable quantity of a commodity *vis-à-vis* a fixed amount of money, can hardly ever be found in a clear-cut, polarized, uncomplicated form;[16] usually, the two tend to coexist. This tendency, however, can readily be observed throughout Europe, over many centuries, and even on the very threshold of the "metric" era—that is, at the turn of the eighteenth and nineteenth centuries, it was so strong as to constitute a major obstacle to the introduction of the metric system in practice.

The principle of variable measure and fixed price, which we have exemplified above in different periods by discussing the methods of regulating the price of bread, may also be exemplified across space. For, in many cases, when we wish to investigate the geography of prices, it is actually the geography of measures that we should study. To illustrate the problem, let us chart on a map such data as we possess relating to the bushels (or some corresponding measure) in the large and small towns of the Cracow *voivodeship* in the middle of the sixteenth century.[17] Our map shows the area over which a standardized measure was in use: the area encompasses the basin of the Vistula from Cracow to Sandomierz, the district of the lower Dunajec and the Wisłoka, and the area between Cracow and Myślenice in the south, and Będzin and Pilica to the north and northwest. Admittedly, the country town of Pilica is situated on the river Pilica, but at a point where the latter is not yet navigable, hence one would expect it to be associated with the Cracow market. Two areas lying farther away from the Vistula, on her right bank and left bank, namely the districts of Biecz and Jasło, and that of Słomniki, Lelów and Żarnowiec, respectively, use a measure of corn approximately one-third larger than the Cracow bushel. The western, Silesian marches of the Polish state employ a measure of corn that is considerably larger. On the other hand, the southern border region of Podgórze, Nowy Targ, and Nowy Sącz, being an area deficient in food supplies, employs a very small measure. The picture is quite logical. First, the regions engaged in export down river employ a small measure; secondly, those farther away employ a larger one; finally however, the regions less than self-sufficient in respect to food supplies use the smallest measure of all.

In an ideal situation, the price of a given type of grain ought to be the same throughout the market, whereas the measure would vary in different places, its variation masking the costs of transport and the traders' profits. The diversity of measures found should

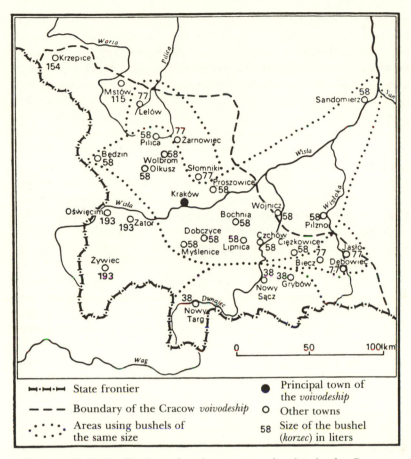

Geographical distribution of grain measures, by size, in the Cracow *voivodeship* in the second half of the 16th century (based on data collected by A. Falniowska-Gradowska)

act as the counterpart to the uniformity of the price. In practice, however, it is difficult to find any large area that exhibits precisely that sort of behavior in prices and measures because too many factors are involved in the emergence and successive changes of measures for the picture to remain undistorted. Here, we wish to demonstrate only that the market-*cum*-transport factor, too, has played its part.

S. Mielczarski writes: "It would seem that, in the second half of the sixteenth century, we have to consider the emergence of differential prices against the background of already differentiated measures."[18] Quite so, indeed. Measures, once they have been differentiated (the determinants of the differences including the aforesaid factor of distance from the market), would naturally tend to persist; at the same time, there might be alterations in the regionalization of production, and prices would certainly undergo changes.

Taking a large view, it is worth mentioning that the Gdańsk measure was apparently the smallest in the lands of Poland; the Wrocław measure was the smallest in Silesia; and in the Grand Duchy of Lithuania the statute of 1677 decreed that "a striked barrel should contain seventy-two gallons. In other municipalities and country towns, however, where lesser markets are held, the measure is to be decreed in proportion to the aforesaid barrel."[19] Interestingly enough, the practice was not seen as strange by contemporaries, and, indeed, it was natural for them to think in precisely such categories. Let us exemplify by citing the treatment of the matter by the inspectors of the *starostships* in the Rawa *voivodeship* in 1564.[20] After inspecting the Regnów manor they state: "Thus the tenant does not sell the rye locally, but takes it to the river to float it to Gdańsk, for men on the royal domain ought to take rye to the river, and whereas 9 Rawa bushels works out at 10 Gdańsk bushels, and then at Gdańsk the rye amounts to 56 bushels, so it pays to send it to Gdańsk, spending nothing on the barge dues."[21] Thus transport to the river port is defrayed by compulsory labor dues, and then the river transport by the difference in measures. An exceptionally "pure" case of this practice has come to our notice in Ducal Prussia.[22] Given the relatively small areas of the provinces and the dominance of Królewiec (Königsberg) as the chief export port, the phenomenon of measures, being the function of distance from the market and increasing with that distance, reveals itself with the greatest clarity (naturally our "distance" is a social category, taking into account transport facilities, and not just so many miles).

The Warsaw measure is smaller than that of Podlasie[23] [the district *ca.* 50 miles northeast of Warsaw—TRANS.], the Poznań measure is small relative to the other measures of Greater Poland[24] [the large region centering on Poznań—TRANS.], and the measures of corn in Podolia increase as we move from the east westward.[25] Again, as we have already remarked, measures become larger as we move from Cracow westward in the direction of the border between Poland and Silesia and then, on the other side of the border, decrease gradually as they near Wrocław. In keeping with the same principle, the Gdańsk bushel is small—indeed, very small. It is interesting that this fact was taken into account by the inspectors of the *starostships* as early as their very first inspection (*lustracja*). Their report on the village of Pożarów (Ożarów) in 1565 notes that its fields yield 1,380 bushels of rye and goes on: "Yet it is not the *starosta*'s wont to sell the rye at the river bank but to float it to Gdańsk, and then a deal more of the rye is counted by the Gdańsk measure, since 8 Warsaw bushels works out at 9 Gdańsk bushels, the rye for sale thus adding up to 172½ bushels. Part of it will go to pay for the barge dues, and part may be sold at the manor."

The geography of markets would thus seem to coincide to some degree with the geography of measures. In the heart of the market the measure was smallest, and then it grew as the radial distance away from the center increased. The boundaries between the tracts of land belonging to the different markets would constitute "barriers" of sorts, made up of maximum measures. Such boundaries would, now and then, coincide with the watersheds, since the waterways played a major part in the transport of grains.

To return, then, to our point of departure in the present chapter: the price as a mechanism reducing to a common denominator all the factors playing some part in a given transaction is a relatively recent phenomenon. In the mentality of the preindustrial society it was the measure that, to a large extent, played that role. The student of the measures of pre-colonial West Africa, Niangoran-Bouah, has demonstrated that in the civilization he has investigated there were in use in selling, feminine or "weak" measures, and masculine, "strong" ones in buying; that in the countryside measures would be larger than those in towns, the difference between them masking the cost of transport and the merchant's profit; that loans would be made and then repaid using different measures, the difference serving to conceal the element of interest; and so forth. He terms the modern European system of a variable price for an unchanging product "the purely European intellectual way." He does not know that it was not very long ago that the European metrological thinking was no different from that he describes operating in pre-colonial West Africa.[26]

· 15 ·

THE INERTIA OF MEASURES
AND THEIR VARIABILITY

"The persistence of measures is closely bound up with the questions of communal memory (*mémoire collectif*)," maintained M. Bloch,[1] using Durkheim's terminology. The following hypothesis may be ventured: that the tremendous diversity of measures coexisting at any given point in time in the pre-capitalist epoch, coexisting indeed in neighboring villages or within the estates of a single landowner or monastery, would be matched by a persistence of measures that was at times nothing short of astonishing.[2] A French student, for instance, demonstrated the absence of change in the land measures in a single parish in Normandy, according to evidence dating from 1049, 1282, and 1792.[3] The Carolingian royal foot was a measure that was not, after all, newly introduced by Charlemagne but one that he found already in existence and merely reaffirmed by royal guarantee, extending the area of its use within the framework of his unifying activity. Allegedly, it underwent some change in 1667 but, in fact, seems to have stayed basically unchanged until the French Revolution.[4] Similarly, A. Gilewicz's enquiries into former Polish measures led him to conclude that "the size of early measures remained unchanged, in principle, until the eighteenth century. The fact that this was the case from medieval times until the nineteenth century has been brought home to the present writer in the course of tracing the development of a number of units."[5]

The "inertia tendency" of certain measures over the centuries has been attested by numerous students, but has also been questioned by others in many instances. This disagreement among scholars pertains to the measures used in Belgian towns from the fifteenth to the nineteenth century.[6] The Florentine *staio* appears to have remained unchanged from the early fourteenth to the beginning of the nineteenth century.[7] The Gdańsk bushel kept the same size from the sixteenth to the eighteenth century.[8] L. Musioł found measures unchanged in three Silesian villages over a period of 400 years,[9] and he was also led by the evidence from Silesia to conclude that the measures in which the church tithes were assessed were particularly persistent and immutable.[10] In France, as far as the Paris region was concerned, Y. Bezard found that the measures

111

applied to the fields did not change from the sixteenth to the eighteenth century,[11] while G. Fourquin went even further in postulating his thesis of their fixity from the beginning of the fourteenth century until the French Revolution.[12] Again, A. Machabey concluded that many French weights underwent no change from the thirteenth or fourteenth to the eighteenth century.[13] And in West Africa, measures in use today (other than metric) coincide with archaeological findings dating from the medieval period.[14] In Tunisia[15] and in India, too,[16] the staying power of traditional measures has been well attested. Traditional measures "ought to" persist forever in their sameness. That was what was expected of them, immutability and antiquity. Thus, for instance, due to the continuity of the central authority, the Capitoline measure was apparently still in use in the Rome of the popes.[17] Moreover, such measures endured unaltered because their survival was aided by conflicts of interest and by the prevailing ideology.

All that notwithstanding, the nature of traditional measures inevitably renders them changeable. If a surface measure of cultivable land is also the measure of the time required to work it, then technical improvements in the yoke, by raising productivity, must inevitably bring about an enlargement of the unit of land measurement. If the area in question is measured by the "input" of the sower, then changes in the seeding intervals must bring about either an enlargement or a diminution of the unit of measurement. If, finally, the measure of area is equivalent to the amount of land needed to afford fair and square nourishment to one peasant family, then an increase in the yield of land must bring about a decrease in the unit in question. If the unit of weight is the pannier for the pack-ass, then an improvement in breeding through good fortune in selection cannot but increase that unit.

Above all, in both medieval and modern times in Europe, a dichotomy has been discernible between urban and rural measures. The balance between the forces of change and the forces of persistence has been different for the former and for the latter. Thus, in noting the convincing demonstration by H. van der Wee of the immutability of the measures in Flemish towns over many centuries,[18] we must remember that he is considering free cities, whose position enabled them to safeguard that immutability; we shall see, below, that it was not so in the towns of Poland, with their *szlachta* masters. The relatively greater stability of the measures in towns was thus a result of their freedom and independence from feudal seigneurs, as well as of their more pronounced internal social differentiation, the latter being responsible for the resistance of some interested group or other to every attempt at changing the measures.

These matters have been far different in the context of rural relations. Here, we find as a rule that "feudal might was right" in the dependence, in varying degrees, of the peasant upon the lord, and the blatant inequality of their strength. Insofar as the land measures, aided and abetted by the virtually sacred ritual of the permanence of the boundary landmarks, were able to resist attempts to alter them, this resistance was reasonably effective in Western Europe, where the subservience of peasants to lords was less marked than in Poland. But dry measures, the measures of corn in which the peasants rendered the lords their dues in kind, almost universally reveal an upward trend in their dimensions. This trend tends to be halted only in those regions that experienced the gradual elimination of the dues in kind. It is the opinion of many French historians that by the eighteenth century, measures in France, although most strikingly differentiated from district to district, were stable within the districts and did not change. Agreed—if we confine our attention to the urban markets. But if we go beyond the walls of towns, out into the countryside, then even in the less-advanced provinces, where the payment of dues in kind was of most significance (Brittany, the west, and the southwest), we shall observe changes in measures until the eve of the Great Revolution. Let us just mention here in passing yet another factor of relevance to some changes in the dimensions of urban measures—to wit, commercial relations.

The representational and functional nature of traditional measures invests them with an unavoidable variability. Yet they undergo changes over time primarily because social forces committed to the guarding of their invariability are of unequal strength. Hence, the utter inflexibility to change of some measures across the centuries coexists with great changeability—unidirectional in time—of others. These tendencies only appear to be contradictory: for it follows from the very social essence of traditional measures that some of them, at some points in time, cannot but resist change, while other measures, at other times, cannot but change. It is precisely the study of the coexistence of these two tendencies that forms the major task of historical metrology—if we accept as its objective the investigation of measures as a social phenomenon and do not restrict it to studying the performance by measures of narrowly technical service functions. In the history of traditional measures, there is indeed a dialectic of the changing and the unchangeable.

·16·

THE TENDENCIES TOWARDS
STANDARDIZATION

Thus, in the long period, traditional measures oscillate between inertia and change. At the same time, however, other factors cause the standardization of measures to spread over larger areas. Basically, there are two such factors: commercial ties and the will of the state.

It is, by and large, usual that imported goods are measured by the standards of the exporter. Accordingly, in Flanders one measure was applied to the home-grown corn and another to corn imported from overseas;[1] salt was measured by the measures of its place of origin;[2] the measures of Flanders and of Cologne coexisted, and which one was employed depended on the seller.[3] This is hardly surprising. After all, the salt would be barrelled in the country of its origin, and it would be a troublesome and expensive procedure to arrange its purchase by some other measure (although, in fact, this was done occasionally at the Wieliczka saltmines in Poland). In the textile trade, the width of cloth was determined by that of the looms in the country of production, and the length by the custom prevailing there. It was quite unlikely that those dimensions would be the ones current in the purchasing country.[4]

The more valuable the commodity was, the more important for the buyer; the fewer the counties producing it, the stronger was the position of the merchant bringing it to market, and his position of strength enabled him to dictate to the buyers the measures to be used in the transaction. Yet there were also instances when the stronger position was actually enjoyed by a foreign merchant engaged in exporting a certain product from some particular country. Possibly this was the case for the Baltic last; or to give an example from a different region: wherever Arab traders reached in their search for gold, they would succeed in imposing the Arabic system of weights, especially small weights.[5]

Practices of this type, while making for the standardization of measures for a given product in markets that were geographically far apart, would yet increase the diversity of measures coexisting in particular places. Not until the rise of closer commercial ties between regions, provinces, and countries, ties that embraced various articles, would the processes of standardization come to be effective. The fairs of Champagne in their heyday popularized the Champagne measures (in particular, those of Troyes) over a large area.[6] Did their use endure? Certainly not. If this popularity did not endure, it nevertheless lasted at least as long as did the renown of the fairs of Champagne. S. Hoszowski has, accordingly, proposed the following generalization: "The equalization and standardization of weights and measures stand in direct relation to the range of exchange relations (commerce) between given territories."[7] Although this formulation has had its critics,[8] they do not convince, for their reservations are of a kind that can be readily urged against any and every historical generalization.

The second factor favoring standardization of measures was that of administrative reforms. Which of the two was the more important? This is a hard question to answer. In practice, the influence of the fairs of Champagne was surely stronger than that of many a royal reform. Ultimately, the will of the state would win through, but not until much change in economic life, and in the nature of the state itself, had taken place.

Finally, the seigneurial system played an ambiguous role as a factor in the standardization of measures. Feudal seigneurs determinedly sought standardization of measures within their domains, and not infrequently, this meant extensive areas.[9] Nonetheless, the standardization within a particular demesne enhanced the separatism within the territorial state as a whole.

The ancients bequeathed us some relevant legends—of the standardizing efforts of Philip of Macedon and Alexander the Great, of the Capitoline measure, and of the standardization reform of Justinian. We are not competent to pronounce judgment on the kernel of truth these legends may contain. Yet the very facts of their survival and their insistence on regarding such reforms as highly creditable to the rulers in question speak for themselves.

Since the end of the ancient world, Europe has experienced three major waves of activity in quest of metrological standardization—all within wider unification attempts: the Carolingian, the Renaissance (or absolutist), and the Enlightenment (enlightened absolutism). The climax of these attempts was eventually to come with the beginnings of capitalism—the fundamental reform in this field, the introduction of the metric system in 1791 (completed in 1799) by revolutionary France. Henceforth, the metric system was to conquer the world. On 14 October 1918 it was established in the Soviet Union by decree of the Council of the People's Commissars. Japan adopted it in 1958. Among today's major powers, the Anglo-Saxon countries alone do not employ the metric system, albeit having long since accomplished within their borders a great deal of metrological standardization. Naturally, even in the countries that adopted the metric system, traditional measures continued in use for a long time; they are still to be found today, notably in retail trade and in farming. Nonetheless, the proposition may be ventured that their continued survival is proportional to the economic backwardness of the country. Indeed, attempts to standardize weights and measures in societies that have felt no need of such a development have proved quite strikingly ineffectual.

The effectiveness of the Carolingian reform [early ninth century—Trans.] has, without doubt, been much exaggerated by many historians.[10] Possibly, metrological reforms met with more success in medieval England, which was already then a more efficiently administered kingdom; arguably, however, their effectiveness was the result of fundamental compromise: the monarchy would decree standardization, but would respect and tolerate manorial measures.[11]

In the efficiently administered East Prussian state of the Teutonic Knights, the standardization reform was carried out by Ulrich von Jungingen, familiar to us in another context.* This was a partial reform, concerned with agrarian measures, the standardization of which was sought as a prelude to the standardization of the feudal dues,[12] and the old rod of Chełmno was recognized by Jungingen in 1392 as the official unit throughout the Order's territory.[13] Czechoslovakia saw an attempt to put through a standardizing reform by Premysl Ottokar in 1268.[14] Spain, for another example, experienced during the Middle Ages at least five major attempts to standardize measures—by Alfonso X in 1261, by Alfonso XI in 1348, by John II in 1435, by Ferdinand and Isabella in 1488, and, finally, by Philip II in 1568.[15] The repetition bears witness to their ineffectuality. The New World saw some interesting consequences of the diversity of measures when the Spaniards from different regions of Spain imported different measures, which brought about endless misunderstandings, facilitating abuses and hindering the work of the state authorities.[16] In the end, the reform of Philip II gained a measure of success—although limited, of course.[17]

This brings us to the second, Renaissance series of reforms aimed

* A reference to his crushing defeat by the Poles in 1410 at Grunwald/Tannenberg—Trans.

at standardized measures. In Spain, there was the aforementioned one by Philip II; earlier still, there was apparently one in France, in the reign of Francis I;[18] again, there was one in Lombardy in 1597, formally revoked in 1605,[19] and another in the Savoy in 1613.[20]

In Russia, analogous attempts were undertaken by Ivan the Terrible in the sixteenth century. One of his lieutenants, the German von Staden, in writing the Tsar's encomium states: "The present great King has seen to it that throughout the lands of Russia, all over his realm, there should be one faith, one weight, one measure."[21] However, this sentence clearly tells us nothing but that Ivan did desire metrological unification. In 1624, the government, seeking to eliminate local measures, standardized the *tchetvyert*, to be used henceforth in assessing inland customs dues. But local measures continued in use notwithstanding; obviously, the official Treasury "quarter" was smaller than the local ones.[22] In practice, for centuries yet, metrological diversity continued in Russia.

Similar developments are discernible in Poland in the sixteenth century. Two dates are of particular relevance—1507 and 1565. In 1507 Sigismund the Old integrated the Cracow and the Poznań measures, unequivocally confirming, however, at the same time, the distinctiveness of the Lvov and Lublin measures. On the other hand, the statute passed by the Piotrków *seym* in 1565 standardized the weights and measures throughout the kingdom; the bushel (*korzec*) alone was legally confirmed in its variety, and its standardization henceforth was lawful only inside the boundaries of any *voivodeship*, in keeping with the standard of its main town. It is well known, however, that Sigismund's attempts at metrological unification in Poland were unsuccessful. Economic regression, the weakening of the authority of the state, the postwar chaos and the decentralized governance by local dietines—all these aided and abetted the growing decentralization of measures.[23]

The eighteenth century saw the third wave of reforms aimed at standardization. East-central Europe was in the van, through the efforts of Austria, Russia, and Prussia. Their enlightened absolutist monarchs, in building and streamlining vast administrative and bureaucratic apparatuses, would time and again encounter obstacles arising from the variety of measures employed in different parts of the same country. A "disorder" of this kind just was not to be suffered by absolute monarchs in their inquisitive efforts to know all and to inventory as much as possible of their possessions. The eighteenth century was an age of "political arithmetic." And how was it to be practiced, how were the wanted sums to be obtained and the four arithmetical functions put to use at all when the measures differed from region to region? The new need overlapped the age-old, traditional one: that since measures were an attribute of sovereignty, it was meet for them to extend territorially wherever the King's writ ran. Common measures would bind together the different imperial provinces situated within historically determined frontiers, though not infrequently separated by hundreds of kilometers. Common measures would also underline the territorial distinctiveness of the empire as a political entity. The pre-metric standardizations of the eighteenth century did not seek universality; on the contrary, the objective was, as it had been since the Carolingians, that measures should be identified with the ruler and be coextensive with his rule. Universal, mankind-wide objectives were yet to make their appearance—as a concomitant of the metric reform.

This quest by the political authority coincided with the conclusions of interested scholars. The progress of mathematics in the seventeenth century, and of the natural sciences in the eighteenth, led numerous scholars towards the contemplation of a fixed, unchanging, and exact standard of measurement. As in many other walks of life, so in metrology the state authority of enlightened absolutism had little difficulty in securing the cooperation of men of learning.

And what of the common people? The common man, on the one hand, from generation to generation, eternally cherished the dream of the "just" measure; yet, on the other, his interest was narrowly defined: his particular village, the nearest country town, the parish with its demand for the tithe, or the local lord exacting his dues, restricted his vision. It was only in the eighteenth century in Western Europe, and later still apparently in her Eastern reaches,[24] that the common folk began to sense a need for a standardization of measures of greater geographical range. Besides, the absolute monarchy, while ceaselessly insisting that all its policies sought the good of the people, did not see fit to elicit their views.

The reform came earliest, in 1705, in Austria. The evidence from Silesia[25] enables us to ascertain in detail the nature of the reform. It was not markedly successful. The authorities dealt severely with the uncooperative municipalities, while it was, in truth, the large landed estates that sabotaged the reform. There were renewed attempts at standardization—in 1715, 1740, 1750, and yet again later.[26] They were sabotaged by the Hungarian noblemen in Slovakia, particularly in places where local measures were larger than those decreed by the Habsburgs.[27] Despite considerable exertions, the achievements of the 1750–1751 administration were modest. A hundred years later, many parishes were still collecting the Mass tax by "our forefathers' measure"; this, despite the fact that the authority, learning from experience to respect tradition and commercial ties, decreed, in spite of the loss of Silesia, that the standardization of measures in the Teschen part of Silesia should follow the measures of Wrocław.[28] The reform, launched in 1775, was conceived as a long-term one and was carefully planned. Yet it, too, failed to run its full planned course, for history did not allow the absolute monarchy enough time to complete it. We shall return to this reform again, with reference to Lombardy, in our final chapter.

In Russia, too, successive attempts to carry through standardization of weights and measures were made during the eighteenth century, from Peter to Catherine.[29] Peter the Great's standardization *ukase* was promulgated in 1724. In 1736 the Commission for Weights and Measures was set up, and it proved very active— especially in its early years. That entire period of reform was concluded with the fundamental *ukase* of 1797. In Prussia, partial standardization was attempted by Frederick the Great *ca*. 1750. In The Grand Duchy of Tuscany measures were standardized as of 13 March 1781.[30] And in Poland, the Coronation Seym decreed standardization in 1764 (extending it to Lithuania from 1766).

Yet, in the period of the European Enlightenment, multifarious reforms undertaken by enlightened absolutism, in association with the same manifestations of the evolution of the forces of production and progress of science, and utilizing at times the studies of those same savants, for all the pressure exercised by the enormous bureaucratic and administrative apparatus, were crowned, at best, by partial success; whereas the metric reform, born of the French Revolution, was destined to conquer the world of the future.

·17·

SOCIAL CONDITIONS AND THE EMERGENCE OF CONVENTIONAL MEASURES

To reiterate, early measures, roughly until the beginnings of capitalism, partook of substantive character, "signified" or represented something, expressed something human relating to man's personality or the conditions of his existence. Modern measures, however, have no meaning other than that of sheer convention; what matters is the acceptance of the system, and not the magnitude of the basic unit, which might equally well be large or small. The units of conventional measures are defined in terms of physics or astronomy (the weight of a certain volume of water at a specified temperature and pressure, or a stated part of a meridian), but they have no social significance. This has been most strikingly obvious in the history of the meter,[1] which was initially accepted as 1/40,000,000th of the earth's meridian (*metre vrai et définitif!*) by the French statute dated 22 June 1799 and remained unaltered as a unit of measure by resolution of the International Conference of 1870–1872, even though ever more accurate measurements kept altering our estimates of the true dimension of the meridian (a process that doubtless will continue). Until 1960, the unit of measure, in countries adhering to the metric system, was not the 1/40,000,000th part of the meridian, but a designated bar of metal, kept in a Parisian assay office.

What then is the social meaning of the transition from representational measures, with their human associations, to abstract measures of convention, "signifying" nothing? The transition points up the need for common, intersubjective, verifiable standards, independent of human individuality. The Protagorean maxim that "man is the measure of all things" has had to yield, in our quest for the objective and immutable. How difficult, though, it is to find immutable elements in a world in which nothing stays unchanging! The picture of attempts to find such elements, such fixed points of reference, is therefore full of social values. It is not surprising, then, that in his quest for some immutable magnitude, man should opt ultimately for a finite feature of the planet he inhabits, his sole constant base. But then it turned out that he could not, as yet, quite accurately ascertain the dimensions of his planet, and, moreover, that even it was not wholly immutable. At that point, there was nothing left but sheer convention.

But the history of the meter does not end there. The Sèvres standard, made of an alloy of platinum and iridium, was internationally binding from the First International Conference on Weights and Measures in 1889, and after seventy years' service was retired—or, more accurately, put to rest in a museum. For it had two shortcomings. First, its exactitude: at the end of the nineteenth century, its accuracy to 1/10,000,000th of the meter represented a great achievement, but today it no longer satisfies the requirements of precision engineering. For example, in the production of remote-control rockets, the tolerance may be no more than one ten-millionth of the millimeter, not the meter, and thus calls for a thousandfold increase in accuracy. Secondly, too much risk attached to the situation wherein the functioning of a worldwide system of weights and measures depended upon a single standard piece—however carefully it was tended and guarded. The new meter had to be more accurate and capable of being reproduced. On 14 October 1896, the meter of Sèvres was "dethroned."

This was no mere change of the standard piece. It was, for one thing, an abolition of the standard as such. Today, the standard meter may be reproduced in any suitably equipped laboratory the world over. It was, moreover, a change in concept and definition. The official definition of the meter now reads: "a length equal to 1,650,763.37 wave lengths of the orange light emitted by the Krypton atom of mass 86 *in vacuo*." People to whom this makes sense are few and far between. It is no longer for "every schoolboy" to grasp what the meter is. We have gone a long way from the measures of the feudal epoch that meant so much in human terms. The "dehumanization" of the tool so intimately bound up with everyone's daily existence that is the meter has reached a kind of *ne plus ultra*. Yet, at the same time, the process has taken us very far along the road of more effective and fruitful international understanding and cooperation.

That the standardization of measures is a historical process, parallel to the widening of the marketplace, goes without saying. We have noted, above, that as Polish nobility grew commerce-minded, it engaged in a struggle for the standardization of those measures that were relevant to commerce—namely the units of wholesale trade—while simultaneously battling to retain the right to determine those measures that were relevant to their dealings in the countryside. In regions where the bourgeoisie was relatively stronger, and it—rather than the gentry—dominated the wholesale trade, it was the bourgeoisie that would champion uniform measures.[2] As the market becomes nationwide, within the context of the nation-state, the standardization of measures becomes symbolic of the changes already wrought, as well as a factor accelerating further change. Goethe saw this clearly when, in speaking of the future unification of Germany, he insisted that "Deutschland sei eins in Mass und Gewicht."[3] In France, the First Republic, with its centralizing tendencies, saw this as another area in which to exercise much influence. Nearer to our own time, India since Independence has had to combat local separatism and, in the course of so doing, duly introduced the metric system in 1961.

As early as the introduction of the metric system in France, however, far-sighted men saw in it an international institution, hoping—rightly, as it turned out—that other countries would adopt it, too. In the grandiloquent phraseology of the French Revolution, here was a measure "for all peoples and for all time." There is no practical social meaning in the meter today. It is but a convention. What does matter a great deal is the social agreement to use the meter as the common unit of measurement. It signifies no less and no more than a great stride of mankind along the road leading to the goal of a "common language" of mutual understanding and collaboration.[4]

However, for a society to be able to adopt measures of pure convention,[5] two important conditions have first to be satisfied: there must prevail a *de facto* equality of men before the law, and there must be accomplished the process of alienation of the commodity.[6] There is no scope for measures of convention alone, and thus for the metric system, in communities where different categories of people are subject to different laws; or, conversely, the presence of different laws for different people means different measures for different people. Inequality before the law implies unequal laws or rights in relation to measures: some people decree them, others have to put up with them; everyone has a measure of his own, the strong imposing theirs upon the weak. "The measure" is not impersonal but rather "human"; it belongs to some, it does not belong to others, and it is dependent upon the will of whoever has the power to enforce it. Not until there is equality before the law can there be metrological equality. The measure will then cease to depend on anyone's will, and this will symbolize its "kinship" with the dimensions of the terrestrial globe—for, after all, no one can exercise any influence over that.

Finally, the measure has to undergo alienation, just as commodities have. The craftsman's handmade product, turned out to the order of some particular customer, bears a dual human stamp: of the particular producer and of the particular consumer. Each craftsman has a "style" of his own, each buyer has requirements peculiar to him. Among the many qualitative features of the craftsman's product, features that are quite particular and unrepeatable, there is its "measure." One will have a "fair" measure, another—"unfair." It is a measure arrived at by way of direct personal contact that entails dispute, bargaining, and compromise. However, a mass-produced commodity cannot but be devoid of that dual human stamp. There is nothing about it that will tell us who made it, nor who it was made for. The measure given to it is unrelated to its maker or its user. The mass production of commodities is intended for vast, remote, and diverse markets. Each of them will have a measure of its own. Commodities cannot bear the measure of the country of origin if it be unintelligible to the buyers, nor of the country of destination, for those will be many and varied. The dimensions of such a product cannot be expressed in any measure that "belongs" to some particular locality or nation. No measure can enjoy favored treatment in a market situation. Each must be abstract, just as market value is—or rather, conform to the abstract character of market value. The dimensions of the human body are given numerical values, and the need to measure them on every occasion is then eliminated. There is no further call for compromises. The measure ceases to be "fair" or "unfair." It becomes morally neutral because nobody can influence it any longer. Its existence has become "extra-human." Now, finally, and only now, can the measure be just one of convention.

Traditional measures were "human" in many respects. They were expressive of man and his work; they depended at times upon his will, which in turn depended upon his character and attitude toward fellow humans. Yet, at the same time, traditional measures offered endless opportunities for abuses and acts of injustice, for violence of the strong towards the weak. The meter, in "dehumanizing" measures, in rendering them independent of man and "objective" in their interrelationship with man as well as morally neutral, has also transformed an instrument of "man's inhumanity to man" into a means of understanding and cooperation for mankind. But, to repeat, for the meter's final victory, two conditions had to be satisfied: the equality of men before the law, and the alienation of the commodity.

PART TWO

·18·

CLASS STRUGGLE IN THE POLISH COUNTRYSIDE FROM THE SIXTEENTH TO THE EIGHTEENTH CENTURY

Whichever preindustrial society we consider, particularly in its rural aspects, our immediate impression is one of metrological chaos, since the situation is so utterly remote from what we are used to. And, indeed, there are many historical works that bewail the "chaos," and leave it at that. Yet the chaos is only apparent and "there is method in this madness." Feudal measures are not as inaccurate as they may at times appear to us, and the differences in them, as well as the coexistence of different methods of measuring, have a profound social significance. More than that, the rules are strict, offering no scope for arbitrary behavior, and a binding code of morality—albeit, usually unwritten—does exist. Every transgression of rules amounts to a challenge to the prevailing social norms and meets with stiff resistance. We therefore have no hesitation in applying the concept of class struggle to the matters to be presently considered.

Admittedly, the metrological arrangements we find in the Polish countryside in the sixteenth century are extremely complicated, but all is governed by firm and clear rules; we shall be, however, quite unable to grasp their character unless we bear in mind the determining factors behind Poland's particular agrarian system of the period. The system had evolved from the Middle Ages, when Polish landowners had sought to attract settlers by offering them holdings protected by Germanic and Polish law. The situation, however, was modified under the impact of an aggressive system of manorial demesne farming [the *folwark*] at the end of the fifteenth century and in the sixteenth. The aggression of the *folwark* added labor services, which from the sixteenth century were assessed in terms of time-work, that is, number of days due, rather than piece-work, that is, amount of work due. The amount of land accruing to the demesne, at any rate in that period, was increased by buying up the village jurisdictions and putting wastelands under the plow and, to a lesser extent, through diminishing the size of the existing farmsteads owned by peasants. Given a situation like that, it was inevitable that the strife surrounding measures should focus pri-

127

marily on those used in the assessment of dues in kind that the local lord demanded.

The records bequeathed by the first *lustracja* are most particularly illuminating on this matter in view of the large quantity and the uniformity of their evidence, but in relying upon them, we must bear in mind two issues. First, the date of the first *lustracja* was determined institutionally and was associated with certain steps taken by the state treasury; thus, the inspections did not commence spontaneously in the second half of the sixteenth century, but we simply do not possess a source of similar scope and productivity for the earlier period; secondly, careful reading of the records of the inspectors is enough to make us see the presence in them of antecedents, replete with social content, while the state of affairs actually described by them was a codification of many former contentions, struggles, compromises, and agreements.

What is particularly striking in the records of the first *lustracja* is their precise formulation: no room is left for arbitrary or ambiguous statements. For all that the population of any particular locality obviously knew well the kind of bushel (*korzec*) in local use, the inspectors never refer to "the bushel" in general, but always state the type of bushel in question. True, different bushels might be used in neighboring villages—but there was never any doubt as to what kind they were. Thus, although there were diverse bushels, the relations between them were fixed and known. Naturally, within a certain finite geographical compass, this would apply to the relation in which the bushel in a particular village stood to the bushel in a neighboring village, or at the landing-place of river-borne trade, or in the traditional selling place for produce, or even in the provincial capital. Matters to be codified almost always included the method of measurement—"striked" or "heaped." The relevant nomenclature is extensive; we know of "heaped" or "topped-up" measures, "semi-topped-up," those with a "head" or "small head," and measures "brimming over" or "spilling over," and so forth. Nor were the dimensions of the "heap"—where indicated—left freely to the parties to a given transaction. Set customs would be strictly observed, and various inspectors, in different parts of the country, always reckoned the "heap" of wheat, rye, and barley at one-third, and the "heap" of oats at a half, of the bushel. In some instances, albeit infrequently, mention is made of the approved way of pouring the grain into the bushel—to wit, that it should be "from dropped-arm height,"[1] and should not be "pressed down."[2] The question arises why those particular issues should be specified, and why in some cases, but not in most? Is it too bold to surmise that the inspectors noted the matter in those villages where their attention was drawn to it, where their informants stressed the issue, and that this was presumably more likely to be the case in places where, in fact, the way of pouring grain into the measured container had once led to clashes, resulting in a compromise by the time of the inspection?

That the existing arrangements were highly complicated and, simultaneously, gave rise to the strictest principles of conduct, is attested by yet another practice. In a number of places, there co-existed different methods of measuring. This was so quite often in the case of milling arrangements; to give an example, the mill at Kłodawa would pay its dues in "heaped" measures of malt but "striked" measures of oats.[3] It was not unknown, besides, that the peasants' dues owing to the manor in one and the same village were paid partly in heaped and partly in striked bushels.[4] Other contemporary sources also reveal similar practices in other places. It is, moreover, interesting that when the *starosta* or his officials failed to inform the inspectors in detail how the tax grain was measured (was it "from dropped-arm height"?), or accepted (by "heaped" or "striked" measure?), the inspectors would not fail to make it quite clear that the omission was morally reprehensible, and they would unequivocally suspect a desire to diminish the unit of payment.[5] The conviction ruled that in matters of this kind complete clarity should prevail and all arrangements should be free of any ambiguities; no less was demanded by social morality.

Since, therefore, considerable store was clearly set by these matters; since in addition to respect for the existing practice, the greatest possible unequivocality was being sought; since certain issues would be emphasized in some cases but not in others; since on occasion the selfsame peasants paid the same lord a part of their dues one way and part another—surely, then, we are entitled to see in all this the outcome of a lengthy evolution through quarrels, conflicts, and compromises accumulated over many generations. The exactitude of the inspectors shows their anxiety about injuring either party and rekindling old feuds.

Our representation of the state of affairs revealed by the inspections may seem rather static. In fact, this was not the case. Certain time-hallowed customs, guarded and preserved by conflicting interests, were after all constantly attacked—either openly or clandestinely—both before and after the inspections. Again, the sources that have come down to us threaten to delude the historian.

It is true that sources relating to sixteenth-century metrological disputes in the countryside are few and far between, but then it was not until 1582 that the office of the Crown Referendary was instituted. That disputes similar to those described in its records had also occurred earlier is clear from the decree of Sigismund Augustus dated 23 June 1569 relating to the onerous dues of the inhabitants of the villages of Jerzmanowice and Ciołkowice, wherein we read that they "ought to pay up the *osep** in corn too . . . according to ancient custom, by the oldtime measure and bushel, on the eve of St. Martin's." Also, in the decree of King Stephen Bathory of 24 May 1577, dealing with the burdens of the villages of Krościenko and Iskrzynia, we read: "They complained, too, that the *osep* due was collected from them using a larger bushel than formerly was the custom. If this is true, as your serfs have told us, then it was greatly injurious to them, and in no way could we approve of your way. . . . You are to collect the *osep* due from them by the former bushel and no larger."[6]

On the ecclesiastical estates things were no different; we might cite in support a 1576 document, the "Ordinance for the Kielce Complex of Estates."[7] In it, we read: "*Item*, as far as the ordinary *osep* is concerned—regarding which we have received complaints that it is measured, when collected from the peasants, by a larger measure than used to be the case—it has now been decided that none other but the Cracow measure be employed . . . , and in order to secure this practice henceforth, the quarter standards were prepared and certified by the Cracow measure and sent off to be kept at Kielce and to be used from now on every time any *osep* due is handed in and striked, with no heap at all." Let us parenthetically note that the document cited suggests that the Cracow measure, established by the Piotrków statute of 1565, served for a time as a sort of "neutral" unit, referred to whenever disputes arose. Presumably, it did not discharge this function for long.

Beginning at the close of the sixteenth century, we have at our disposal a rapidly growing quantity of information upon the issues we are concerned with here, and it kept on growing until the fall of the independent Republic of Poland-and-Lithuania [at the end of the eighteenth century—TRANS.]. This was not due to any rapid growth in the importance of the practices in question, but to the changing nature of our sources: after the first inspection, many

---

* The *osep*, or *sep*, was the basic contribution in kind, namely, in corn or grain, which the lord of the manor levied from his serfs—TRANS.

others were carried out during the following century, the Crown Referendary's Court intensified its activity, and manorial archives extended their scope and contents, as did the files of rural courts of law. Nonetheless, it would appear that the growing amount of relevant source information from the seventeenth and eighteenth centuries was partly occasioned by the growth in the substantive importance of metrological questions in rural life, as well as of metrological conflicts in the struggle between the village and the manor. The gradually ossifying system of feudal relations and labor dues frequently provided occasions for metrological disputes—and they were taking place in a context of increasing reliance upon the written word, with "old" documents such as the inspectors' or the Referendary's records being increasingly invoked by the disputants. It was becoming ever harder to reduce the number of rods of land in the peasants' holdings, and therefore, to reduce their size, recourse would be had to clandestine efforts to diminish the rod. It was even harder to increase the number of bushels due from the peasants as the *osep*; so, the only way to enhance their dues now was by increasing the size of the bushel. Similarly, the peasants' respect for the old customs made it hard for them to ask that the number of bushels they owed as the *osep* be cut, but there was nothing to stop them trying to prove that "of yore" the bushel had been smaller.[8]

In the records of the inspections, the rigor of metrological definitions grew strikingly greater. We find the occasional insistence that the *osep* be measured out without "pressing down,"[9] that malt be given in amounts according to "the measure commonly used in brewers' purchases," and the *osep* "according to the common striked measure townsmen use,"[10] while sometimes it falls to the inspectors to decide between various measures—the old and the current, the castle's and the community's. Significantly and most characteristically, in cases like that, the old measure would always be smaller than the new one, and the castle's measure larger than that of the parish.[11] In the village of Rokitno, where "the castle quarter is larger than the market quarter, his Reverence the Abbot lodged the complaint that despite the decree of King Casimir . . . of 1460 . . . the *oseps* are extracted from those villages using the castle and not the market quarter. *In contrarium*, the representative of the *starosta* produced documents *ex actis revisionis* of 1564 and of the inspections of 1569 and 1617, which demonstrated that such *oseps* as well as other dues had long since been rendered to the castle at Międzyrzec, just as they are now, measured by the castle quarter."[12] So we are dealing here with a "historical metrological" investigation going back over a century and a half! At Pyzdry, on the other hand, the castle measure was allegedly "larger," on account of a change of shape. We read: "The burghers here have complained that, in the matter between them and the castle, the castle measured dues in kind by a broader quarter than formerly established; a second quarter [!] goes to fill the castle's broad measure. And, therefore, seeking to avoid further difficulties in this matter, we submit the quarter with our stamps to collect with the castle's dues . . . and the malt is to be supplied by the old measure, and neither pressed nor raked about."[13] Towns would jealously guard their standards of old measures, and they were better placed than the peasants to preserve them. As far as the doings in the villages on the royal domains were concerned, our information comes, above all, from the files of the Referendary's Court. Z. Ćwiek, who researched them with respect to the seventeenth century, counted seventy complaints from the peasants about the increase of the *osep* measures by the intendants of the royal domains.[14] Surely, then, the abuses in private landlords' villages, where the peasant had no access to properly constituted courts of law, must have been far more widespread?

From the mid-eighteenth century on, we have in the surviving sources hundreds of items of information relating to disputes about measures. We may assume that such high incidence bears witness not only to the growth in the quantity and quality of sources, but also to the exacerbation of metrological issues in the Polish village. The scope for openly increasing the serfs' dues was reduced— whether the increases had been the result of larger numbers of bushels of corn yielded up to the manor, or larger numbers of days of unpaid labor on the demesne—perhaps because of more commercial transactions entailing higher demands upon the village to help with transport and increasing the participation of the peasants in the market economy. But when it became less feasible to demand from the peasants four bushels instead of three as previously, the temptation grew to tamper on the sly with the size of the bushel. When it became less feasible to demand four days of corvée labor in place of the former three, a transition would be commonly made to the so-called *wydział* corvée, based on the amount of labor rather than on the number of days of work, and this could only be enlarged by increasing the unit, i.e. the rod used in the measuring of the field constituting a day's work. Again, the peasant's transport dues could be increased by increasing the load per cart. The growing mobility of the peasants, however, widened their horizons and emboldened them to resist. As the ownership of land was being concentrated in the hands of the magnates and their latifundia became even vaster, increasing numbers of peasants on private estates, even though they as yet had no access to the state courts of law, were gaining access to some other court of law and the right of appeal from manorial decisions to the magnates' judicial institutions. The administration of the latifundia, including the parts that dealt with the peasants' complaints, grew more efficient. Consequently, more cases were being dealt with and fuller records were kept—coming down to us in due course.

Let us begin with the disputes concerning dry measures, which illuminate two issues: the quest by the manor to intensify the exploitation of the peasant, and the strife for shares in the fruits of this exploitation between the manor and other interested parties. The peasants complained, for example, that the officials "assess the *osep* with a measure so large that it doubles the amount due and insist that it be heaped and pressed down. . . . They take our rye to the Vistula using this huge measure of their own, larger even than the manorial or the demesne one."[15] The Referendary's Court is fully cognizant of such manorial moves. It therefore instructs its commissioners to carry out investigations as to whether the manor had not managed surreptitiously to alter the measure. In 1781, the commissioner being despatched to the village of Zederman in the *starostship* of Rabsztyn is instructed to "establish his jurisdiction upon arrival and request that the *osep* measure be brought and presented to him, and if it is the true old one and not increased, . . . then he should calculate it in gallons (*garniec*) and convert it, striked, into the current Warsaw measure."[16] A similar case with respect to the barrel for *dziakiel* [corn tribute in Lithuania—TRANS.] is reported in 1772 from the Suraż *starostship*.[17] Again, we have the statement settling the dispute in 1781 between the village of Kaszów and the convent of the Tyniec abbey, resulting from the peasants' complaints about an underhand increase in the size of the *osep* containers.[18]

Just as in the disputes about the measures of length, the Referendary showed here, too, a tendency to establish some guarantees of the fixity of measures. For example, when the inhabitants of the villages of Radzikowo and Kębłowice experienced a difference with their owners, Onufry Bromirski and Jakub Miszewski, the Referendary decided in 1779, in setting the magnitude of the bushel quarter to be used in the assessment of the *osep*, that "the standard of the said quarter is to be kept in the manor and an identical one, equal in all respects to the former, including metal fittings and an iron bar on top as well as a sealed bottom, shall be made for the headman at Radzikowo and kept by him for the Radzikowo and Kębłowice communities."[19]

In the supplications that we have of the peasants on the primate's estates, the same question crops up time and again. We learn that "Squire Korycki, his lieutenant and also a hard man to the people and village of Ostrówka, takes oats for next to nothing, and fills the bushel pressed and heaped."[20] The "enlightened" magnate, in this case the primate, acts in the same way as the monarchs of the Enlightenment period, and the administration of the primate's estates follows the same principles as those of the Crown Referendary's Court: it seeks a compromise, and it looks to prevention of further disputes. The only difference is that whereas the Referendary, in translating the assessment of feudal dues into the new Warsaw measure (in principle tacitly, but sometimes explicitly), permits the use of traditional measures in daily practice; the administration of the primate's estates, on the other hand, follows no set principles, now resorting to former measures, now converting them into the newly binding ones. The administrators would always, of course, see to it that no loss resulted to the landowner.

Let us quote the verdict of inspector Lachmański in the case of the peasants' complaints in the Żnin complex of estates: "The *osep* measure used in the collection of corn from the villages . . . has to be just, being the royal Warsaw measure of 32 gallons of the same constitution, not pressed down but accepted smoothly striked [*sic!*] . . . the oats to be poured into the quarter with a shovel and not with the hand, the same applying to the barley, albeit without levelling it when measured. However, in view of the measure being duly striked but not pressed down, to ensure that the tenant incurs no loss, there should be added, per each quarter, two gallons against loss through drying [*sic!*], mice, and so forth, the said quarter to be measured by myself in the presence of the villagers and assessed inclusive of the said gallons [thus no longer in keeping with the law—W.K.], so that the quarter shall contain 10 gallons and have then my seal affixed thereto."[21] In yet another disagreement, the Rev. Szwaynert, a high official, pronounced that the peasants of Wielenin "shall yield up the *osep* not heaped, as hitherto, but hand-meted, by my own measure, properly stamped."[22] An analogous decision was arrived at administratively in the village of Cisów,[23] and further examples could readily be supplied.

Inspection records tell us of scores of metrological disputes, even during the inspection of 1789, a full quarter-century from the reform of 1764 and therefore at a time when all measures and methods of measuring should long since, according to the letter of the law, have been standardized nationwide, leaving no room for alteration, altercation, duplicity, or injustice. Yet, as late as 1789, the village community of Czubrów complained: "Outsize measures of *osep* have been extracted from us . . . though when we were sold by W. Wytyszkiewicz to W. Radwański we immediately stressed that our old measure was to be kept." And in the subsequent petition they wrote: "We ask . . . that the *osep* be collected from us . . . by the proper Olkusz measure and no other, according to the inspector's decision; and we are prepared to give evidence before the law how they beat us on the tree stumps at the manor and insisted that we deliver the *osep* using a measure that was larger than the just one, and now they keep the just one at the manor, though they should return it to us"[24]—all of which the accused denied. The village assembly of Zederman in the Rabsztyn *starostship* was particularly unyielding in metrological clashes and brought before the inspectors the following accusations: "According to the decree, the castle should take from us the *osep*, oats and rye, by the royal bushel striked, but they take it heaped. They have seized our communal bushel and will not let us have it back."[25] The villagers of Kosmalów, Zimnodoły, and other places complain that they have to supply the *osep* "heaped up and pressed down as hard as it will go, being told to pour the grain from shoulder height,[26] with the overseer raking it about on top, so that each bushel truly has five and not four quarters, which, by means of such excesses, they take from us improperly."[27] Incidentally, in the "Complaints Form" of those communities we read that they were expected to supply the *osep* "by the Olkusz measure, heaped up, according to the inspection,"[28] although the use of the strickle had been legally binding since 1764.

Things were no different in the *voivodeship* of Sandomierz. The villagers of Tczów complain that "the manor demands wheat and rye heaped, and they top up the oats until they spill on the ground"—the manor, naturally, denying the truth of the charge.[29] At Bartodzieje, too, the people complain that the bushels are taken from them heaped and not striked.[30] Again, the complaint from the village of Gzowice reads: "According to the inspection, we should supply 36 Warsaw bushels, but they take 40 from us, heaped up, to boot."[31] And the community of Rawica petitions thus: "Our landlord collects the *osep* using a large measure, heaped, and we have, in effect, to give him 2 bushels where 1½ are due."[32] Similar examples are plentiful.

Certainly, metrological arrangements did not present a uniform picture throughout Poland. It would seem that disputes about weights and measures were less common in Greater Poland and Pomerania, the striked measure being applied there almost universally (save in the brewers' measure of malt). Nonetheless, our examples above, drawn from western Lesser Poland, are sufficient to make us see that such disputes were frequent in the country.

Just as the *szlachta* were interested in increasing the units of dry measures and of weight (in particular, as we shall see, the smaller ones), so, too, they would seek changes in the measures of length, and thereby of area. Here, however, the interrelations varied, and it would appear that, in different periods, different tendencies prevailed. In the sixteenth and early seventeenth centuries, when the tendency was marked for the demesne to grow in size at the expense of the peasants' holdings, it would appear that the prevalent procedure was an attempt to diminish acreages through reducing the unit measures of length. On this pretext—if a pretext was needed—the peasants' shares were decreased. In the second half of the eighteenth century, however, when it was becoming difficult to reduce the size of the peasants' holdings, and at the same time the system of *wydział* corvée was spreading wider and wider, the practice of lengthening the "rod" or "yardstick" was, apparently, favored as a method of extracting more labor from the serfs. In both periods, however, these two contradictory tendencies were active simultaneously, with conflict resulting at times. Also, at the end of the eighteenth century the "rod" was decreased to aid the process of decreasing the peasants' holdings. In 1785, the primate's villages of Sędziejowice and Żaglin complained that the recent tenant, one Podgórski, "had acted most injuriously towards us, so much so that even in the measuring of the fields he had used a very small rod, which reduced us so badly, we no longer know what to do and how to support ourselves."[33]

Yet in practice, it would seem, the stronger tendency among the nobles was to extend the customary measures of length, especially so in the period of the expansion of the *wydział* corvée, for it was these measures that served to assess the amount of labor due per serf-day. This, at any rate, transpires from numerous eighteenth-century transfers. Thus, the communities of villagers in the Ryty *starostship* accused before the Referendary's Court the *starosta*, Onufry Kicki, and his officials of "decreeing sticks of excessive length for plowing."[34] Again, the villagers of Turów and Kurów complained that, as their landlord, the mayor of the town of Wieluń—or rather his tenant—"applied a four-ell stick to the area for plowing."[35] Similarly, several villages including Dobrzec, complained that the municipality of Kalisz and its named landlords "force them to plow fields measured out with any stick they please, not the one of lawful dimensions."[36] The primate's serfs at Bobrowniki complained in 1785 that the area for plowing was measured out by a nine-ell stick and not one of seven and a half ells as formerly ("they chopped it off only last week"); the tenant naturally denied this.[37] During the 1789 inspection, the people of Tczów complained: "They give us huge areas to plow, overlong strips (*stajo*)* and large patches,"[38] and at Gzowice the villagers said: "The land we have to plow and to hoe they do not measure by the stick . . . of seven and a half ells but by a longer one the landlord made for himself, and none of us can manage to complete the work in a day; we have to finish it the following day."[39] If the good men of Gzowice are to be trusted, then the old stick had been truly "just," that is, functional, in that the day's work measured by it was within a man's capability.

Several villages on the same estates recorded similar complaints: "They have now made two strips out of three, and when they have it in for someone, then they will count as many as 50 rods to one strip."[40] The community at Błędów, likewise the primate's, writes under the same date: "The ground is measured by a stick other than the one they got from the treasury, and on top of that, the steward adds to it, so that more or less three extra rods per worker's plot are assessed."[41] Similarly, the serfs of Wrzeczko insisted that the perch used to measure out a day's plowing was unfair.[42]

Naturally, the manor would only make sure that the peasants would not diminish the "stick" used in the assessment of labor dues. In the inventories and in written instructions there are copious references to the rod or stick "marked by the squire,"[43] in the safekeeping of the "governor" or the "village administrator." The village, as the evidence cited shows, accused the manor of secretly lengthening the rods and sticks. The Crown Referendary's Court, in endeavoring to ensure the observance of the 1764 statute concerned with the standardization of measures, and also in seeking to mediate between the serfs on the royal domains and their lords,

* A *stajo* is an old Polish measure of length, affirmed by the statute of 1764, relating to cultivable land, and equal to 134 meters—TRANS.

established the following norms: (1) the length of the "rod" shall be standardized nationwide at 7½ ells; (2) the rods shall be stamped by an *ad hoc* commission appointed by the Referendary; (3) standards, duly stamped, shall be produced in pairs, one to be kept at the manor and one by each village community. So, for instance, when a dispute arose in 1788 between a number of villages and the nunnery of Franciscan Sisters at Cracow, the Referendary's verdict ran: "We recommend that the stick or rod to measure the land shall be made, in time for the next executory commission, of the length of 7½ ells, stamped, and one provided for the manor and for each village community."[44]

The phrasing was somewhat fuller and more interesting, however, in another case, possibly of a more acrimonious metrological dispute. In adjudicating upon the difference between the villagers of Mogiła and the local monastery, the Referendary decided: "Two rods shall be measured up and stamped, one to be kept by the village headman and the other at the manor, *the headman to have the right to check the latter against the former on every necessary occasion* [italics added—W.K.], and if an injustice should be perpetrated, then, on an appeal being made to the manor against the overseer, compensation is to be meted out to the injured parties without any further investigation."[45] Of course, the verdict does appoint the lord of the manor to be *iudex in causa sua.* Yet it is, in our view, very important that the village headman was, by this decision, entitled to check "on every occasion" whether the manor's rod had not been secretly lengthened; it seems possible to see here a major success scored by the village community in question in the metrological aspect of class struggle.

Instances of the measures of length being decreased are few and far between in the eighteenth century, whereas instances of increasing them are legion. For example, when the villagers of Bobrowniki in the estates of the primate were dissatisfied with the lengthening of the rod, the inspector questioned the tenant who "answered, admitting in all conscience, that he had not told the steward, who was a local man, to measure out large sizes and confronted him with witnesses who had been present on the occasion that led to their complaints about the steward: that he had ordered the rod to be cut a measured half-foot longer; seeing this, the peasant who has been the instigator of today's accusation, produced the offending piece of length, which he had chopped off and hidden, and applied it to the rod at this point."[46] In the dispute with the peasants of the Żnin estates, the verdict of the primate's inspector reads: "And since the village communities complain about the calculation of the length of strips, he rode out into the fields, ordered remeasuring with the true rod in his presence, saw that the lengths varied from 22, 27, 28, 35, 37, to 38, and, therefore, in order to avoid further differences, made the following decision: that the amount of land to be plowed in a day must be neither less nor more than 300 cross [i.e. square] rods . . . the rod itself being no longer than 7½ mercantile Warsaw ells."[47]

As one would expect, conflict situations relating to the measures of length did occur, affecting both the manor and the village. The peasant assembly at Goszczanów, if the evidence of the inspector of the primate's estates is to be trusted, placed itself in a tight situation. The inspector writes: "The received inventory states rent for vacant plots (wasteland) at one złoty per stick, adding up to 30 sticks in length by one stick of width. The stick should be 8 feet long, but during the last inspection by squire Kozietulski the peasants complained that it was too long, so it was shortened to 7½ feet. And now the peasants complain that for measuring the wasteland, this stick is too short. It was remeasured, found to be 7½ Warsaw ells long, and was therefore approved."[48] It would appear that the Goszczanów peasants had gone too far. They fought for a short stick with which to measure their labor dues, and then wanted a long one to be applied to the wastelands they held on lease. They could hardly have it both ways! The matter was settled by arbitration in favor of the official Warsaw measure.

Lesser units of length (e.g. the span) were also increased in efforts to increase the textile yarn dues collected from the peasants.[49] On one occasion, in the estates of the primate, the inspector concluded that "the village community quite justly complains in the matter of yarn because the ell applied to their dues was longer than the common norm." In this case it was, in fact, the ell that was being called into question. The manor preferred a longer ell and, accordingly, a longer measuring stick when assessing the peasants' *wydział* corvée, but a short one when the wastelands were being leased for rent to the peasants, and again, a longer one when collecting the skeins of yarn from them. The peasants obviously wanted things the other way round.

On some occasions, the yarn was collected by weight instead of by the ell-length. Again, disputes would arise associated with this method.[50] What was consistently demanded was that things should be done in the open: the steward was to mete or weigh in the presence of peasant witnesses. Yarn presented significant problems of its own in the peasant economy, which was still primarily a subsistence economy. This is clearly revealed by a petition from the villages of Gorzyce, Dochonowo, and others: "In the year 1785, landlord Sutkowski issued distaffs for spinning, but his weight was unjust, and when our wives and female tenants completed the spinning and we delivered the yarn to the manor at Jaroszew, the weight was short, so we had to give up some of our own yarn; and so ruthlessly did he insist upon this, though the times were hard, that now not only we, but our children, too, will have to go without shirts."[51] There we have it: if "our own yarn"—be it wool or flax—had to be foregone, then so did one's own shirt; the question of purchasing it for money did not arise.

The manor would also try to enlarge the bushel in assessing labor dues at the time of threshing. The primate's village of Smardzew, for example, complained: "Once they would give no more than half of a quarter to beat per day; now they give a large quarter that we have to beat and thresh."[52]

A second fertile field for disputes associated with measures of capacity arose from the endeavors of the feudal landowners to secure their own gains from other potential rivals in the exploitation of the peasant. The landowners' efforts here were particularly aimed at gaining control over municipal measures. As far as the situation in the countryside was concerned, the chief issue at stake was bound up with the manor's battles against the "abuses" perpetrated by the millers and innkeepers. In the stewards' manual of instruction, we frequently encounter reminders to "have in the mills no measures other than stamped ones."[53] The Referendary, doubtless prompted by the peasants' accusations, offers similar admonitions: "Millers must charge fair stamped measures per bushel, to wit, no more than a sixteenth part, or two gallons."[54] Many similar suggestions may also be found in the tendency of the manor to oppose the alterations of measures by the innkeepers, who would, of course, reduce the measures of beverages, thereby raising their prices, just as the millers would enlarge dry measures in which their charges for grinding were calculated. Huge numbers of relevant injunctions turn up in the sources.[55] In such situations, the manor might safely assume that it would enjoy the support of the village. For example, the steward's manuals command him "to call up the villagers and recommend them to see to it that the publicans pour their liquor using stamped measures"; to enable the peasants to see to this, it is further demanded that "all tin measures in the inn be riveted to the wall and stamped in accordance with the measures instituted by the Commission."[56]

The requests are sometimes amusing. For instance, that addressed to the tax clerk of the Rogów estates in 1779 states: "In serving liquor and beer, just measures have to be used . . . and prices charged in proportion; particularly, drinks in glasses should have little wooden labels attached by string, stamped and signed by the scribe, and registered, and these measures must always be wholly accurate; and whosoever breaks or damages the vessel has to pay for the damage."[57] And so, when a customer raised his elbow, a wooden plank, duly signed, would dangle from his cup. When a dispute between the villagers and the innkeeper was irreconcilable, the latter would be sacrificed as a scapegoat.[58]

The practices of piece-workers in diminishing measures of capacity have also been noticed in the history of Polish manufactories. Wagon drivers taking charcoal to the works would often engage in such practices; therefore, the managers, according to Osiński, would "stretch the bushel upwards, vertically, by inserting three bars to prevent its narrowing by the filler."[59]

In the continuous, agelong struggle focussing on metrological questions, the village, as we have made clear, "stood firmly" by the measure it considered fair. Was it also considered fair in internal transactions among the villagers? This facet of the question has produced the least evidence, for obvious reasons. In the law court files of the village of Kasina Wielka, under the date of 1605, there is a highly relevant passage: "The community has declared that none dare sell by any measure other than the Wieliczka measure, and should any one transgress, then he shall give up sixty pieces to the lord and receive twenty strokes of the lash in the stocks."[60] The procedure must have been approved by the manor (convent, in this case), since the monetary part of the penalty was to go there. But the extremely harsh corporal punishment suggests to us that the parish as a whole desired retribution, particularly since the reference is to the sale of grain and not to any feudal dues. We have many proofs that their old bushels were carefully kept by village communities. One of them, belonging to the Gniezno chapter, must have fallen upon hard times indeed early in the eighteenth century because it had to give in pawn to the neighboring village the privilege of possessing the traditional measure.[61]

As generations passed, the metrological conflict grew more acrimonious. The possession of land was concentrated in fewer and fewer magnates' hands, their latifundia became more vast and required more complex administrative machinery. There were ever more clerks and they cost more and more (both in absolute figures and relative to the lord's income), and—now rightly, now wrongly—they were increasingly suspected of feathering their own nests. Therefore, as far as the exploitation of the peasant was concerned, the magnate's situation was different from that of the middle-grade noble owning one or two villages. The latter wished simply to employ a larger bushel in the assessment of the *osep* due to him and, whenever possible, did so. But the magnate, on the one hand, desired a larger bushel too, but, on the other, was anxious lest the gain should fail to enhance his wealth and the heart of his estate—to wit, the peasant farmsteads owing him a variety of dues—be enervated, while the gain would surreptitiously accrue to someone or other in the administrative hierarchy.

Given the apparent antagonism between the manor and its administrative personnel and their seeming conflict of interests, keenly perceived and well attested in the sources, there was yet beneath the surface a deep solidarity between them. Inevitably, this bond is but rarely revealed by the sources. Even the contemporaries were not always aware of it. There are many references in the sources testifying to the manor's efforts to check the millers' abuses;[62] in fact, however, the manor not infrequently winked at them, since a larger income for the miller made it feasible to extract a larger sum for his lease of the mill.[63] The landlord of the village of Czubrowce in the Ojców *starostship* in 1789 clearly forewarned the inspectors that, should they allow the peasants their wish to have their dues assessed by a smaller measure, the end result would be to the disadvantage of the state treasury,[64] and the tenant would recoup his losses from the owner.

Hence, in the manuals of estate management, which constitute our best evidence of the desires, intentions, plans, hopes—in fact, the overall economic policy of the great landowners (the one-village lord had, after all, nobody to write manuals for)—as far as measures were concerned, the stand taken was initially equivocal; but, as time went on, the manuals came increasingly to favor stable measures and set measuring methods. The crystallization of their attitude was a lengthy and complex process—after all, they had to take into account deep-rooted customs and conflicting interests. For not only—indeed, not most importantly—was it the *osep* that was at stake, but the measurement of the corn of the demesne farm itself (the *folwark*)—at threshing, at receipt by the granary, and yet again when distributing it therefrom.

In the manual Piotr Michałowski issued to the stewards of his estates in 1774, we read: "Every kind of corn being moved from the barn to the granary should be measured, as a rule, not heaped up but striked, with a top."[65] The instruction is, indeed, none too clear. Presumably, it would be clarified for the stewards for whom it was intended by the inclusion of the phrase "as a rule," namely, the rule of custom. And yet it was precisely in the lack of precision of the written instructions that there lay opportunities for abuses by the management. Hence, in an instruction for the barn scribe in 1787, the phrasing was clearer: "The barn scribe shall take it as his duty to see that all corn moving from the barn to the granary be measured striked, in the presence of the overseer, with a small top amounting to one eighth on top of the bushel or quarter [clearly, the different grades of officials were being set to watch each other—W.K.]. . . . As far as the overseer is concerned, he should not, immediately after measuring, distribute the grain from the barn, but only from the granary, the scribe again using the strickle, leaving a small top."[66] Here, it is custom that seeks to codify and calculate, the guarantee against abuses being the mutual watchfulness of the scribe and the overseer. And so, for all that a desire to collect corn using a large measure and to give it out using a small one may be evident, it may be better and safer to use but one and the same measure throughout.

"The overseers," so we read in the instruction for the manager of the manorial farms at Buczemla and Oleszkowice, "ought to distribute grain from the barn using the strickle and employ the same method in collecting it for the granaries whence it will go."[67] The matter is clearer still in the instruction for the manager of the Stanków estate: "The receipt of the grains at the granary as well as their distribution has to proceed every time by the Lvov measure, which is at present settled as striked, with no top whatever, as it is easier for the auditor to accept a reasonable deficiency from mice and rats than to incorporate in his accounts some unusual excess arising from topping-up."[68] Was, then, the bookkeeping convenience the motivation here? Surely not. The "excess"—and not only in the eighteenth century—suggests that there was here a potential scope for malpractice. The instruction manual of 1791 for the Opatów estates insists on similar procedures;[69] that at Tworyczów goes still further in explicitly referring to the need to prevent stealing;[70] while that at Boćki, as late as 1800, actually threatens the tax clerk with having to compensate any party harmed by some dishonest practice, if found out.[71] There were similar ambivalent views of measures other than dry measures. Sometimes the attitude was purely traditional, for example, in the manual for the manager of the manorial farms of Endrychowice, Bobłowo, and Zubowszczyzna in 1759, wherein we read: "The *morgs** have to be measured out without burdening the serfs, and taking time into account."[72] We have here a case of the traditional stance—but one that was simultaneously obsolete; for the idea was that the time needed for work would determine the measure rather than *vice versa*! Consequently, the measure would have to be altered if it turned out that a day's labor was insufficient to accomplish the task so assessed. Elsewhere, however, we note a quest for an "indifferent" or neutral measuring unit—as at Tworyczów in 1763: "Let the sticks by which the serfs' land for plowing is measured stay neither larger nor smaller."[73]

It would appear that in the eighteenth century efforts to prevent abuses by officials had to be intensifed. Concern about them had long existed—as long, indeed, as very large estates themselves, with their serf-worked demesne farms (*folwarks*). Let us cite Gostomski's pertinent, vigorous and cynical observations: "The lord or superintendent must show the overseer, in his own presence and the presence of the villagers, how to measure from the barn, just as the measuring should be done for the market or for the barge or at every distribution—to wit, so as to leave nothing in the bushel. Herein the loss is considerable: because in taking from the barn and to the market heaped, and then selling levelled flat, the official levies at least a tithe on his master . . . which practice must not be condemned when it serves the lord's purse instead of the thieving official's."[74]

The owners of the latifundia, consciously or unwittingly modelling their conduct upon that of the sovereign rulers of the Enlightenment, tried, like the latter, to standardize measures throughout their estates. The inspection records from the sixteenth and early seventeenth centuries reveal that, now and then, measures varied within the same complex of royal estates; and eighteenth-century administrative manuals produced by magnates tell us that they assiduously sought to standardize units of measure within their "states." However, the standardization of measures within the lands of one particular owner meant that they would differ at least by some mark or stamp, let alone by their magnitude and nomenclature, from those binding elsewhere.

* Unit of land measure, ca. 5,600 sq. meters—TRANS.

We have already noticed that landowners would try to impose their "particular" measures upon their own officials, peasants, millers, and innkeepers. Naturally, they tried to do the same in relation to their towns. There is good evidence for this in the numerous instructions issued to the governor and other officials at Opatów.[75] Time and again they are advised that the magnate's employees are expected "to check frequently that just measures are in use." Upon the magnate's estates, the lord himself would determine the measures, and the men acting on his behalf would count among their functions the control of measures, arbitration in disputes among the serfs, and also judicial action upon metrological transgressions.

Princess Anna Jabłonowska, in her manual on urban administration, assumes the role of defender of the burgher-consumer against faked measures, as well as defender of the morality of burgher-sellers; she imposes control of measures through appointing a "watchman of good order" and an obligation to make use of town measures ("so that every object shall be weighed in the public scales at the town hall").[76] As a curiosity, we may add here that in the instructions for the scribe at the forge where Jabłonowska herself is the supplier, as a seller of copper—the buyers being most commonly Jewish merchants—her order is that the weights (she goes into much detail regarding their safekeeping) "have their denominations spelt out on them in oil paint in Hebrew, so that each Jewish buyer shall truly know that cheating is quite out of the question here."[77]

The measures determined by the great landowners for their own estates were controlled by their employees. In order to facilitate the work, and simultaneously underscore the lord's sovereign power, the measures were marked with the lord's stamp—not infrequently, the family crest.[78] Just as in the royal towns measures would be marked by the deputy *voivode* with his official stamp, so in the royal villages the inspectors or the Commissioners of the Crown Referendary applied theirs. A name would follow the stamp. In the eighteenth century, in White Russia, the term "Radziwiłł measure" was well known, signifying a measure binding throughout the vast estates of the Radziwiłł family.[79]

The right to determine measures was an attribute and a symbol of the noble's authority in his lands, just as it had been a symbol of Charlemagne's sovereignty in his empire. Our sources do not suggest that the *szlachta* felt any real obligation to observe the metrological laws enacted by their own Diets—neither that of 1565 nor of 1764.[80] Let us, therefore, cease to wonder at finding but one citizen (the one "just man" to the historian of metrology two centuries later!) who, as early as 1767, ordered that the royal Warsaw measure, decreed by the statute of 1764, be used in his estates of Knoryda, Romaszko, and Boćki.

Once more, it was left to Gostomski to express with infallible accuracy the consciousness of the Polish noble—and he a *voivode* himself, writing 23 years after the Piotrków statute! He writes: "The sower's bushel for all foodstuffs, animal feeds, and malt must be the same on every lord's demesne, but *it need not correspond to that of the market or of the voivodeship: at home on his own ground he can decide howsoever he pleases.*"[81]

· 19 ·

THE STRUGGLE OF THE NOBILITY
WITH THE BOURGEOISIE
IN THE URBAN MARKET

The centuries-long struggle for the control of measures in the urban market in Poland is of the greatest interest. It forms a chapter in the history of metrology, and at the same time, it reveals the multifarious conflicts that were tearing apart Polish society at the time. For the market—although attracting some groups more than others—was the meeting place for all: the peasant and the burgher, the noble and the cleric. It was in the market that the goods produced would be invested with value, and shifts in the distribution of the national income would take place. To be able to influence the market meant being able to guide those shifts in certain directions rather than others. The *voivodeship* price tariffs[1] constituted the most basic means of exerting influence; their workings would accord with the interests of those in power—to wit, the nobles. There also existed very considerable opportunities for breaking, circumventing, or sabotaging the tariffs by those against whom they were designed. In practice, it would seem, they were circumvented at least as often by decreasing or increasing measures as by increasing or decreasing prices. The researches of students of the history of prices have established the discrepancies between the stated tariffs and prices that obtained in fact, and have thus set the lower limits of those deviations. However, their other direction generally remains beyond our apprehension, since we are unable to establish precisely the quantitites to which those *de facto* prices referred. The fact that the opposition to the tariffs took the form of altering measures brought about the extension of *voivodeship* controls to that area, too. For, lacking the power to control the measures, it was impossible to affect the shifts taking place in the distribution of the national income through the intermediary of the market.

However, within the urban market everybody appears in at least two roles: the noble and the peasant as sellers of corn and buyers of alcoholic beverages and manufactured goods, and the burgher, the reverse. Each of them is anxious that some measures should not grow larger, while ready to welcome increases in others. Such

147

feelings produce certain solidarities and antagonisms. For example, the noble defends the peasant, on whose well-being he depends; yet if the peasant is to be exploited, then the noble wishes to enjoy the monopoly of so doing. But the noble owner of the town hides the metrological swindles of his burghers (while complaining about swindles being perpetrated in other towns), for he too wants no competition in exploiting them. Affluent urban merchants, after all, do not view in the same way the measure of the grain they buy and the measure of bread.

Moreover, does not the town live in the closest relationship with its rural "hinterland"? It cannot be emphasized too strongly here that neither the *voivodeship* tariffs nor the Diet's attempts to bring about standardization referred to the measures the gentry had in use on their own estates, but only and exclusively to those used in town markets, a fact that frequently went unnoticed. The Piotrków Diet statute of 1565, which remained Poland's fundamental metrological statute for two centuries (until the Coronation Diet statute of 1764), stated unequivocally: "And furthermore all *oseps* and dues, howsoever titled, all the serfs shall give unto Us and unto their Lords, and unto whomsoever they owe them, according to the old measure and relation."[2] The determination of the standard and the safeguarding of measures were attributes of authority, and no one—certainly the diet of the nobles least of all—would dare impugn that authority. This is a matter of particular importance in forming a proper estimate, as we shall try to do later, of the boldness of the 1764 reform—on paper, at any rate.

The *szlachta*'s attempt to stabilize measures used in town markets was coupled with their quest to increase them in their own estates. But the divergence between the two could not grow indefinitely. In the final analysis, urban institutions would submit to the pressure from rural institutions, and this is scarcely surprising: for one thing, towns were like islands in the rural sea; and, again, this move served the interest of the most influential stratum of the urban population. The bushel would grow larger in the countryside, and after some delay it would also do so in the town.

Let us support our general statements with a handful of examples drawn from the resolutions of the dietines held in the southeastern marches of the Polish Republic, namely, the *voivodeships* of Lvov, Przemyśl, Sanok, and Halicz. The phrasing of the resolutions arrived at by the gentry in question reveal their objectives and desires. The monotony with which they are repeated bears witness to their ineffectiveness, and thus, in our view, they were needed not because (at any rate, not only because) of the weakness of the executive, but because of the inconsistency of the nobles, who sought to increase measures in the countryside while stabilizing them in towns. In the resolutions, we find within the period from 1616 to 1764 some scores of provisions touching upon the urban market measures. The phraseology of the resolutions is unequivocal. They speak of the bushel "containing no more than . . . ,"[3] or that "the *półmiarek** must not contain more than . . .";[4] they demand that summons to court be issued to those "who dared employ in towns, large and small, oversize measures, to the disadvantage of the poor" (this injunction, however, does not preclude, in the very same resolution, a demand that "the pound must be no less than . . .").[5] The proclamations insist that burghers "may not dream up measures *in super*"[6] and that municipal authorities "must carefully revise the *półmiarek* and get rid of the oversize ones."[7] Again, it is stated that "the *półmiarek* must be no larger than . . .";[8] or, more clearly, that townsmen enlarge measures, and "when a seller comes to use a measure, they foist on him the false one,"[9] or, indeed, quite plainly, that they do so with measures "which *pro libitu suo ementes* [*sic!*] employ in calculating the grain, whereby we ourselves and our serfs are greatly annoyed."[10] Or that the burghers "substitute, in buying various grains, *ex arbitrio suo półmiarek*, and decline to acquire stamped measures for use in purchasing corn";[11] "they displease nobles and their serfs by enlarging some measures beyond what is customary, and then the clergy, levying the Mass tax, also try to adopt such measures in collecting from the serfs."[12] And so on, *ad nauseam*.

The meaning of it all brooks no misinterpretation: the seller's anxiety is quite clear in every instance. Taking the language of these pronouncements literally (and it is their literal sense that shows most clearly the intentions of the *szlachta*), it is not the standardization of measures that is being decreed, but "maximum measures" of sorts; and the seller would not object at all to the application of smaller measures. After all, the phrase that "the measure must be no more than" is repeated time and time again. A certain anxiety about the serf's "fair deal" is discernible, too, an anxiety lest the burgher usurp too large a share of the value created by him.[13] But the burgher, in his turn, is also subject to a feudal dependence; just as the owner of a village seeks to ensure that his peasant shall be exploited by no one else, so the owner of a town, in seeking to

---

* An obsolete Polish unit of dry measure, literally the "half-measure"—TRANS.

exploit his burghers, may well defend alterations of the measures that they perpetrate surreptitiously. That this position was often adopted we learn from the following *seymik* resolution, dated 15 September 1716: "And should the squires, landowners, or governors of towns and country towns place obstacles in the way of the institution of just measures, in order to safeguard the burghers' interests and that *ad quodvis iudicium pro irrogandis super ipsis poenis*, the deputy *voivodes* will be entitled to summon them" (cautiously, nothing is being said about the magnitude of the penalty).[14] Likewise, the landowners are not to hinder the collection of fines imposed upon towns for metrological offenses,[15] the attempt at prevention presumably implying here that such cases had occurred. Finally, with the 1764 reform making its mark, allowing the various parties to transactions to save their faces, as it were, and give up reciprocal deceits, the *szlachta* of the Lvov territory declared: "Let us hold ourselves obliged, one and all, that each of us shall in his towns, country towns, and locations, both hereditary and held by lease, without any delay determine weights, measures, and ells, and have the goodness to adhere to them."[16] Obviously, whatever came to pass in the countryside was of no concern to the signatories. And yet it was they whose influence was decisive.

We have earlier quoted numerous attempts to regulate the size of the bushel (and *półmiarek*), in which the emphasis lay upon the phrase "no more than," since our aim was to draw attention to the defining of *maximum* measures. Let us now consider the question of precisely what "no more than" meant in terms of gallons (*garniec*) to the *półmiarek* in successive periods. Apparently until 1674 the ever-present "no more than" referred to 24 gallons.[17] In 1677, we find the figure of 26 gallons,[18] but the norm of 24 gallons was again in use in 1683.[19] However, ten years later, in 1693, it was 30 gallons,[20] a figure that there was, it seems, an attempt to discard a dozen years or so later—but in vain, for the March 1707 *seymik* retained the norm of 30 gallons for measuring spring corn, while trying to return to 24 gallons for the winter crop which was of prime concern to the *szlachta*.[21] Yet two months later, the maximum of 30 gallons had to be accepted "for both the winter and spring corn."[22] Even this did not last long. By 1716 the figure of 32 gallons was agreed upon,[23] to stay until the end.[24] Within half a century, from 1674 to 1716, the measure—assuming, of course, that the gallon's dimensions did not change—was increased by a third.

Given our present state of knowledge, the mechanics of the change in the norm are not easy to determine. It would seem that, on one side, an important factor was the steady continuing rise of nominal prices. The burgher, compelled thereby to pay ever higher prices for his corn, tried at least to retard the rise in price by increasing the measure. In the end, during the period 1674 to 1716, the price of corn in the markets of the major towns rose enormously.[25] On the other hand, it seems arguable that this development was possible solely because the bushel used in the countryside was simultaneously growing in size. This would be effected by the lord, seeking to enhance his dues in kind. The enlarged bushel would, despite the peasants' resistance, come into common use, enabling the burghers to increase the urban measures. The peasants, numerically dominating the market, might well have refrained from objecting, since they were already accustomed to a larger bushel in their village, and perhaps even perceived it as normal and "just." In conclusion, let us note that the seller's anxiety about the developing situation is discernible in three concerns of the relevant *seymik* legislation: the question of the "heap," the question of whether the seller or the buyer should perform the measuring and apply the strickle, and the question of the availability of the standard.

As mentioned in the preceding chapter, the nobles, in collecting their dues and *oseps*, always demanded that they be rendered "heaped." In the town market, however, acting as sellers, they demanded "striked" measures.[26] But it was scarcely feasible, in the long run, to have it both ways. Nonetheless, the gentry from generation to generation persistently hoped to do so, demanding in addition that the strickles be stamped.[27] At times, they would at least seek a compromise in keeping with the spirit of the times and agree that the cheaper grains might be sold "striked." Thus a *seymik* decree of 20 May 1707 lays down "that the *półmiarek* contain 30 gallons of either spring or winter crop, striked, but excepting oats which should be filled therein heaped,"[28] and in another district it was declared in 1733 that the burghers "shall measure rye, wheat, and peas striked, but other grains heaped."[29] Until the partition of Poland and even later, further attempts were made time and again to enforce striked measures—but without success.

It was the constant endeavor on the part of the nobles to establish the custom that the seller—that is, the noble—and not the buyer, be entrusted with the actual task of measurement. The importance of this task has been discussed in chapter 7, above, with reference to its impact on the final outcome. Again, the attitude of the nobles was inconsistent: they wanted to measure the corn themselves when they collected it as dues from their peasants, and they wanted also to measure it themselves when they sold it to the townsfolk. The fact that they had repeatedly to try to establish this dual practice provides the clearest proof of their limited success.

Lastly, the *szlachta*, while jealously guarding standards in their own possession, would yet demand that in urban markets standards be publicly displayed and available for use without payment, for here their position as sellers was vulnerable. In demanding free use of standards, the *szlachta* would again pose as defenders of the peasants: "Whosoever shall dare take howsoever little per bushel or per measure from our serfs, let him pay as penalty one hundred marks."[30] Time and time again, the demand would be reaffirmed that measures, that is, standards, should be displayed for public use and made of durable material (stone). Doubtless, every case where "just" measures were introduced in a town would involve as its concomitant the destruction of the measures previously used, as "unjust"; the loss thereby to our museums and to the historian today has been irreparable. But in the urban market the noble, and the peasant, too—his interest in this instance protected by the noble—were buyers as well as sellers. The *seymik* resolutions appear to imply that they purchased only textiles and alcoholic beverages. Hence the amusing attempts to fix maximum dimensions for bushels, and minimum ones for ells, quarts, and pounds.

The texts of the *voivodeship* price tariffs present matters similarly, if somewhat more circumspectly. The magnitude of the bushel is not defined in them explicitly as a maximum, but it is firmly stressed that "he who sells corn may strike it, and he who buys it may not stop him."[31] And it is stated clearly, with respect to lengths of cloth, that they must not be shorter or narrower than 30 ells and 2 ells respectively.[32] Nothing could be more typical of the situation than switching from maximum to minimum definitions of measures! We witness here an attempt by its leaders to further the interest of a social group that for some commodities acts the part of seller (corn), and for others, that of buyer (textiles). As buyer, what they sought was that there should be no price rises camouflaged by a diminution of measures; as seller, that there should be no lowering of prices masked by the buyer's crafty substitution of an enlarged measure. To achieve those objectives, however, was by no means a simple task because, for one thing, the attitude of the *szlachta* could not but be inconsistent and, secondly, the burghers had at their disposal some effective means of passive resistance and of sabotaging their social superiors' rules and regulations.

To repeat, the principal reason why the *szlachta* could not help acting inconsistently was that their position in their own villages dictated different ends and means of their conduct; they favored a larger bushel in the country, while in towns they did not want it to grow in size; again, in the countryside they were in favor of the measuring being carried out by the collector of the produce, whereas in towns they favored the supplier. They fulminated against the metrological "swindles" perpetrated by the burghers in towns belonging to other landowners, but were prepared to defend similar "cheats" in their own towns. They were indignant when the quart was made smaller in the inns on the estates of others, but concealed like manipulations by their own innkeepers.[33] Finally, the *szlachta*'s attitude lacked consistency because while it might be within their powers to order about their own villages or nearby towns regarding weights and measures, they appreciated that they were in no position to lay down the law in Gdańsk. Naturally, however, there was nothing to stop the landowner from applying one bushel to the dues he received from his serfs, and another to the grain he dispatched to Gdańsk; indeed, such gentry "theorists" as Haur or Kluk advised precisely this course of action.[34]

Ideally, what the landowners wanted was simultaneous employment of three different bushels—one in their own villages, another in the town, and yet a third one for their river-borne trade. But the explicit phraseology was all about a single measure—unchangeable, "immemorial," and "just"—and their objective was unattainable. It was not, of course, just the landowner, but the peasant, too, who sold grain in the town marketplace, the buyer being the townsman. Numerically (though not necessarily as far as the amount of produce for sale was concerned), the peasants preponderated. In spite of this, within the configuration in question, the peasant's position was the weakest: the townsman would be able to pay him the lowest price because of both the peasant's weakness and his "no costs" production. The price would settle within a range of bargaining between that offered by the townsman from a position of organized strength (he was aided by the coercive institutions of the municipal legislature and police) and that asked by the peasant from an incomparably weaker position. The transactions entered into by burghers and peasants would thus threaten to depress the market price, and this could not but affect the price that the nobles could obtain in the same market. True, the nobles were stronger, but their position was adversely affected by the fact that they sold in larger, indeed "wholesale," amounts. In these circumstances, it was in the nobles' interest, as we have already noted, to bolster up the peasant's position. The affluence of the two was thus to a degree shared, while, in addition, the nobles would hold on to their "monopoly" of exploiting the peasants—hence the direction, remarked above, of the endeavors of those who represented the nobles as an estate.

We have also alluded to the fact that the burghers held the upper hand in market contacts between themselves and the peasants—contacts that were fundamental to the existence of towns as well as to the evolution of the commodity market. The burghers' territorial concentration, the fact that the peasants arrived in town as unorganized individual suppliers from a multiplicity of places, the organization of the urban government, the coercive institutions at its disposal, their better education (e.g. in matters of arithmetic), their superior material resources (e.g. the possession of stocks of goods), and their access in certain instances to the state authority, which might offer a ready ear to potential creditors—all these strengthened the burghers' hand. But in their contacts with the nobles, they formed the weaker side. True, the great free towns of the Hanse in Germany, being the most powerful party in the market, also enjoyed control of the measures,[35] and the same situation obtained in the Italian towns in the heyday of the free communes. In Polish towns, however, formally it was the *szlachta* who had the decisive say in metrological matters through the legislative agencies of the *seym, seymiks,* and the *voivode*'s office. In legal terms, municipal authorities were reduced to an executive role: they were to ensure that measures approved by the state organs were duly used within their boundaries; but in practice, the towns were not defenseless in this contest. Unlike the free German cities, they could not promulgate laws and issue regulations, but they could—and, it seems, effectively did—sabotage the laws and regulations the *szlachta* made.

A different situation existed in an independent town, which was not subject to the will and whim of an owner or the *starosta* in the manner described, which was the rule in the Polish towns. In thinking of a town that was independent due to its political system and particularly its geographical and economic position, we have in mind, of course, the city of Gdańsk, where a triangular relationship was played out in the market. The actors were the Polish noble—or rather his agent—the local merchant-intermediary, and the Western European export merchant. The Gdańsk merchant held a privileged position, for, as the intermediary, he enjoyed a monopoly. He had to consider the Western European exporter—after all, the latter had free choice to buy his grain at Gdańsk or at some other Baltic port—but he did not have to worry about the Polish landowner-supplier, for the Vistula would take his grain nowhere but to Gdańsk. Every time a transaction was concluded, the latter was in a no-option situation: he could not return home with his grain and he had to sell it. Moreover, their situation, as well as the economic organization of the municipality of Gdańsk, enabled the local merchants to be efficiently organized, for how, otherwise, could they have taken advantage of their monopolistic position? Foreign buyers represented organized powers. Polish sellers, however, not only lacked organization, but competed among themselves, and there was no powerful state in the background ready to defend their interests. It was of no great consequence, but it typified the whole situation, that semi-overt metrological swindles continued over the centuries.[36] While regularly cheating their own peasants in the matter of measures, here in Gdańsk the *szlachta* more than met their match, and themselves suffered greatly from dishonest practices of the same sort.

The measure, therefore, is an instrument for influencing the market in a direction favorable to whoever is in the position of strength. And that party will be the social class which, at a particular point in time, holds, or shares, or participates in power. The control of measures by the town is a manifestation of the exploitation of the countryside by the town, for that control constitutes an instrument of that exploitation. The control of urban measures by the nobles is an expression of the domination of towns by the *szlachta* landowners and, in turn, of the exploitation of the towns by them. In the concrete situation of Poland under the sway of the *szlachta*, both elements of exploitation coexisted, albeit in unequal strength.

· 20 ·

THE HISTORY OF THE STANDARDIZATION
OF MEASURES IN POLAND FROM THE
SIXTEENTH TO THE TWENTIETHTH CENTURY*

[This chapter opens with a discussion of the 1507 reform.] The sweeping reform of 1565 did not affect the measures being used in the countryside. Evidence suggests some attempt to implement the reform in the towns. In the seventeenth century, regional assemblies of nobles were criticized for their relevant resolutions by political commentators, who pointed to the proliferation of measures, continuing as though the 1565 reform had never been passed, and who called for more uniformity. These demands multiplied in the second half of the eighteenth century, leading finally to the statute of 1764, which, for its time, was highly advanced. This was the first Polish metrological legislation to be enforced along the prescribed lines, with equal stringency in the countryside. However, although the regional inspectors calculated peasants' dues using the new units of measure, they failed to prevent the use, in day-to-day village life, of the traditional measures prohibited by the new law. A quarter of a century after the reform of 1764, on the eve of the final partition of Poland, uniformity of measures was still little in evidence.

The Prussian edict of 1796 established control of measures, in the territories annexed by Prussia, by a compromise that accepted the Wrocław units as measures of length and weight and the Warsaw units for grain and beverages. In Galicia, annexed by Austria, Francis II ordained the use of the Lvov measures. [The metrological enquiry set up by the administration of the Duchy of Warsaw in 1811–1812 is next discussed, and then the work of the Society of the Friends of Science, which paved the way for metrological reform.] The 1818 reform of measures in the Congress Kingdom of Poland (under Russian sway) produced the so-called "new Polish measures," a system based on the Parisian meter with one line equal

---

* Since this chapter is concerned entirely with developments in Poland and demands some knowledge of Polish history and historical geography, the author agreed that it should not be included in the English version of the book; the present summary is a translation (supplied by Adam J. Szreter) of the French summary by the author in the original—Trans.

156

to two millimeters, and preserving the traditional nomenclature and divisions. [The reform of measures in the free town of Cracow (Austrian part) is next discussed.] The use of the new measures on large estates was optional, and this amounted to a step backward from the principles of 1764. In places where the new measures were larger, the nobles would introduce them without conversion. The introduction of Russian measures in the Congress Kingdom took place in 1849, and the proposition to return to the "new Polish measures" in 1861 was rejected by Tsar Alexander II. [Spontaneous adoption of metric measures during World War I in areas under German occupation is noted, and the retention of Russian measures in the Austrian parts of Poland.] The decree by the head of state in the newly independent Poland in 1919 introduced the metric system, a measure that was facilitated by popular resentment of Russian measures as a symbol of tsarist domination.

PART THREE

· 2 1 ·

ATTEMPTS TO STANDARDIZE
MEASURES IN FRANCE
FROM 789 TO 1789
AND THEIR FAILURE

In Part One of the present work, we tried to indicate the many-sided social involvement of traditional weights and measures, their associations with man, with his work, beliefs, and values. In Part Two, we outlined the metrological vicissitudes in Polish lands from the beginning of the sixteenth century until the early twentieth, and the introduction of the metric system in a Poland restored to independence after 1918. Our next task is to examine the manner in which the new, modern, and indeed revolutionary metric system was "thought up" and introduced into daily life with high hopes that it would, at one stroke, resolve and reconcile the convoluted mass of clashing interests and different conceptions bound up with traditional measures. It was hoped, indeed, that with the new system would come equality for all, not only before the law, but also before the abstract idea of the meter that was embodied in the small bar of metal—always under guard, never seen by any one.

Let us, then, consider France, the birthplace of the metric system. The dramatic history of French measures will be presented here, as befits a drama, in three acts. Act I—pessimistic; in it, we shall unroll the story of how, over a thousand years, increasingly modest attempts at standardization followed one another, all to no avail, until even the most progressive men lost hope. Act II—optimistic; in it, we shall discuss the astounding outbreak of popular longings for a uniform "just" measure, of the faith of the masses in its feasibility, their condemnation of traditional measures that aided and abetted those cheating them, all of which rolled, like a mighty wave, right across France during the period of the preparation of the cahiers de doléances in 1789. Act III—realistic; in which we shall see how, at long last, this seemingly simple reform proved in practice most difficult to effect, and how it encountered passive and even active resistance from the very parties that had but recently vociferously pressed for it.

Legend has it that it was Charlemagne who "invented" the school, and that he standardized measures throughout his empire. As re-

161

gards the latter, we possess evidence in a few surviving passages from normative declarations of Charlemagne and his successors.[1] Let us cite the most important of them:

[A.D. 789] Volumus ut aequales mensures et rectas, pondera iusta et aequalia omnes habeant, sive in civitatibus, sive in monasteriis, sive ad dandum in illis, sive ad accipiendum sicut in lege Domini praescriptum habemus;[2]

[A.D. 800] Volumus ut unusquisque iudex in suo ministerio mensuram mediorum, sextariorum et siculas per sextaria, octo et carborum et tenere habeat, sicut et in Palatio habemus;[3]

[A.D. 813] Volumus ut pondera vel mensurae ubique aequalia sint et iusta;[4]

[A.D. 864] Mandamus et expresse praecipimus ut comes et Reipublicae ministri ac caeteri fides nostri provideant quaternus iustus modius aequusque sextarius, secumdum sacram Scripturam et Capitale praedecessorum nostrorum, in civitatibus et in vicis et in villis ad vendendum et emendum fiat; et mensuram secundum antiquam consuetudinam de Palatio nostro accipiens et non pro hac occasione a mensuariis vel ab his cui censum debent, maior modius nisi sicut consuetudo fuit, exigatur.[5]

Ordinances like those, repeated over the best part of a century, give much food for thought. The Crown's program in metrological matters, which is set out in them, was not to change in a thousand years. It rests upon the recognition of the royal prerogatives in the field of weights and measures as an attribute of royal sovereignty. The standard kept in the palace would hold sway over the entire area where the royal writ ran; the standardizing plans therefore applied to the state's entire territory. At the same time we have here some reliable evidence, bearing witness as early as the Carolingian period to the duality of metrological practice that, too, was to last a thousand years: namely, different measures for buying and for selling, for collecting and for distributing, and, also, the increasing of measures employed in collecting taxes in kind.

We are not competent to analyze the Carolingian reform, of which both the genesis and consequences would be of interest. It would be interesting, on the one hand, to know to what extent those reforms constituted attempts to spread throughout the empire measures that were in use in some part of it, probably in its heartland, for they could scarcely have been newly "invented" measures. It is not out of the question that a compromise was being sought between two coexisting systems, as would appear from a mixed duodecimal-vigesimal system of division. On the other hand, it would be of interest to know how far the Carolingian system was, in fact, introduced into daily life—its geographical reach and its social penetration. That it was employed in the administration of the imperial demesnes is undeniable; furthermore, it is certain that for a long time it applied to coinage (by weight), but it seems doubtful that the people of the empire at large, particularly those living some distance from its center, adopted it for practical purposes.[6] There is linguistic evidence that the vigesimal system must have pre-dated the reign of Charlemagne, namely in the daily usage of "the sixty," "the seventy," "the eighty," "the ninety," and a little earlier, "the hundred-and-twenty," too—all based on the unit of "twenty." However, in Belgium, which formed part of the Carolingian empire, people referred to the "settente," "ottente," and "nonente." Should we see in this usage some evidence of a differential reception of the Carolingian reform by people whose system was based on a number other than twenty?

The legend about the Carolingian "standardization of measures" was destined to live for a thousand years in the consciousness of both the rulers and the ruled. Statesmen and rulers, in their repeated attempts at standardization, would often invoke the Carolingian precedent; while the masses, in their demands for standardization, would keep reminding successive monarchs of the Carolingian reform as a model to be followed, and of metrological powers lost as a Crown prerogative during the centuries of the "Gothic obscurity." Exactly one thousand years from the date of the first Carolingian capitulary—that is, at the time of the preparation of the *cahiers de doléances*—the exemplar of Charlemagne was frequently invoked. At the close of the eighteenth century, in keeping with the spirit of the social contract, the voice of the people spoke of the uniformity of measures "in the beginning," and of their differentiation afterwards. The trend they bespoke was from uniformity to multiplicity, not from diversity to uniformity.

"All measures were equal in the days of our earliest kings,"[7] writes Paucton in 1780, and he goes on to explain that the chaos in the realm of measures "ensued, it would seem, as a concomitant of rents and other seigneurial entitlements." Some seigneurs increased, and others reduced, measures because some wanted larger taxes from their vassals and others wished to attract new settlers to their lands. This, according to Paucton, was the source of everlasting dispute between the Crown and the seigneurs, touching metrological rights.

Although some successes were, in fact, won by the Carolingian policy seeking uniform measures, yet as time went on and the royal authority weakened *vis-à-vis* that of the regional rulers and local feudal lords, there is no doubt that those successes ceased to count.[8] For centuries, the object of the struggle was to determine the level within the system of feudal decentralization at which weights and measures would be controlled. There were important issues at stake for the lords: first, income from the control over weights and measures; second, the amount of rent, wherever it was paid wholly or partly in kind; and, third, a vital form of social control over their feudal dependants that affected their daily doings and dealings.[9]

It is not difficult to imagine the numberless and endless disputes arising from these issues and demanding practical solutions based on precedent. Such practices and usages were eventually codified, and they are set out in the *coutumes* of the various provinces of France. The *coutumes* are well acknowledged as a source difficult to interpret: even when a particular collection is easy to date, it may yet be difficult to date some of the rulings it contains as some go farther back in time than others. For the *coutumes* were not only collections of customary law: at times they introduced new legal rules, too.[10] It is, furthermore, known that the *coutumes* attach a great deal of importance to the distribution of competences among the occupiers of the different rungs on the ladder of the feudal hierarchy. Thus, they seek strictly to differentiate various rights among those entitled to the high, middle, and low assessment of justice (in French: *haute, moyenne,* and *base justice*) with reference to three matters, namely, the right to determine the standard, the right to pronounce judgment upon metrological transgressions, and the right to derive direct economic gains from the monopoly of measures through the leasing of public weights and measures, the appointment of land surveyors, the sale or lease of their office itself, etc.[11] The *moyen justicier,* that is, the judge whose competence it was to mete out justice at the middle level, played the central role in the whole system.[12]

Numerous formulations in the *coutumes* show without any ambiguity that they tried to settle disputes that had arisen time and again in the past, and make the substance of the dispute quite plain. For example, the *coutume* of the region of Auxerre states that "the possession and leasing of the standards of weights and measures falls within the competence of the high judge, and the passing of verdicts in cases about transgression of the said measures falls within the competence of the middle-level judge, up to the sum of 60 sous." It then goes on: "He to whom the middle assessment of justice applies shall consider disputes regarding measures, up to and including 60 sous, these measures to be obtained from a seigneur to whom the high justice applies; and should some measures be adjudged as deserving to be burnt, then the said high judge shall consider such cases with no limit on penalties, and the said middle judge, cases up to 60 sous."[13] Elsewhere, again: "The said middle-level judges hold the right to lease from the seigneur, who has transferred to them his judicial rights, measures relating to corn and wine";[14] and "Trying cases about measures is within the competence of the middle judge, and he is to receive the measures at issue from the seigneur of the high judge. Should it be established that the said measures are false and ought to be destroyed, then the high judge shall be competent to impose penalties without limit and the middle-level judge up to 60 sous of Touraine. . . . The possession and lease of standards of weights and measures is a function of the high judge."[15]

In some cases, the autonomous rights of the municipalities in the area of weights and measures were reserved. For instance: "Linen cloth shall be measured by the Clermont *aune*, which is considerably smaller, as is shown by the iron standard in the upper market hall of Clermont";[16] "The consuls are in charge of the scales, in their own right as well as on behalf of the urban commune, and they are outside the castle gates";[17] or again, the ruling that in the whole judicial district of Orléans, wine should be measured by the "large Orléans measure."[18]

Sometimes municipal privilege would take the form of an exclusive exemption from some payment, for example, "The burghers of this town [Pernes] are entitled to take the goods home and to measure them there, at no payment to the tenant."[19] Sometimes similar exemptions would be less specific, for example, "excepting those localities of the said district which had very long since had their own measures."[20]

In one case, in Normandy, the metrological privilege was raised to a higher level: "The control of weights and measures throughout (toute . . . la seigneurie des mesures et de poix) belongs to the duke," and in this case—unique, in our knowledge—it is also clearly stated that the lord in question, to wit, the Duke of Normandy, may lawfully alter measures.[21] Here we have, expressly stated, the supreme instance of metrological privilege. Frequently, however, the *coutumes* bear witness to the attempts to stabilize measures. Obviously, they aimed above all to prevent alterations of measures by seigneurs who were vassals of lords enjoying the metrological privilege stemming from the principle that "he, who is authorized to mete out the middle assessment of justice . . . is competent also to decide matters of low justice, and he, who is authorized to mete out high assessment of justice, has rights at the middle and lower levels. The one to whom the middle level applies is competent to control measures. And should someone use a measure other than that of his overlord, or an unstamped measure, he shall be fined 60 sous."[22] Elsewhere, even more clearly, it is said: "If a lesser lord, who is beholden to fix his measure according to the standard of his overlord, should use a larger or a smaller one, then he shall forfeit and have taken from him the right of leasing measures, and may be fined any amount at the judge's discretion."[23]

Still more important, time and again we come across the proviso that the privileged party himself may not alter the standard: "The seigneurs are bound to keep at home standards that they have no right to alter"[24]—although this does not make it clear just who is to ensure that no alterations are made. The matter remains unclarified even when the threat of removing the privilege is unequivocal, as in the following instance from near Tours: "The seigneur, who enjoys the right of measures, be he a baron or a castellan or whatever, is not entitled to have more than one standard, which he may not enlarge or diminish, but must enjoy his aforesaid right just as it has always been done. If he acts otherwise, then he forfeits the said right and he shall be deprived thereof."[25] What is of special interest in the above passage is, of course, the phrase, "not entitled to have more than one standard." The realistic nature of this phrase is beyond doubt. It is there precisely because experience had shown that the seigneurs would keep more than one standard; they would then employ one when buying or collecting their dues in kind, and a different one when selling. This kind of practice is also attested by other sources to which we shall refer. Parenthetically, it may be mentioned here that the *coutumes*, in their normative endeavors, tried also to regulate many other, presumably contentious, matters. Some would set forth the duty of displaying standards in public places, or would deal with the manner of measuring, by striked or heaped measure (*ras* or *comble*), or again, would fix the ratios of the depth to the diameter of the *boisseau*.

Let us, indeed stop to consider the question of proportions. In Anjou, for instance, it was laid down that the depth of the *boisseau* should equal one-third of its width;[26] in Touraine, the ratio was virtually identical.[27] In the countryside, the normal eighteenth-century ratio was 3:1.[28] However, in Paris, although the diameter of the *boisseau* was again larger than its height, its shape was yet markedly less flattened.[29] The standard of the 1767 Nantes bushel, preserved in the museum Conservatoire des Arts et Métiers, exhibits actually a slightly larger height than diameter.[30] These proportions provide food for thought. The bushel's height—as we remark elsewhere—is, for practical purposes, subject to a limit: it cannot be greater than the distance between a man's dropped hand and the floor, for, otherwise, the filling of the bushel would be unduly labor-consuming. But to flatten the proportion to the ratio of 3:1 considerably exceeds technical requirements, as the examples of the urban *boisseaux* that we have cited prove. If, therefore, at the time of assembling the *coutumes*, the measure under consideration has a very flattened form, then it must have evolved so over a lengthy period. Naturally, the shape of the standard does not affect its net capacity, but certainly does affect the magnitude of the "heap" it is feasible for it to carry. If the party in possession of the metrological privilege is, at the same time, authorized to collect taxes in kind from his serfs, and it is customary for them to be rendered "heaped," then circumstances favor the shallower standard. This practice is attested also by evidence relating to the Polish countryside in feudal times. The French urban authorities, however, had no need to resort to a shallower *boisseau*, and hence in towns its height remained markedly greater. However, the emphasis in the *coutumes* on the flattened shape of standards seems to suggest that the intention was to stabilize that shape, which had evolved over a long period, beyond which it is not technically possible to go much farther, although attempts would be made to do so.[31]

The French monarchy tried, on many occasions, to intervene in metrological matters in order—as elsewhere—to restrict the seigneurs' sovereign claims and to enforce royal institutions throughout the kingdom, thus advancing the cause of its unification. Now and then, metrological disputes would present the Crown with opportunities for intervening and playing upon various social antagonisms in the quest to strengthen its authority. The Carolingian reform apart, such efforts grew increasingly frequent from the beginning of the sixteenth century, and their monotonous repetition over nearly three hundred years certainly testifies to their ineffectiveness. The catalogue (leaving aside a minor attempt by

Louis XII in 1508) opens with an exceedingly interesting decree issued by Francis I in 1540: "It is the view of many seigneurs in France that their seigneurial and judicial powers, attested by title deeds, incorporate the rights of weight, of the ell, and of other measures. But the supreme authority of the King incorporates the right to standardize all measures throughout his kingdom, both in the public interest and for the sake of promoting commerce, among his subjects and with foreigners."[32] Here, we have the nub of the contention. Did the King, in fact, possess that right? Certainly, Francis I and his successors were to insist upon it—some more successfully than others. The Crown's course of action would remain constant: to respect those privileges, which produced income for the lords (such as the leasing of standards, the right of measurement in public, or fines for metrological offenses), but to insist that they observe the royal standard. To win the latter point would be clearly most worthwhile for the Crown, for then, whenever disputes arose, the defendents would have no option but to compare their measures with the royal standard. The decree that we have cited makes the matter crystal clear: "We wish that all abuses, swindles, peculations, and mistakes made because of the diversity of measures, ells, and methods of measuring should cease in all cities, towns, regions, lands, and estates."[33]

But then, only three years later, the edict of 1543 signalled a retreat from that ambitious attempt. Again, Henry II renewed the royal quest. His decree of 1557 states:

> Weights and measures shall be reduced to clearly defined forms and shall bear the appellation of royal weights and measures. Since in all the duchies, marquisates, counties, viscountcies, baronies, castellanies, cities and lands observing the laws of our kingdom, weights and measures are of diverse names and dimensions, wherefore many of them do not correspond to their designations; and often, indeed, there coexist two weights and two measures of different sizes, the smaller one being used in selling and the other in buying, whence innumerable dishonest deals arise; we now designate persons of standing to see to it that overlarge weights and measures be reduced, both ours and those used by dukes, marquesses, prelates, counts, viscounts, barons, castellans, and other holders of the rights of weights and measures, in relation to wheat or other cereals, wines, salt, goods and produce, which are, according to custom, sold by weight or measure. And since our city of Paris is the

chief and principal one in our kingdom, being the main seat
of our parliaments and courts of law, we desire that the said
reduction of weights and measures be first carried out in Paris.
All lords in the city, suburbs and environs, who hold the privilege of measures, are obliged to furnish the names of their
weights and measures, from the largest to the smallest. . . . A
similar reduction will thereafter be carried out in other cities
and provinces of the kingdom. . . . Old standards, not har-
monizing with our measures, shall be broken up. . . . All shall
be required to regulate their measures according to ours.[34]

The following year, 1558, the above ruling was extended "to all
towns, settlements, lands, estates and judicial districts within the
competence of the Parlement of Paris."[35]

So many fine words, but to little effect. It could scarcely have
been otherwise, when even the central organs did not dare, in
practice, to undermine the seigneurial monopoly of measures, this
being precisely the exclusive right to determine their dimensions.
To appreciate this, it is enough to quote the decision of the "Grand
Conseil" in 1565, according to which "the serfs must render to their
lord the taxes and rent due to him according to the measure which
he holds in his estate, even if some other measure were in existence
in that locality."[36] On 6 February 1563, the Parlement of Paris
decided that "measures employed in assessments of inheritance,
unless determined by a contract of sale, shall be those binding in
the area where the inheritance in question is located, and not those
of the area where the contract was entered into."[37] These two decisions are sufficient evidence for us to maintain that despite the
legislation of the period 1540–1558, seigneurial monopoly of determining the dimensions of standards continued to be respected,
and with it, regional diversity of measures, too. In view of these
facts, a great deal of importance should not be attached to the royal
declaration of 1575, which, harking back to the decrees of 1557
and 1558 and even to Louis XII's edict of 10 October 1508, maintains that "weights and measures in the kingdom ought to be standardized";[38] nor to another decree of 1579, which declared that "the
nobility, clergy, and other persons henceforth shall no longer be
able to collect or receive their rightful dues other than by just
measures, on pain of loss of rights and a fine."[39]

Of the prevailing social mood, of the tensions and ambitions
behind it, of the various pressure groups that brought about both
the legislation we have discussed and the resistance that so effec-

tively blocked it, we know but little. It has come to our attention that as early as 1510 the three estates of Auvergne put forward a postulate in favor of standardization,[40] and that in 1560 similar proposals were put forward by the estates of Orléans and Blois, which pressed for them yet again in 1576 and 1588.[41] A counsellor of Henry III and president of the law court of the Parlement of Paris, Messir Barnabé Brisson, presented an amply substantiated case against the existing metrological chaos in 1576.[42] Brisson referred to former legislation and emphasized that it had been honored only in the breach. He stressed the chaotic nature of the situation brought about by seigneurial metrological monopoly. Indeed, he went much further in writing: "Only when there is but one weight and one measure for the whole kingdom, will particular lords, who enjoy the right of measures, be no longer able to stamp them with their marks, and be no longer competent to act as judges in metrological disputes."[43] This view constituted a blow at the judicial side of incomes derived from metrological privileges, particularly so if one bears in mind the stated principle that "the jurisdiction in questions of weights and measures is the prerogative of the Crown of France, and it is only by the King's grace that particular persons availing themselves of certain powers may do so."[44] It was precisely this that was at issue; the Crown of France had never relinquished the right in question, neither had it succeeded in having that right acknowledged by the feudal lords.

Admittedly, the proposal to standardize all weights and measures on the Parisian model was revived at the meeting of the States General in 1614,[45] but the subsequent lapse of that body, right until 1789, meant that the proper forum wherein to consider the matter was unavailable. There remained only the provincial estates, or the Parlements, which preferred to settle metrological disputes—which concealed social conflicts—with the aid of prejudication. Thus, the Parlement of Rouen, in 1678 and again in 1680, recognized the seigneur's standard of measure as correct in assessing the wine tax, yet did not preclude the possibility of employing the vassal's standard if he was able to produce documents affirming that he had held such a right of old.[46] Likewise, the Parlement of Bordeaux in 1692 decided that the term *boisseau* signified a striked *boisseau*, but only if extant documents did not spell it out otherwise.[47] A more far-reaching attempt was that of the edict of November 1705, whereby copies of all standards in use were to be deposited in the custody of local courts of justice to permit recourse to them whenever necessary.[48] Doubtless, here was an attempt on the part of the state authority to lay its hands on the "exhibits," which were crucial in metrological disputes. It would not seem, however, that this attempt gained wide currency in practice.

Again, the Paris Parlement declared on 18 August 1710 that "ground rent should continue to be paid according to the measure originally employed in its assessment, and to this end there should be provided, at a shared cost, two measures, one for the owner and the other for the payer."[49] All traditional rights, subsuming a variety of measures and methods of measuring, were accepted as normal and fair. The request that two identical copies of standards be made was only for the purpose of avoiding disputes at collection time. Again, nothing else was aimed at by the Chambre des Enquêtes in laying down on 11 April 1715 that "for the receipt of seigneurial taxes, the measure is to be checked either in the neighboring royal court of justice or in the seigneur's court, for he held the right of keeping the *boisseau*."[50] This was a very weak test, since it accepted the standard not only of the King's court but that of the seigneur, too.

Less is known of the attitudes of the Church and various grades of clergy towards standardization. In the *Memoires du Clerge* of 1716, it was stated that "the measure used in the payment of the corn tax to the parish priest ought to be either the local public, or the royal, measure." This might at first glance appear as an expression of strongly anti-seigneurial sentiment, but the following explanation suggests otherwise: "The reason for this is that it is not proper to accept measures other than those checked in the royal law court or in a seigneurial court, the seigneur holding the right of possession of the *boisseau* and the right of coercing those subject to his jurisdiction to use that *boisseau* when buying or selling."[51] True, only the parish priests have been specified here; but we shall see more than once that the higher clergy were ready to defend metrological privileges. The parish priests, in fact, often held equally fast to them; witness, for example, the 1768 letter from the procurator-general Dudon to the mayor and consuls of the town of Boulasac, wherein he chastises the local priest for refusing to permit a verification of his measures: "There is no need to look into the details of the case to make it plain to you how utterly unjustifiable is the stand taken by the vicar in refusing to submit to your jurisdiction in a matter of this kind. You should persuade him to present at the town hall all the measures he uses in order to verify them by the standards."[52]

Centuries of strife and compromise produced a situation of apparent chaos in the country, and since every custom-hallowed law and usage was safeguarded by conflicting interests, it is not surprising that the authorities were doomed to inaction. Even in Paris[53] and its surrounding district, utter confusion ruled. Standards did exist, but they were dispersed and in poor condition. The *toise*, made of iron, was kept in the Grand Châtelet; its twin copy, however, was in the chamber of the Academy of Sciences at the old Louvre. The standard of the *aune* was in the care of the guild of the *Marchands Marcier* in the rue Quinquempoix; standards of measures of capacity, both dry and fluid, were kept at the Paris town hall; prototypes of weights were kept at the Mint and at the Grand Châtelet. Furthermore, in the city market there were in use, concurrently, the weights of Versailles and those of St. Denis, even though this practice was banned by the royal edict dated 12 September 1778. It seems, however, that the edict of August 1669 had proved more effective in announcing standardization of the *arpent* within the competence of the Department of Waters and Forests; for although binding only in the management of the state woodlands, this practice applied, in fact, more widely.

Parisian grain measures were made of metal. The *boisseau* was nine *pouces* in diameter and eight *pouces* and five *lignes* in height. The height was thus only a little less than the diameter, and this was almost universally the case as far as urban measures were concerned. In 1640 Marsenne calculated that a striked *boisseau* of this type would hold 172,000 grains of wheat, while a heaped one would hold 220,160.[54] In official transactions, the "heap" was abolished in the seventeenth century, and the dimensions of the *boisseau* correspondingly increased, but in unofficial transactions this step opened up opportunities for even larger "heaps." According to La Condamine,[55] the old, pre-1668 foot had been set equal to the length of the half-arch of the inside door in the great pavilion of the Louvre on the side of the rue Fromenteau. The new standard was set into the wall of the Grand Châtelet, but its iron workmanship was poor; it grew rusty and was, in fact, very inaccurate. La Condamine, therefore, suggested that the *toise de Perou* should be recognized as the standard, against which his fellow scientist, de Mairan, championed the length of the second's pendulum. This scholarly dispute was to continue for a long time, but it remained confined to their studies.

On venturing outside the city tollgates, however, one enters the domain of pure feudalism, in which there were only two avenues whereby the monarchy could seek to secure for itself any right of intervention: by counteracting the attempts to increase the lords' measures;[56] and by imposing the royal measure in places where peasants presented grievances about the seigneurial measure, and the seigneur was unable to produce any documents wherewith to buttress his stand.[57] The issue was one of setting some limits to exploitation; as Jacquet, a source not unsympathetic to the lords, wrote in 1764: "The idea that the seigneur-castellan may determine measures as large as he sees fit by his *fiat* is absurd and incomprehensible.[58]

Metrological disorder naturally greatly aggravated the problems of administration under a monarchical régime of enlightened absolutism,[59] notably in the areas of greatest sensitivity: fiscal policy and food supplies. Control of measures formed an integral part of the control of prices,[60] for comparability of measures was a precondition for control of the market, for comparability of data in official price listings (the *mercuriales*), and for the regulation of food supplies. It was essential to the administration that relations between measures in the particular markets of the kingdom and Paris measures should be fixed and observed. The first attempt to do so was undertaken by the Comptroller-General of Finances in 1754.[61] Apparently it was not successful because ten years later the administration had to take up the matter again.

On 22 April 1764, the Comptroller-General of Finances enquired of the provincial *intendants* "whether it might be of advantage to commerce to reduce all different weights and measures recognized in the kingdom to a single common standard. Intendant Berry Dodart's reply was positive; but some years later, during the period of the Revolution, his colleague Mercandiez of Orléans complained that "the feudal powers-that-be, who could have lost thereby, and the cunning merchants, to whom the idea of standardization did not appeal, set their faces against the experiment."[62]

Clearly, Paris insisted no longer—no longer, even though it knew, or could readily guess, what went on in the country. In 1724 the questionnaire called "Enquête Dodun" contained, among its numerous questions, this telltale one: "In cases where the *étalons* exist, do you find their dimensions and shape convenient to use, and have those who have been entrusted with the keeping of them been guilty of abusing their privilege?"[63] The phrase "those entrusted with the keeping of them" may refer equally well to feudal lords and their underlings or to the lessees of metrological rights.

In 1747, the Comptroller-General of Finances ordered the issue of a questionnaire concerning local measures of grain, a procedure that was repeated in 1754–1755, and yet again in 1766–1767.[64] A new circular from the same source, dated 10 September 1766, demanded only the determination of local measures in keeping with Parisian units. Simultaneously, Tillett was given the task of working out the relations between French and foreign measures.[65] These were important questions, bound up as they were with that of free trade in grain. To stress their importance, a royal *ordonnance* was promulgated.

In the face of mounting difficulties for standardization and the remembrance of the failure of so many successive attempts in that field, the royal declaration of 16 May 1766 set the seal upon the Crown's surrender. We read in it: "Although it would be most desirable commercially to establish a uniformity of weights and measures and thus ensure good faith between buyers and sellers— which trust, indeed, is the spirit of all active trade—the unsuccessful attempts made in that direction on various occasions have given rise to doubts as to the likely success of yet another such effort." The King, therefore, decided to be content with a limited, indeed a puny, measure: "It seems to us that it would be at least of some advantage in facilitating and safeguarding commercial operations and in the diminution of major inconveniences stemming from the diversity of measures, to prepare accurate tables wherein everyone could find the ratios and relations between all weights and measures currently in use in the kingdom."[66] The frame of reference would be "l'once et la livre poids de marc, la toise de six pied-de-roi et l'aune mesure de Paris" [the ounce, the weight of a true pound, the *toise* of six royal feet, and the Parisian *aune*], these being most widely known and applied. To have even this modest undertaking carried out proved no easy matter—witness the repeated circulars pertinent to the matter, of 6 May and 4 December 1767;[67] nevertheless the tables were prepared in many regions of France.[68] With this attempt, there terminated, ruefully and modestly, all efforts of a thousand years of the French monarchy's quest to standardize weights and measures.

Meanwhile, however, as a result of centuries of evolution and fragmentation of control over measures among thousands of greater and lesser feudal lords, the metrological chaos prevailing in France was unimaginable. Far more *boisseaux* than parishes were found in 74 parishes around Angoulême, as some of the parishes used two, three, or even four different *boisseaux* concurrently.[69] The situation was no different as far as land measures were concerned. The future *département* of Oise had in use 175 different land measures.[70] Even in the Paris region, heavily influenced as it was by the capital city's market, every parish used measures peculiar to itself,[71] and, in the veritable jungle of them, only now and then are the tendencies that we have pointed to discernible: if, for example, the *boisseau* of the abbey of St. Denis was larger than the Paris *boisseau* in the ratio of 1.95 to 1.30 liters,[72] then it is not impossible that the explanation for this situation is to be sought in the abbey's right to collect taxes in kind. The confusion yielded illimitable opportunities for fraudulent practices, and it is not surprising that there were confidence tricksters about who roamed the country with stolen *boisseaux* and claimed to be verifying officials.[73] Again, we know of rogue lords who would introduce in the areas of their jurisdiction new measures in order to levy new taxes on the populace.[74] Some instances of this kind were amusing: Mme Talon, the mother of the procurator accompanied the Paris Parlement law court on a visiting session to Clermont in 1665 where she sought to verify and standardize local measures; her attempt failed—the old virago, needless to say, having failed to appreciate that the intention of judicial excursions of that type was to intimidate the nobles and not the burghers.[75]

Arthur Young, appalled by what he saw, left us the following description penned on the eve of the Revolution:

> In France, the unending proliferation of measures is quite beyond imagination. They differ not only in every province, but in every canton, too, and almost in every town; the differences drive people to despair, they affect the nomenclature and the dimensions, too, of the units of both area and volume. Add to this the peasants' universal ignorance. . . . The French peasant knows nothing outside his farm and his market. He sees no journal or periodical more than once in a lifetime, or else the diversity of measures in the kingdom would certainly attract his attention.[76]

We shall soon see that Young was very much mistaken in his appraisal of the French peasant, but this in no way detracts from the soundness of his view of the metrological chaos. Yet, discernible in that apparent chaos, some iron sociological rules held.

The efforts aiming at standardization on the part of the French monarchy over many centuries were—so it turns out—doomed to failure. It is astounding that it should have accepted defeat in this area, although it was able to impose its will upon the feudal lords in numerous, more fundamental matters. But a recalcitrant baron

could be subdued; it was far harder to overcome the passive resistance of the feudal estate at large on an issue which, in the last
resort, did not seem fundamental. A model that fits this situation
well is one that explains that the absolute monarchy was able to
restrict the political powers of the feudal lords without encountering a truly bitter resistance precisely because it respected, even
assured, their material privileges. The monopoly of measures was
an integral part of the latter.

Even the most brilliant of men lost confidence after so many
unsuccessful attempts. Montesquieu writes:

> There are certain ideas of uniformity that sometimes attract
> great minds (for they even affected Charlemagne) but infallibly
> make an impression on little souls. They discover therein a
> kind of perfection, which they recognize because it is impos-
> sible for them not to see it: the same authorized weights, the
> same measures for commerce, the same laws in the state, the
> same religion in all its parts. [We shall come across the same
> assortment of problems many a time yet—W.K.] But is this
> always right and without exception? Is the evil of changing
> constantly less than that of suffering? And does not a greatness
> of genius consist rather in distinguishing between cases in
> which uniformity is requisite, and those in which there is ne-
> cessity for differences? . . . If the people observe the laws, what
> signifies it whether these laws are the same?[77]

This ambiguous passage, with a question mark prevailing over an
assertion and with its resigned tone, may be rooted equally well in
adherence to a principle as in disbelief in success; the context suggests that the author rather favored pro-seigneurial tendencies,
although his phrasing hints at some anxiety about the arbitrary
character of the central authority.

However, twenty years later Necker, too, although his intentions
are clear, evinces hesitation. The report he submitted to the King
in 1778 states:

> I have taken the trouble to investigate also the means that
> would have to be applied to the task of standardizing weights
> and measures throughout the kingdom, but I have as yet
> doubts as to whether the advantages thereof would outweigh
> the diverse difficulties in altering the measures in countless
> contracts, acts of sale, acts of feudal obligations, and many
> others. However, I have not yet quite given up.[78]

Only some armchair theorists of the Great Encyclopaedia retained faith in the possibility of standardizing measures within the framework of the absolute monarchy. The entry *Mesure* maintains: "It is easy to see that nations will never agree to the extent of adopting the same weights and measures together. But this can be achieved within one country under a single ruler." The author then goes on to argue against the opponents of standardization.[79]

Yet, in another volume of the same Encyclopaedia, under the entry *Poids*, the opposite view is voiced:

> The diversity of weights is a most tiresome issue in commerce, but it is an inconvenience that cannot be helped. For not only is it quite impossible to reduce the weights of all nations to a single common one, but there is no possibility of achieving this even within the boundaries of a single country. That this is so, we have a proof in the fruitless endeavors to standardize the weights in France in the reigns of Charlemagne, Philip the Long, Louis XII, Francis I, Henry II, Charles IX, Henry III, and Louis XIV.[80]

There may yet have been a handful of dreamers undaunted by these failures. In 1745 in the Academy of Sciences La Condamine presented a blueprint for a "universal and invariable measure based upon the length of the second pendulum at the equator."[81] At the same time, Camus was busying himself with detailed calculations of the Parisian *aune*.[82] In 1775, men of the calibre of Turgot and Condorcet were contemplating the possibility of establishing an invariable measure "derived from nature."[83] This is not the place to analyze their work and opinions. Some of them were primarily laboratory-oriented, for the existing measures were not equal to the needs of scientific research: they were not exact enough and were too diverse. Other investigators favored standardized, invariable measures as a facet of the socio-economic system—for example, to promote free trade—while the armchair theorists hardly ever bothered their heads about the social obstacles in the way of the realization of their ideas. Rare, indeed, were views such as that of the Agrarian Society of Limousin (1 August 1761 and 20 April 1765), which combined a critique of the metrological chaos with criticism of the privileges of the nobles.[84]

Practical men, especially those directly involved, saw the matter differently. When the Assemblée Provinciale of the Ile-de-France asked the lower-level Assemblées d'Élection for their opinions on the standardization of measures in January 1788, this is what the Assembly of Sens had to say:

> If the countless variants of measures are a bad thing, then we would venture to say that today more than ever it is impossible effectively to do anything about it, even in respect of corn measures, and even in that respect, only in the market. True, an attempt so circumscribed would avoid innumerable difficulties, such as the necessity of preparing new inventories of landed estates, new obligations of rent tenants, etc.—these, above all, being issues that have rendered it impossible over eleven centuries to establish a uniform system of measures in the entire kingdom. But a uniformity like that, even if confined just to the marketplace, would still give rise to numerous difficulties: first, it would signify a substantive diminution in the scope of the competences of the high justice (*haute justice*), based upon the provisions in all the *coutumes*; second, it would hinder the exercise of privilege with regard to public weights, and this would provoke powerful protests from the holders thereof, simultaneously forcing them to incur heavy expenditure in exchanging existing market measures for new ones. The execution of the plan would put a partial brake on commercial activity and bring about rising grain prices, a development all the more to be feared because the extent of the rises would be unpredictable. Lastly, the differences in measures, which exist everywhere, are a most attractive incentive to engage in commerce, between cities or between provinces. A large number of traders, particularly those known as *blatiers*, are attracted to their type of enterprise solely by the diversity of measures . . . nicely in accord with the intentions of the edict of 7 June 1787. The differences in measures are such that they defray the costs of transport and charges for the use of the market scales. But to introduce uniform measures would ruin that type of commerce, destroying in the process numberless small markets, whose sole base subsists in those differences of measures, and which—unimportant as they are—do after all provide a service by supplying local inhabitants with all they need as consumers. The security and ease of the grain trade, which the proposed uniformity of measures may be meant to enhance, suffer but little from their diversity because that alleged security and ease depend above all on the reliability of measures. All the merchants and dealers in grain know now-

adays how to perform the necessary arithmetical operations to reduce measures of all sorts to the Parisian standard . . . their self-interest guards them against surprises . . . we therefore think that it would be no inconvenience at all to leave things alone, while it would be dangerous to tamper with custom, which, even if it be less than fair, has been such for a very long time and is so in all countries.[85]

Could anything be more explicit? At the outset, the most important issue, feudal metrological privilege in the area of seigneurial jurisdiction, was put out of court. The question was confined to measures in urban markets, and as far as these were concerned, the argument incontrovertibly set out the impracticality of the proposed reform and the social dangers it entailed.[86]

When we go outside the city walls, we enter the domain of omnipotent feudal arbitrariness. A manual of seigneurial jurisdiction, published in 1764, devoted a special chapter to procedures arising from the possession by the lords of judicial powers in the area of weights and measures.[87] The picture presented in the manual leaves nothing in need of elucidation. On the royal demesnes—and only there—jurisdiction and control over matters metrological lay within the competence of the local royal judge. Elsewhere they were within the ambit of the seigneurial judiciary, and all that had to be determined was the boundary, dividing the competences and incomes accruing from them between the high and middle levels of justice. The manual paid due attention to variations (mostly minor) among the provinces in the manner of the still-binding *coutumes*. All the powers of middle-level justice derived here from authority delegated by the executant of "high" justice; "the middle judge . . . is obligated to adjust his measure to the standard of the seigneur, his superior." Accordingly, when a verdict ordered that measures be destroyed as false, the higher judge had to be informed. The author of the manual tells us that some *coutumes* required the castellans to submit, once in a lifetime, the standard they used to the municipal authorities of the nearest town. Not much, however—indeed nothing at all—existed by way of sanctions when death claimed them before they had discharged that duty. The decision quoted in our manual as prejudication in a case of that very nature, given out by the investigating magistrate on 12 August 1758, determined that, despite his failure to submit the standard in question, "the seigneur may continue to employ various measures in the collection of rents, even though the rent-payers' grievances were concerned precisely with the enlargement of measures and demanded that they be reduced to the royal measure, the king being the immediate overlord of the law court district of Oe,"[88] that is, of the seigneur the complaint was about. Thus, even in a case when the seigneur had failed to observe the appropriate formality, his metrological authority in his estates was under no threat. Only when the castellans and higher judges "did not possess the original measure, commonly known as the *étalon*, may their rent-tenants and persons within their jurisdiction pay the rents due according to the King's measures"; in cases like this, "the seigneur has no right . . . to demand that his rents be paid by a measure larger than the *étalon*—if he has the right to keep it—or larger than the King's measure."[89] This stipulation, however, presented no threat to the seigneurs, for they had the means to see to it that the standard, were it true or false, was always in their keeping. At any rate, Jacquet cannot bring himself even to hint that the lords might falsify measures to their own advantage. He merely warns against "bad faith of tax farmers, interested tenants, and also the ignorance of the rent payers, which must never be ascribed to the seigneurs."[90]

We are at liberty, however, to enquire why the middle-grade judge was bound to recognize the standard of his superior colleague, whereas the latter was not obliged to recognize, in turn, the standard of his overlord, the King? Jacquet tries adroitly to evade this dilemma. He concedes that the supreme authority over weights and measures is the King's; that the King's edicts have a higher standing than those of his vassals; that, therefore, the King "might decree that all weights and measures be assimilated to the royal weights and measures"; yet the King, although he has the authority, does not do so, in his gracious acknowledgment of the seigneurs. On the contrary "all edicts and decrees touching this matter emphasize particularly that His Royal Highness has no intention of harming the seigneurs who enjoy the right of weights and measures in their lands."[91] This statement was doubtless correct.

In his heart of hearts, the Parisian lawyer Jacquet is here divided. He respects the seigneurs; if some metrological abuses occur, surely it is never their fault, but that of the administrative personnel on their estates, or of the tax farmers. He respects royal rights, too— praising, at the same time, the King's gracious desistance from using them. To him, the villains of the piece are the merchants, smartly sabotaging the introduction, for practical purposes, of long-standing decrees of standardization.

The rascally merchants . . . have obstructed the execution [of those decrees] on the pretext of [a variety of measures] being of service to commerce.[92] The resistance of the seigneurs holding the privilege of measures (some larger, others smaller than the King's measures) applicable to the corn-rents from their tenants could be appeased by assimilating all their measures to the King's standard, and correspondingly increasing or decreasing in the new inventories the number of measures of corn due to them. And the tenants might well accept this gladly, being assured, once and for all, that the measures would not be increased again on the pretext of a new control.[93]

Now, the last part of the above quotation is quite an eye-opener, for Jacquet unwittingly reveals that he realizes well the danger of such practices.

Where seigneurial officials did enlarge the measure used in the collection of taxes in kind, their offense remained unlawful and, even though of long standing, did not entitle the seigneur to collect more than was his due. Nor was he permitted "to have more than one standard, because if he did have several, then the collectors and tax farmers would find it easier to cheat the tenants; for they would surely not fail to demand that the dues be paid using a large measure, as long as the tenants noticed nothing, while using the proper measure if some brighter ones did notice."[94]

Jacquet offers an interesting interpretation of the prohibition in respect of not just enlarging but also reducing the standard: "The prohibition imposed upon the castellan not to diminish the measure certainly does not concern his relations with his tenants, for they could only gain thereby; it is directed at the seigneurs of lower rank, who, since they take their measures from the castellan, would thus lose some of their revenue." The regulation banning the enlarging of measures by the seigneur, however, is a different matter: "Transgressions of this kind result in irrecoverable losses to the tenants, who are burdened enough as it is; therefore, there can never be too much care taken to prevent abuses like that."[95] Clearly, the author was not unaware of abuses "like that" taking place.

Finally, Jacquet maintains that, in some cases, the lords would be deprived of their metrological rights because of the abuses perpetrated. He feels this to be fair in relation to a seigneur "who took advantage of a privilege granted him solely for the purpose of counteracting attempts to falsify measures."[96] Here, he offers a functional defense of feudal privilege; had Jacquet lived to see the

Revolution, he would surely have been a Girondin. Examples of law courts' verdicts in metrological suits, which are published in an appendix to his manual, include many abuses that were to the disadvantage of the peasants. In particular, the seigneurs of Oe showed a predilection for increasing their income in that way, and they provide some interesting examples. If we have devoted over-much space to the views of the lawyer Jacquet, that is because we attach much importance to his manual—on the one hand, because it is based not only on the analysis of normative texts but also on knowledge of relevant judicial practice, and on the other, because his work was used as a textbook among practitioners of the seigneurial judicature.

His work, however, was not unique. Edme de la Poix de Frémonville, author of a book of instructions for compilers of estate inventories, writes in a similar vein: "All measures of the seigneurs should be presumed to accord with the measures in use in the nearest market town, unless there exist documents containing statements to the contrary," but "if such documents do exist, then the seigneurs should avail themselves of them."[97] So much for the philosophy of lawyers in matters metrological. The way of thinking of the police was somewhat different.

Delamare, in his treatise on the police,[98] assures us, in a phrase that we shall encounter time and time again in almost identical wording, that "all measures were equal in France under our earliest kings." Alas, as early as "towards the close of the reign of Charlemagne, and still more markedly under his grandson, Charles the Bald, flaws appeared in the equality." Meanwhile, however, to Delamare, uniformity of measures constitutes a proof of the state's might, or even of divine commandment, as the Hebrew example he cites testifies—at least in his view. The philosophy of the police must ever be the philosophy of the authority of the state. Successive attempts at standardization by the kings were aimed at the public good, but "particular seigneurs set against royal wishes their own privileges, granted by some of our kings or derived from immemorial custom. They held firmly to the view that, according to feudal law, the administration of justice was theirs by patrimonial title, that the policing of weights and measures was an integral and inseparable part of that justice, and adduced, finally, the evidence of the *coutumes*." The *coutumes*, according to Delamare "had been initially common custom, but in due course assumed the character of laws, sanctioned by public authority; it would be difficult to question the laws that they affirm in the absence of any ordinances to the contrary from our kings."[99]

Delamare is aware of metrological offenses: "Lust for profit has prompted some to provide themselves with overnormal or undernormal weights, depending on whether they intended to act as buyers or sellers in commercial dealings."[100] His *Traité de la Police*—while the author is in no position to question the legal force of the *coutumes*—admires the legal force of the *coutumes* for Tours and for Poitou, which at least require local seigneurs to deposit standards of measures in the nearest town or the nearest royal court of justice. Here, he presents what may be considered a minimalist police program, and so, the police philosopher would seem more progressive than the Parisian lawyer Jacquet. But although both accept, on paper, the Crown's competence to settle metrological disputes, neither of them thought—it was indeed beyond their conception—of challenging the absolute metrological authority of the lord on his estates. The peasant was obliged, in theory, to employ measures determined by the judge of the middle level of justice, who in his turn had to employ those handed down by the high judge, while the latter was not at all obliged to employ those of *his* overlord, the King.

In practice, until the Revolution three quarters of Frenchmen were, in metrological matters, subject to justice as meted out by the seigneur, who thus often acted as *iudex in causa sua*. The view, advanced by many historians, that in the eighteenth century units of measure in France varied between places but remained constant over time holds water only as long as we do not venture outside the city walls. The attempts of the royal judiciary to encroach upon the areas of traditional competence of seigneurial justice most frequently ended in failure.[101]

In the Revolutionary era, when the need arose to translate measures in actual use into new metric ones, even experts in the field would, not infrequently, be surprised. A few questionnaires that were issued to help the central authorities become acquainted with the problems (in 1791, and in years III and IV of the Republic) reveal a desperately complex picture. Historians investigating the history of prices have underestimated this complexity—and with reason. After all, their evidence came from the urban markets, mostly from the larger cities; had they ventured beyond these at all, they would have been confronted with a vastly different situation. Moreover, not only were the measures there diverse, but they kept changing—more often than not growing larger—as the wave of seigneurial reaction rolled across the country on the eve of the Revolution.

A silent struggle was meanwhile being fought in France over the nobles' endeavors to increase rents. Hundreds of *feudistes* were searching through seigneurial and communal archives for documents that would enable lords to revive rights long since consigned to oblivion.[102] It was only natural that former measures should attract their interest, too. Whatever the legal validity of the arguments derived from them, the peasants would automatically oppose them, as they did every kind of attempt to increase the dues. They would protest, but in the end, anxious lest under the existing institutional arrangements they should lose their farmsteads, they would give in. But the injustice rankled. As we shall see later, the moment the existing institutions faltered, they sought redress. Meanwhile, the defenders of tradition were on the offensive and tried to base their stance on theoretical arguments. V. R. Marquis de Mirabeau writes: 'Our present legislators, without going to the roots of the diversity in this field among the provinces of our vast country, would wish to abolish all variety of customs and usages and *coutumes* of weights and measures, but they erect their constructions upon shifting sands."[103]

So, Montesquieu does not believe the practicality of metrological standardization, Necker too has doubts, and Mirabeau condemns it on moral grounds. The provincial assemblies of nobles were hostile to it; the law courts respected existing privileges all the more readily in that this enabled them to avoid clashes with the seigneurs; the police were powerless, since the seigneurs were sovereign in their lands; and the King, having relinquished the ambitions of his ancestors, confined his own ambitions to a modest attempt to construct conversion tables.[104] Exactly one thousand years had passed between Charlemagne's first metrological capitulary in 789 and the outbreak of the French Revolution. The same disputes about the same problems had continued for a thousand years, and for a thousand years the same abuses and deceptions were practiced, while successive attempts at standardization ended in the same kinds of failures. On the eve of the Revolution, at the most a handful of committed theorists retained faith in the feasibility of standardization.

Then, like a bolt from the blue, comes the summer of 1789—and throughout the length and breadth of France, the preparation of the *cahiers de doléances* is unanimously accompanied by a popular demand for metrological standardization and the abolition of the absolute authority of seigneurs in the realm of weights and measures.

· 22 ·

"ONE KING, ONE LAW, ONE WEIGHT, ONE MEASURE!"

Ten years after Necker's report, and a year away from the diatribe of the nobles' assembly of Sens at the time of the preparation of the *cahiers de doléances*, it turned out that the entire nation wanted standardization of weights and measures, believed it to be attainable, was convinced that it was indispensable and even that it would be relatively easy to carry out. As de Tocqueville was to write: "The *cahiers de doléances* will stand as the testament of the French society of the ancien régime, as the most perfect manifestation of its desires, and as an authentic expression of its last will."[1] Was "the testament," however, truly "authentic"? The debate concerning the *cahiers de doléances* is almost as old as they themselves. It is not our intention, here, either to summarize or to continue it; but since it is our intention to make extensive use of the *cahiers* as a source relevant to this study, and since contrary to the present scholarly "fashion" we are firmly convinced of their value, we must pay attention to the controversy surrounding them.

In the earliest studies and publications, interested historians confined their attention to the *cahiers des bailliages et sénéchaussées*, whose small number facilitated their analysis. Before long, however, critical voices arose, pointing out that those *cahiers*, which in principle were supposed to synthesize the preliminary *cahiers* from the parishes and corporations, in actual fact diverged from them in numerous matters of import. They were edited by regional Notables, who passed over in silence numerous grievances and wishes that expressed the needs of the common people. They blunted the sharp edge of these demands, while they added demands that mattered to the higher ranks and showed indifference to the lower ranks of the third estate.[2] That this was so is confirmed in our analysis.

In due course, the historians' interests duly shifted towards the preliminary *cahiers* of the parishes and craftsmen's corporations. This shift was favored by many students, Jaurès being the foremost among them. Large-scale publication of the documents in question eventually began,[3] and the six or seven decades separating that initiative from our own day resulted in numerous large volumes. Of the vast mass of forty thousand surviving *cahiers*, however, many

185

inevitably remain in manuscript. They were, after all, the product of the most extensive investigation of public opinion in Europe until our own century! Despite the realization that many extant *cahiers* would never be published, the publication of the first volumes heralded a new stage in their critique.[4] At its heart was the classic, perennial problem of historical textual analysis, that of authorship; in the case of the *cahiers*, the fundamental issues are questions of "models" and of editing.

There is no doubt that at the time of the summoning of the States General in the Spring of 1789, France was flooded with the "models" of the *cahiers de doléances*.[5] Their provenance was diverse: some emanated from the government agencies; some from the opposition (the Duke of Orléans);[6] some were produced by the thinkers or reformers of the day.[7] We know for a fact that countless "models" circulated all over France. But, as Picard says, "the models that were in fact taken up were those that harmonized with popular sentiments. There were also others, reactionary and conservative, but they fell upon stony ground."[8] Furthermore, the "models" were often used merely to help in the formulation of preambles to the *cahiers*, or for rhetorical decoration.

It is scarcely surprising that the "models" were readily accepted by the people at large. After all, as the masses were now being called upon to advance their wishes, they found it difficult to formulate them. They lacked the relevant experience, let alone the fact that the majority of the participants in the parish assemblies were illiterate. Let us cite a poignant example: at one meeting, the embarrassment of the flummoxed parishioners was settled by a neat suggestion from the clerk to copy the text that was supplied the evening before by a chance acquaintance at the inn; and the "chance acquaintance" was none other than Mably himself![9] Surely, many similar cases must have occurred, though the relevant evidence has not come down to us.

We do possess direct knowledge of numerous models, while indirectly we can learn even more through extracting the identically worded sections as we analyze the *cahiers*. The method is actually the same as the comparative analysis of several versions of the life of some saint, whereby the medieval historian succeeds in bringing to light the initially unknown editor of the earliest source. There is no doubt that since a model would be passed from hand to hand in the region, dozens of parishes copied the same one; occasionally, it is even possible to track its progress on the map. Yet to accept that this was a common practice is by no means tantamount to accepting that the *cahiers de doléances* are of little value as a source of information about popular anxieties.

The first task of the critical scholar is to extract all original or uncopied *cahiers*, and some editors have attempted to do so. Thus, A. Le Moy, investigating the region of Angers, appraised as original 51 out of a total of 178 parish *cahiers*;[10] P. Bois, investigating the region of Château-au-du-Lois, arrives at the figures of 29 out of 39;[11] and both writers adjudge the "guided" as well as the literally copied *cahiers* as less than true primary sources. However, even if attempts of this type should leave us with only half or a third of all *cahiers* as "original," then out of some 40,000 extant, we would still possess a great deal of valuable evidence. But to proceed in this way would be overly critical. The crucial question is, are we entitled to consider, *a priori*, the copying of a model as a passive, uncritical, and unthinking act? Might it not, at times at any rate, signify an active assumption of a considered stance of approval? Might not the model's author be expressing in it the desires and aspirations of the masses and, in fact, merely formulating for them what they might otherwise have been unable to formulate? These propositions may be rejected out of hand as lacking any substantiation: it may well be said that the oft-voiced demand of the *cahiers* for the freedom of the press was a concern of the *philosophes* but not of the illiterate members of parish assemblies. But an argument of some weight may be put forward in support of our suggestions, too. The point is that although we are aware of hundreds of *cahiers* clearly having been copied, only a small number of them had been copied with no changes at all, neither deletions nor, more often, additions. It is this that, in our opinion, ought to reverse completely the common negative evaluation of the "models" as a source. For, if an assembly, in copying a model, introduced any alterations at all, then this implies that the copying was not a passive act; while the fact that other passages had been copied, word for word, should be seen as evidence of their acceptability,[12] given that, to repeat, complete absence of any alterations is extremely rare. Even more telling for the student of metrology is the evidence like that in the *cahier* of the parish of La Meignanne (near Angers) where the only alteration was the addition of a demand for standardization of measures, written in the hand of the local clerk.[13] To conclude, we are not disposed to gainsay the value of the *cahiers* as evidence, even in the instances when they have been copied.[14]

In recent years critics of the *cahiers* have been pressing their case yet farther. According to them, even the uncopied *cahiers* repeat

the same demands, which were mainly in the interest of the Notables alone. They were indeed, at any rate in rural areas, usually
prepared by the Notables and were tacitly agreed to by the parish
assemblies, who saw no harm in them. They reflected ideas that
were "in the air" at the time. They contained ideas current among
the higher strata of the third estate, even in—and only in—the
"original" *cahiers*.[15] That such *cahiers* were indeed prepared by local
Notables cannot be doubted at all; this much is obvious, if only
from our knowledge of the extent of illiteracy in France at the time.
What we know little about, however, is who precisely those Notables
were. The analysis of style in the original *cahiers* shows rough phrasing more often than it does any pretentious grandiloquence drawing upon Latin tags, literary sources, or the writings of the *philosophes*. The present writer would wish for more analysis of
handwriting, a task beyond the users of published sources, while
those he has seen in manuscript[16] had clearly been written by men
unaccustomed to daily practice of the difficult art of writing.[17]

But to establish the identity of the editor does not exhaust the
question of authorship, not even when his name and social standing
are known to us. For the crucial question still remains unanswered:
to whose thoughts did he give expression? Did he merely clothe in
literary garb grievances made by others? Did he perhaps accept
dictation?[18] There is, in our quest, no substitute for social analysis
of content, phraseology, argumentation, description, and rhetoric.
Time and again they are banal, run-of-the-mill, and threadbare,
yet often they tell us in concrete, rich detail of people's sufferings,
embarrassments, and feelings of injustice. If the complaint we find
there is about a false *boisseau* used by the seigneur in the assessment
of his corn-rent, it matters little that the *cahier* was put together by
a lawyer in the nearby town, for this would scarcely have been *his*
grievance! And why attach much significance to the identity of the
"pen pusher" serving the parish assembly if his writing renders
faithfully the differences argued and the compromise reached at
the meeting? We shall come across instances of this kind time and
time again below.[19]

There are some indeed, even among the original *cahiers*, written
in the pretentious style of the provincial *litterateur*,[20] but far more
frequently do we find them written in an awkward, even incorrect,
manner.[21] Clichés and imitations abound, yet even in the very
clichés, characteristic differences are discernible.[22] Metrological
postulates may be found in *cahiers* that had been copied throughout,
but also in *cahiers* deemed by their editors as mostly original.[23] Why

should we find a particular metrological postulate heading the list of grievances[24] in two quite independently produced *cahiers*? Or, contrariwise, we find a demand of that type added right at the end, seemingly after the *cahier* had been completed[25]—or later still, apparently after it had been sent in to the deputies.[26] Authenticity is not to be gainsaid when one parish demands "that there be but one measure and one weight in the entire kingdom," and another copies the words but adds "and a uniform measure for wine, at least in each province."[27] When we read the complaint that "the unhappy tenants are overburdened not only by the excessively large measure but also by the ruthlessness with which the payment in kind for poultry is levied on them,"[28] we cannot but discern, in this clumsily tagged-on reference to poultry, a reflection of crudely but authentically remarked associations. And when it is stated by a certain parish: "Let all landowners, be they nobles or bourgeoisie or common people, be henceforth obliged to accept all dues owing to them by the royal measure . . . , and by the measure of the marquis de Châteaugrison of Rennes, which has the unqualified approval of all the vassals,"[29] it quite movingly echoes, as it were, many debates, quarrels, and compromises. To those examples, we may add many phrases, shot through with local detail and color, that refer to the "large measures" awaiting vassals bringing their grain dues to the seigneur's granary;[30] or record a demand that the King's measure be applied, and couple this with another that the dues should not have to be delivered farther away than to "the nearest seaport";[31] or complain of false measure at the mill, let alone the trouble of reaching it by the roundabout route "because there is no road and no bridge to enable us to cross the river";[32] or describe contentious matters in nearby villages and country towns;[33] or, finally, echo the quarrels at the seigneur's granary, in the market place, or at the abbey, when the tithe was being delivered. I know not who wrote the *cahiers* from which these examples are drawn; I do know, however, who guided the writer's hand.

It is true that, as well as such live and palpable passages, there are many banal ones, but even the latter are not without meaning for the historian. The repetition of banalities is historical fact no less than is the occurrence of original remarks, and may even be more important, depending on the sense of the banalities in question. For every society has the banalities it deserves, and, moreover, they vary over time. To cite a specific example, in the *bailliage de Troyes* we can compare the *cahiers* of 1789 with those prepared for the meeting of the States General in 1614, and metrological issues can be clearly seen to be quite unimportant in the early seventeenth century in the life of the villages of Champagne.[34]

Let us finally dwell for a moment on the issue of "exaggerations," that is of the *cahiers* allegedly painting an unfairly black picture of the period. This issue was raised as early as September 1789 at the National Assembly, when the question of the binding power of the *cahiers* was debated at the same time as that of the royal veto. It was then that the significant words were uttered, stating that when the *cahiers* had been compiled by the parishes, the nation was far from being free, that "it had been like a slave, who expects no restoration of his liberty and would be quite contented to have his daily hardships merely alleviated [even though] he would readily accept full freedom."[35] These words obviously do not signify much, but they are enough to draw attention to the obverse side of the matter: yes, indeed, sometimes the *cahiers* do exaggerate and "lay it on thickly" in black, but, on other occasions, their demands are all too realistically modest.

Textual criticism and ascertainment of authorship, as well as the detection of links between texts, are all major achievements of the historical method. The relevant techniques would seem to have attained perfection since the foundation of the École des Chartes and the launching of the publication of the *Monumenta Germaniae Historica*. But the objective of textual criticism is not to drain all life from the text but rather to bring it to life, to salvage it for the purposes of research, and to extract from it as much information as possible in order to enhance our knowledge of the world that had brought it into existence. The *cahiers de doléances* lend themselves admirably to such researches, for, as Jaurès said, "they deserve admiration for their sweep, pulsating clarity, and uniformity . . . [and they constitute] the most excellent collection of national writings that any country possesses . . . the entire rich diversity of social life shows plainly through them."[36]

It is necessary to dwell at some length upon the relevance of the contents of the *cahiers* for the historian of metrology, but, even so, our consideration of them will be rather superficial. To begin with, in the *cahiers* of the third estate, the problems of weights and measures are aired very frequently,[37] at times given much prominence, and often discussed in an eloquent, detailed, and colorful manner. We may quote from a few descriptions of the metrological chaos of the period, which are no whit inferior to that left by Arthur Young. First: "The measures of grain vary from town to town. The bag (*le sac*) of grain contains 4½ *setiers* at Ham, 4 at St. Quentin, and 3 at Péronne and Noyon. The same kind of variety applies to the measures of wine and other liquids, and the measures of land vary enormously, as do weights, too."[38] Next: "The diversity of weights, measures, and ells is tremendous between towns and even within single townships: one has both a large and a small *aune*; another's market has simultaneously in use measures of grain of 64, 68, and even 70 pounds, while in the neighboring one, 84 pounds is the norm, and up to 108 pounds at the fair."[39] Third: "The differences in measures within a province, or even within a parish, provide opportunities for endless abuses that are harmful to commerce."[40] Again: "There are little hamlets where two measures [of land] are in use."[41] "In the countryside, there is not one single weight that can be acknowledged as just."[42] Finally, the inhabitants of a certain parish in Normandy "humbly petition that the *coutumes* weights and measures be made uniform throughout the kingdom, instead of the present labyrinthine choas of laws and *coutumes*, conflicting with one another, which the judges find unbearable and the litigants unreasonably costly."[43]

For practical purposes, the responsibility for that state of affairs devolved upon the seigneurs. Statistics being of little use here, the impression gained is that the complaints in the *cahiers* frequently were particularly aimed at ecclessiastical institutions.[44] The past itself was being apparently charged—the barbaric and dark "Gothic" ages, the old privileges and rights with their incomprehensible Latin; "all seigneurial rights that derive from yesteryear's serfdom (*servitude*)."[45] "We should . . . abolish . . . the ridiculous diversity of discordant customs and usages, the barbaric survivals from centuries of crassness and feudal authority."[46] Again, voices were raised demanding "the abolition of local customs, in favor of the Roman law, [and also] uniformity of weights and measures."[47] The prevailing metrological confusion was seen as the end product of "countless obscurantist, barbaric *coutumes*."[48] Again, some expressed the desire that "there be no more than one *coutume*, one weight, and one measure, without recourse to Roman law or all kinds of local laws that came into being in the days of feudal absolutism which we have toppled."[49] Yet another of the *cahiers* adds that "surely, it is barbarous when the citizens of one and the same state reject one another, look upon one another as aliens, forgetting the ties of brotherhood that bind them together" in a manner "*gothique et révoltante.*"[50]

The men of the Enlightenment did not care for the Gothic style whether in architecture or elsewhere. This is a chapter in the history of tastes and fashions. But it has been said that "the struggle of tastes is a class struggle, too"[51]—in some periods, at any rate. For the choice of styles is not simply a matter of individual free will and aesthetic criteria. Social strata effect aesthetic choices by the structural method, *avant la lettre*. The gothic style is part and parcel of a more comprehensive whole, which has bequeathed us great cathedrals, but also feudal dues and monopolies (the *banalités*), serfdom, and—last but not least—metrological chaos.

The consistent unity of the nation's attitude towards the problem of standardization of weights and measures is astonishing. Everybody wanted it, from peasants in different regions to the craftsmen in diverse trades to municipal leaders of virtually all towns. The point to note above all, however, is that they sought standardization for widely different ends. In a society divided into numberless corporations and parishes, every little social group is in a unique situation, suffers from different anxieties, and exhibits a set of values peculiar to it alone. And yet, out of this chaos, there did emerge certain common discontents and desires, and among the latter, was the quest for standardization of measures. "Comment pouvait-on être Français" at the end of the eighteenth century?[52] Truly, this was no easy matter, but one thing was undeniable: they wished to be Frenchmen. Would a standardization of measures advance this cause? The answer appeared to be yes. Since, however, different social strata sought the requisite reform for different reasons, we ought to consider in some detail the variety of motives behind the drive.

Broadly speaking, there were three sets of motives: anti-seigneurial; commercial; and national. However, in analyzing them, we must bear it in mind that the *cahiers de doléances* were, within a certain finite framework of political conditions and existing institutions, addressed to, and intended for, the King. The phraseology of the *cahiers* may not quite faithfully express the motives and views of their authors. Indeed, at times the chief element of a given formulation was the attempt to hit upon an argument that was "right" primarily in the sense of its power to persuade the recipient. And we shall shortly see that the colorful and detailed descriptions of certain situations that exercised the authors of the *cahiers* may well cause us to place considerable trust in them. Nevertheless, to enhance our certainty and to keep a check on our inferences, an attempt ought to be made to analyze the context in which metrological demands were put forward; to do so, should ensure a more objective view of the significance of weights and measures in the minds and social judgment of the authors of the *cahiers* than do their explicitly stated arguments. However, let us first analyze the latter, beginning with the anti-seigneurial arguments, both because they are the commonest and because their frequently specific detail inspires much confidence.

What the *cahiers de doléances* obviously will not tell us is the degree of bitterness in the metrological quarrels between the peasants and their lords long before 1789. To enquire into this, a different approach is required. After all, in the *cahiers* too, we find here and there references to former disputes.[53] Thus, the parish of Yvignac had been engaged in litigation about measures since 1787 (if not earlier),[54] the parish of Brain from that same year,[55] the parish of Ruffigné since 1780,[56] the city of Caen since 1764,[57] and the parish of La-Neuville-aux-Larris for a long time, too.[58] The enormous complexity of the configurations of dues is also redolent of long-standing conflicts and their compromise resolutions. On occasion, an impasse would be reached, leaving the seigneur with one half of his dues collectable by one measure and the other by another— and both from the selfsame peasants![59]

Accordingly, there was nothing strange about proposals for standardization (especially in the western regions of France) bearing reference to particular seigneurial measures of dues in kind, rather than of a general nature. Let us exemplify: "On all occasions, the measures in which rents are collected ought to be the same throughout the diocese."[60] "Let the measure applicable to the corn-rent, due to the seigneurs and others, be settled in Brittany so as to be one only and the same."[61] "And when the dues are paid, let that be by one measure throughout the kingdom, the measure of His Majesty.[62] "Let all measures which the seigneurs employ in collecting the rents due to them accord with the King's measure."[63] Sometimes, the same sentiments were voiced still more specifically and forcefully, as when the parishioners of Beuzec-Cap-Caval demanded "to pay henceforth their rent by the King's measure alone, striked, and be no longer obliged to take their dues in kind any farther away than to the nearest port."[64] The inhabitants of Esquibien, for their part, demanded that "the seigneurs may not insist upon . . . measures larger than the royal one when rents are being collected,"[65] while the parishioners of Goulien asked "that the measures for grain collection be regulated."[66]

The demands are specific; they show no stylistic borrowings, even in places not far apart, although separate in the *cahiers*. All these features inspire confidence: they tell the truth. Even more so does the demand we have already mentioned—a demand that had clearly resulted from a compromise reached in the assembly—that the royal measure be universally used and the seigneurial ones be abolished, excepting only that of a certain marquis since all his vassals seemed wholly satisfied with it.[67] The seigneurial monopoly of measures, combined with the existence of dues in kind, could not but continually lead to disagreements, a situation that gave rise to incentives for the seigneurs to falsify the standards, and whether they did so or not, the peasants constantly suspected them.

Complaints were countless: "We pay many valuable rents and every seigneur wants to employ his own measure [a lacuna in the text] . . . We would wish there to be but a single measure in France."[68] "When we take the grain to the seigneur's granary, it is assessed with a large standard."[69] "And even the seigneurs' measure is unduly large."[70] "The seigneurs collect oats from us by the large measure."[71] "The measure at Plumaudan near Dinan has grown much larger."[72] "The measure they apply to the grain is arbitrary."[73] "They are making fools of us with their differences in weights and measures."[74] "In this country of ours, every seigneur usually enjoys his own, unique *boisseau*."[75] "Why should the seigneurs or the privileged clergy enjoy the right to please themselves as to what measure of grain is binding on their estates?"[76] "Let the seigneurs' own measures, peculiar to their estates, be henceforth banned."[77] "There are as many different local measures to assess dues in kind by, as there are estates."[78]

Our sources yield also explicit charges of forgery. Here are a few of them: "The nobles' measure waxes larger year by year."[79] "The wheat-rent which our parish has to render is 900 *perrées* by the Lamballe measure, but for some time now this measure has been growing larger and yet it is still 900 to be paid; we beg you, Your Excellency, to reduce the number."[80] "The tax-gatherer of the abbot of Notre-Damme-de-Lisque, of the Premonstratensian order, increased the measure of grain in 1788."[81] On one occasion, a local assembly broke up on this kind of metrological rock when the minority "turned to the seigneur's sister and told her that Lord Lequen had different *boisseaux*: large to collect and small to sell the grain that he collected as dues from the said seigneurie of Plessin-le-Lay."[82] On another occasion, we encounter the following generalization: "The holders of privileges, taking advantage of the various changes and the resultant confusion in the realm of measures, compounded by the medieval Latin (*la basse latinité*) terminology, apply in collecting their incomes whatever pays them best.

In many parishes, the market measure coexists alongside the sei-gneur's measure."[83] Obviously, we have no means of ascertaining whether such complaints were well grounded. But their truthful-ness or otherwise is not of crucial significance to us, nor is the morality of the abbot of Notre-Damme-de-Lisque, or the Lord of Lequen. What alone does matter to us is the utter unlikelihood, in any society, that even slanderous complaints would be made about actions that were entirely improbable—deeds that were just "not done" in a given society. The complaints cited above may in them-selves constitute no valid proof in the particular cases they refer to; nonetheless, in our opinion, they are sociologically very signif-icant.

Sometimes the complaint was phrased so as not to accuse the seigneur but, rather, his underlings. Here are some examples of this practice: "The Lord of Corny extorts almost 600 *hottes* of wine as rent. The amount used to be less, but his functionaries have been very clever at converting the *septiers* into *chaudrons* in such a way that they gained a quart for each *septier*. They cannot claim that this was unintentional because everyone knows that one *septier* of Lorraine has only four quarts, whereas the *chaudron* has five."[84] Another *cahier*, in complaining of extortions, states: "There are almost as many local measures used in assessing seigneurial dues as there are estates. The *boisseau* of each judicial district should be used to verify such measures. The lords who personally supervise the collection of their own corn-rents certainly do not depart from the standards. Yet they avail themselves of the duplicities of their servants who, in turn, seek to recoup from the tenants the high prices paid for the leases. This they do by applying two different measures and using the one that exceeds the standard when the grain is delivered at the lord's granary. The payers are not readily deceived and they complain, but they pay, since their fear of dis-possession is stronger than their justifiable desire to refuse to sub-mit."[85] It is, indeed, difficult to determine in a grievance of this nature whether the complaint was accurately phrased, or whether it was trying not to provoke the lord. At any rate, the point is of little consequence for us. What is important is the expression of helplessness in the face of the lord's monopoly of measures, and the conviction of the injustice suffered.

In some parts of France, meanwhile, more interest was expressed in the stabilization than in the standardization of measures. The crux of the matter was how to prevent alterations of measures by the lords, in particular of dry measures in areas where taxes *in natura* played a major role. The case of the *sénéchaussée d'Angoulême* was an interesting one. Here, demands for standardization of measures were few and far between—but they spoke volumes. Moreover, we have to bear in mind that other sources tell us that a year later no fewer than 80(!) parishes of that region were in dispute with their lords, accusing them of surreptitiously enlarging the bushel![86]

Thus, too, the parish of Blanzac, prominent in similar disputes later, recorded in its *cahier* that "as the *boisseau* of the most noble, distinguished, and beneficent seigneur of Blanzac, so the *boisseaux* of other lords ought to have their size fixed in accordance with the standard deposited at the courthouse of the judiciary district of Angoulême in order to avoid extortions, in particular extortions perpetrated by members of the chapter in Blanzac itself."[87] And the neighboring parish writes: "Let the lords' *boisseaux* be standardized, in keeping with the standard deposited in the judicial office of the capital, and then the same lords' functionaries will no longer be able to harm the unhappy tenants."[88] Again, the lenient treatment of the lords themselves is all too obvious; if anyone was to blame, it was their servants and scribes. This caution is still more striking in the *cahier* of the city of Angoulême,[89] for although the burghers were not really directly concerned, the matter is considered at length, since Angoulême was the capital town of a region in which abuses of measures—or, at any rate, metrological clashes— were uncommonly bitter.

Bearing in mind the endless diversity of seigneurial measures, it is not difficult to appreciate some generalized postulates, such as: "Let the lords whose measures differ, regulate them according to the King's measure";[90] or "Let all seigneurial measures be reduced to the royal measure, so that no lord may usurp the right to have his larger or smaller";[91] or, "The lords dispose of an endless variety of measures; we demand that they all be reduced to the King's measure";[92] and, quite simply, "Let all seigneurial measures be regulated."[93] Countless statements of this nature tell us little by themselves, but when confronted with others of greater particularity and depth, they tell us a great deal. They tell us what it was that troubled both the men behind the words on paper and their fellows who did not find anyone to set it down in writing.

It is worth noting that sometimes a dispute about seigneurial measures, however bitter, did not go beyond the local level. A demand may stay within such modest compass as: "Let the lords be forbidden to collect the corn-rent by their domestic or estate measures, but adjust them to the measure of Guingamp, which amounts to 65 pounds";[94] or "Let all lords be ordered to equalize their measures in accordance with the standard deposited in the judicial office of the region of Angoulême, as the good and noble lord of Banzac has done; thus shall abuses be stopped."[95] At times, the peasants' demand for standardization does not entail getting rid of the lord's measure, as when the men near Angers held that "it is a very important matter that we all have the same rights, that the same measure and the same weight obtain everywhere, and also one and the same *boisseau*, even if it should be used for the payment of rent and nothing else [*sic!*]."[96]

In a great many cases, however, even if no demand for extensive standardization is made (countrywide or for the whole province), the wish is expressed that there should be no further modifications of the standards, preferably guaranteed by the "good King." The popular wishes coincide here with the age-old tendency of the monarchy to act as a conciliatory agent between the lords and their vassals. Thus, in the words of the peasants of Pas-de-Calais: "We should insist on a law that shall order an annual check of all measures the seigneurs use in the collection of their rents, and that the check be carried out in the presence of both parties in the rural court of the seigneurie."[97] Similar sentiments are expressed by the parishioners of Saint-André-de-Blanzac: "All the *boisseaux* of the seigneurs ought to be compared with the standard kept at our country town of Angoulême, so that the functionaries of those seigneurs shall no longer be able to molest the unhappy tenants."[98]

Some demands were very modest, indeed, as when the peasants of Pas-de-Calais wanted merely to be able to see the standards: "We demand that the lords, to whom we are obliged to pay corn tax, be obliged to keep standards at the manor house, properly attested according to the old measure, either large or small, depending on the kind of dues."[99] In another locality, in complaining that even the measure determined by the itinerant parliamentary court was not observed, the parishioners contended: "It is just that all landowners and tenants be obliged to bring along and present their measures so that they may be confronted with the seigneurial standards, which the said seigneurs have the duty to keep."[100]

The demand for laying open the lords' measures makes still more sense in the light of the following elucidation: "The nobles as an estate possess vast wealth, but not only are they unconcerned to help us, but are always on the lookout for new means whereby to oppress and ruin us. . . . they keep the knowledge of weights and measures from us, nor do they disclose the *boisseaux* that they use

197

in levying taxes upon us. One has a *boisseau* of six measures, another of seven, while yet another, of eight. If you but move from one parish to another, you find a different *boisseau*. Neighboring towns, too, have measures peculiar to themselves, and the payers, or even the traders, are confused thereby."[101]

Countless real-life situations and specific troubles and abuses bound up with taxes in kind are reflected in the grievances concerned with weights and measures. Some complain that the lord, but never the vassal, performs the task of measurement: "Concerning the corn-rent that has to be delivered to the lord's granary in each seigneurie, the inhabitants complain of the practice whereby, when the corn is being brought in, the lords themselves wield the measure and apply the strickle. It should not be so, for it is the owner's business to see if the measure shows the right amount."[102] We know of others who, anticipating that they may be asked to pay their rent not in money but in kind, demanded: "If this be so, then the measure of Loheac should be used and not that of Guichen, which is considerably larger."[103] Still others, contrariwise, press that "seigneurial rents be rendered in the future in money, and not in corn, as the measure [lacuna in text] and the measure be made known."[104]

There was even an instance—the only one known to us—when the peasants themselves demanded differentiation of the *boisseaux*. The case, even if exceptional, is illuminating and worth relating: "The majority of seigneurs in this canton collect (on their estates) feudal rent in corn according to the measure of Châteaugrison, most commonly in small oats. Now, small oats here are cheaper by one-tenth than large oats. . . . Yet although rents are owed in small oats, the landlords make no objections to large oats instead, and accept them in the same amounts. Is this not a shocking injustice? Should the seigneurs not keep two *boisseaux*, one for the rents due in large oats, and the other—smaller by a tenth—with which to measure large oats from the vassals whose rents are owed in small oats? And yet those seigneurs will just tell us to pay up in small oats if the rents are owing thus, and they will accept them. A fine argument! Everyone knows that small oats are less useful; should we despoil our land to satisfy the seigneurs' caprice?"[105] The passage is priceless. The idea that different varieties of oats should be measured by different measures is nicely in keeping with the common attitude towards measures by men of the pre-metric and preindustrial times. It did not, apparently, occur to anybody in the case we have cited to fill the normal *boisseaux* with large oats, but to fill it only nine-tenths full! This was unthinkable to the locals since, after all, old inventories had stated that the seigneur should receive his rent by the *boisseau,* and, moreover, it would not have been technically simple, whereas it was natural for them to think in terms of a different container, of smaller dimensions, for the large oats. "Old privileges" were not as a rule questioned, though doubts were sometimes cast upon their very existence. However, it might be proposed that "the seigneurs be obliged to present their title deeds to enable us to ascertain both the value of corn and the size of the measure at the time a given *feudum* had come into being, and to ensure that in the future those same seigneurs shall demand their rents only to the weight and measure binding at the time of origin."[106] Furthermore, it might be demanded that "the seigneur be obliged to present his title deeds, so that we should know the proper capacity of his *boisseau.*"[107]

The demand for the presence of a state law officer as a guarantee of invariability in the seigneurial measures was frequent and coincided with the long-standing policy of the monarchy. Thus we know of a demand in one district[108] for an annual check on seigeurial measures, to be carried out by the royal court of law, and of a similar demand in another,[109] whose collective *cahier* states: "Uniformity of weights and measures [should be binding] throughout the kingdom, and standards ought to be deposited with the clerks of courts or with the police." In Blois, it was demanded that "there should be but one measure and price, fixed every year by the royal court judge on the day the rents are due,"[110] and in Brittany that "weights and measures be standardized throughout the kingdom.[111] Again, an Angoumois source contends that all seigneurial *boisseaux* follow the standard kept at the court of law;[112] a document from the Angers region suggests that "it is desirable that every parish should have a judicial office, which would hold authorized inventories of all rents, calculated in terms of the new *boisseau,* that are payable in the demesnes of the district";[113] and from the area of Châtillon we have the wish that the binding *boisseau* should be remeasured by the state official.[114]

Since, therefore, it used to be requested that the *boisseau* for the collecting of rents be annually verified in the presence of both interested parties,[115] our evidence makes clear the anxiety—presumably based on unhappy experiences—lest during the intervening year the standard should be clandestinely tampered with and enlarged. In the future, it was felt, these matters should be regulated through adjustment of inventories (the *terriers*) that had so

199

exercised the nobles of the region of Sens. One *cahier* from the district of Amont puts it: "The petitioners insist upon uniformity of weights and measures, and that the [new] weights and measures should be used in the inventories, the same as the States General shall adopt."[116]

A great number of disputes, as the *cahiers* imply, were concerned less with measures as such than with methods of measuring (a problem we have already touched upon). Above all, there was the contentious issue of heaped, as against striked, measures. *Ras le bois*, "Let the measure be keenly striked,"[117] was a saying in Brittany, whence also comes the view: "As regards . . . the measure, it would seem proper that it be keenly striked and not somewhat heaped as practiced at Vitre, for thus many quarrels arise."[118] From Brittany, too, comes the complaint that "it is an abuse, craftily introduced into our parishes, that we are made to render our oats-dues alternately by striked and heaped *boisseaux*. . . . The vassals are obliged to pay thus, and yet every one knows that because of the heap the measure cannot ever be accurate and true."[119] And we have further, related examples, such as: "Let heaped measures be no longer at all used, for they lead to uncertainty and to abuses";[120] "Let all varieties of corn everywhere be measured striked";[121] "Let the strickle be always applied by law, so as to eliminate many inconveniences";[122] or "We wish to pay our rent henceforth by the royal measure, striked";[123] and so forth.

Obviously, the issue could not but be bound up with that of the shape of the standard. A *cahier* from Brittany suggests that "the edge or the upper rim of measures that are in use in the collecting of feudal rents should be four or at the most five lines wide, and the height of the measures should be equal to double their width."[124] Now, the reader will recollect that the *coutumes* prescribed a ratio of 3:1 between the diameter and the height of the *boisseau*—and yet here it is 1:2! The latter ratio is demanded, however, because of the absence of any insistence that the measure be striked, whereas the stipulated maximum width of the rim offered opportunities for increasing the "heap."

The true magnitude of a given measure was affected not only by the capacity of the *boisseau* and its more or less flat shape; it could also be affected by the manner in which the grain was poured into it, or even the manner of applying the strickle. The evidence of the *cahiers* is relevant here. Thus we read in one: "In the end, it is in the marketplace that those entrusted with the measuring of grain have to agree, [and they sometimes] pour it in lightly, so that there is no way later of checking the amount there ought to be."[125] Elsewhere, we read that in striking a container of salt "it is right and proper that, after filling, the measure should be striked, and to do that with a roller and not with some devices that are able to take off two or three pounds each time."[126] The matter was also of importance when tax grain was being remeasured; hence the peasants actually demanded that they themselves, and not the lord's clerks, should do the remeasuring and striking.[127] The whole business was complicated indeed, witness the large vocabulary of the methods for measuring: alongside the terms *ras* and *comble* we come across *ras le bois* (for lack of a better phrase, translated here as "keenly striked"), *grain sur bord* ("grain at the rim"), *demi-comble* ("with a half-heap"), or, on one occasion, the far from clear *tro-collée*.[128]

It should be mentioned that for all the countless demands for standardization of measures, a conservative attitude towards them is not infrequently voiced in the *cahiers*. The following is surely a case in point, even though the beginning sounds radical: "Let there be but one measure in the entire kingdom and let all kinds of grain be measured with it; in places where the measure of small grains, such as barley or oats, continues to be larger than that used for wheat, let there be a special measure, and let the measure for wheat be banned in public markets when small grains are being measured; for otherwise the heap that is proper for that measure may turn out higher or lower, depending on whether the wielder of the strickle—and he often a man of little finesse—favors the interest of the buyer or of the seller."[129] In another document a case is mentioned in which the use of the smaller measure was insisted upon for the more valuable variety of oats.[130] Or again, we know of an intriguing instance when the revival of a "top," long since eliminated (presumably by agreement to commute an appropriate number of striked measures), was sought by the seigneur.[131] Apparently, then, even in such a contentious case the peasants did not in principle question the widespread custom of measuring oats "heaped," and the traditional attitude prevailed. It was, indeed, to be still very much in evidence during the period when the metric system was introduced.

The bulk of our evidence about the strength of traditional attitudes comes from western France—from Brittany to Angoumois. This is scarcely surprising, for those relatively backward regions, where rents in kind still prevailed, abounded in particularly bitter and long-standing metrological disputes between lords and peasants. The number and variety of complaints and demands, the realism permeating the descriptions of the abuses perpetrated with impunity, the very wealth of detail—all these features enhance the value of our sources, revealing the extent of the injustice that the peasants suffered. We must, however, reemphasize that in principle the *cahiers* respect the peasants' obligations towards their lords. The very insistence that the latter "present their entitlements" bears this out, while resistance against enlarged measures for dues in kind was aimed against abuses and not against the practice as such. The same applies to demands that the state authority be brought in as a mediator or as guarantor of the invariability of measures. And it applies, too, to the requests for the introduction of new measures in the *terriers*, or for uniform methods of measuring. We have encountered hundreds of references in the *cahiers de doléances* to the need for standardizing weights and measures and for abolishing seigneurial metrological privileges, but we have found none that linked these issues with the levels of rents or other obligations to the lords. Indeed, it is often explicitly emphasized in the *cahiers* that the seigneurs shall lose nothing by any standardizing reforms, and the emphasis would seem aimed at securing their acquiescence.

Let us cite a handful of examples, beginning with more general ones. An extreme and admittedly unique instance has come to us from one of the parishes in the judicial district of Rheims, in a *cahier* of fine literary form and intellectual level, whose authors are simply prepared to forego any standardization of the measures used in rent collecting in order not to fall foul of the landowners. We thus read in it that the measures throughout the kingdom should be standardized "so as in no way to interfere with the weights, measures, and usages that are the lords' due and derive from their rights relative to their vassals and tenants. Our sole desire is that the aforesaid weights and measures be applied to articles that are sold by weight or by measure; we wish to see a ban throughout the kingdom upon the sale or supply of articles by weights or by measures other than those decreed by his Highness."[132] What is at stake here, is an unambiguous standardization only of weights and measures employed in commercial transactions, whereas those used in seigneurial collection of dues are being left severely alone. It is not out of the question that a general statement found in another *cahier* has an analogous meaning, for it demands that measures in the kingdom be standardized 'providing the means, however, to ensure that no losses will result from any innovations and alterations in the existing state of affairs."[133]

The lawyers of Orléans were far more radical in their view that "the King's judges ought to remeasure particular measures and to determine their relations to the royal measure, and thereafter the seigneurs shall be forbidden to use . . . their own measures, either in the market or in collecting their dues.[134] The proposal is rather piquant in that it was conceived by lawyers, whereas hundreds of *cahiers* complain of the numberless litigations brought about by the metrological confusion. But let us leave aside the lawyers' argument, as, indeed, we can afford to, since in numerous other *cahiers* there are also statements evincing a widespread desire to respect the lords' rights. In one instance, we find such an affirmation included in a detailed proposal for a procedure for converting the traditional *boisseau* in which the peasants' dues were calculated;[135] and a similar statement in a complaint against the lord's right of measure, but proposing that it be bought out rather than abolished.[136]

There are many further relevant examples. "Let there be, in the entire kingdom but one single measure, weight, and one ell, and let the existing rents be paid and collected with the number of measures adjusted up or down, according to whether the new ones that we want will be smaller or larger than those used heretofore, so that no losses will be incurred."[137] In another *cahier* we find the statement that "the lords should accept the reform of measures all the more readily in that it will entail no diminution of their rents."[138] In a similar vein, a Pas-de-Calais *cahier* states: "Let the lords whose rents are owing in grain be obligated to keep at their manors standards marked according to the former measure, either large or small, whichever should be applied when the dues are being collected."[139] The contention offers an interesting perspective which is "minimalist" in the extreme: no intervention by the state is asked for, not even a standardization of measures—a larger and a smaller one being stipulated—provided there was no opportunity for the lord to increase their size. A variant, in effect, of this attitude was the demand for "a return to the sources," specifically that "the lords be obliged to present their title deeds to enable us to ascertain the measure at the time a given privilege had come into being."[140] In this case, it is not out of the question that the implications of the demand cut deeper, for who might it be that the lords should "present their title deeds" to? And let us note, in particular, that in the same *cahier* we encounter phraseology unusual for Brittany, quaint yet testifying to a marked development of national and social consciousness: "Friends! Long live God and our King Louis XVI!

May his reign continue forever for the happiness of the nation, for which he is a true father; let us pray to God that He may preserve him and his noble family, too; and may the worthy and esteemed Necker enjoy a long and happy life. We beg them to join with us in checking the abuses being perpetrated by tyrants against that class of citizens, which is kind and considerate and which, until this day, has been unable to present its grievances to the very foot of the throne, and now we call upon the King to mete out justice, and we express a most sincere desire for but one king, one law, one weight, and one measure."[141]

Proposals claiming respect, in principle, for seigneurial rights were frequent; for example, "we wish that weights and measures be standardized in the whole kingdom, but respecting the rights of those who have hitherto held them."[142] Only once have we come across a demand for standardization of measure regardless of the rights of the seigneurs, even though they might oppose the reform.[143] We have, however, come across a great many complaints about measures and ways of measuring that sought to increase the dues. At the root of these lies the reluctance to call into question their traditional amounts. Not once have we come across a demand to decrease the measure—not, at any rate, until late September 1790, when the parish of Guénézan put one forward.[144] We may parenthetically remark here a unique request, connected with metrological proposals, for granting the peasants a freedom to pay either in money or in kind, the implied preference being for commuting the latter to the former.[145] This might lighten the burden for them in seasons of abundant harvests and, consequently, falling prices (a puzzling instance, that, in 1789). And, to cite yet another case of respect for seigneurial rights, the deputies "were expected to seek standardization of weights and measures, while allowing the lords, or other earners of land rents payable in kind, the right to reduce, in the presence of interested parties, the weights and measures that appear in their privileges or inventories to those to be adopted by the States General."[146]

It would, however, be an oversimplification to see in the *cahiers de doléances* only one type of social antagonism focussing on weights and measures, namely that between the peasant and his lay or ecclesiastical lord. For measures were bound up with a great many facets of social life, and the society of pre-revolutionary France was intersected by a great many social antagonisms. Therefore, many other, often long-standing tensions, if not open conflicts of interests, were in evidence, too. We shall consider just three of them:

between the peasant and the state, internecine village antogonisms, and those between the peasant and a lord other than his own.

The charged relationship between peasant and state naturally manifests itself in fiscal matters; that much has been easy to foresee. After all, in the France of Louis XVI almost all the demands of the various social groups were, in one way or another, bound up with questions of taxation. Most frequently, and most contentiously, the disagreements concerned measures of land. Commonplace statements insisting on their standardization occur fairly frequently,[147] but far less frequently than proposals for standardizing dry measures of capacity. This, in my view, entitles us to infer that the problem of agrarian measures was less aggravating in daily rural life, less likely to produce frictions and a sense of injustice, than were the quarrels surrounding the seigneurial *boisseau*. Moreover, it should be emphasized that no single catchphrase gained common currency from the relevant circumstances. Even trite and unsubstantiated proposals vary from case to case and do not imitate. Here, then, is an interesting fact, capable, it seems, of dual interpretation. On the one hand, the originality of the proposals naturally inclines us to set considerable store by them and implies that there were genuinely felt grievances that motivated them. On the other hand, whereas we would have had to consider the presence of "slogans" as significant, we are inclined in the present instance to interpret their absence as due to the fact that the disputes about land measures were less bitter.

Among the more solid proposals, the leading issue is that of the relationship between the agrarian measure and tax assessment. Consider, say, the following: "Since the tax of a royal denarius per measure of land is excessive . . . let the King graciously determine the measure for the future, so that everyone shall be appraised of the area he is liable for."[148] Another parish demands "a remeasurement of all lands (*mesurage et arpentage*) . . . and the remeasurement of all lands in France to be carried out using the standard chain (*chaine*), to wit, the one of twenty royal feet per rod (*verge*)," the objective of the operation being to produce "a just assessment of taxes."[149] Elsewhere, the hope was expressed that, thanks to the standardization of measures, "the tax known as *taille* will no longer be arbitrary because it will be based upon the same unit everywhere, and the assessment will be easier, cheaper, and more intelligible."[150]

Some more advanced and "cerebral" views were voiced, too; for example, the parish of Romigny proposed: "Let there be a general measure fixed by His Highness, binding on all taxpayers in this kingdom. All existing local measures should be assimilated to the royal measure to ensure that taxes are paid equitably."[151] Compared with this general statement, others are more specifically concerned with local injustices.[152] Thus, the parish of Esquelbecq writes: "It behoves us, moreover, to point out that the parish of Esquelbecq is obliged to contribute out of proportion to its area to the requirements of His Highness which are levied from the town and castellany of Berques; its other villages use the Gand measure and their tax burdens are assessed by it, but Esquelbecq is subject to the Artois measure, which is the less by a fifth. For all that our land is no better [were the soil at Esquelbecq better, this might have excused the smaller measure in the eyes of pre-metric man!—W.K.], the parish bears the district's heaviest burden of taxation, given the aforesaid measure. . . . We see it, therefore, as just that a new cadaster should be prepared to put an end to this state of affairs, and the deputies ought to press for that."[153] Views of this kind are numerous.[154] Obviously, with the antagonism between peasant and state focussing on taxation, the question of measures played a far smaller part than it did between peasant and lord. Yet the problem was not absent from this relationship either—not, at any rate, in some districts.

As far as internecine village conflicts were concerned, the most potent animosity was that between the peasants at large and the millers. The matter was highly complicated. On the one hand, the manor's monopoly of grinding was one of the seigneurial privileges (*banalités*), of the same order as the right of measures. We know of cases when the abolition of both was demanded in the same breath; for example, "Let the monopoly of the mill be banned . . . and let there be only one measure for corn."[155] However, it was not the lord enjoying his *banalité* directly, but the miller who actually collected the mill income from the peasants. Would he not, then, swell it by all manner of means so as to increase his own earnings—if only because the lord had raised his lease rent, and the miller wanted to recoup himself? What interests us in the wide spectrum of daily activities is the kind of situation that, with a certain inevitability, made for the rise and exacerbation of peasant-miller hostility.

An obvious and typical example is the following: the parish of Rigny-sur-Arroux insisted that "all millers be obliged to keep on the premises scales and weights so that everyone can be sure that he is dealt with fairly."[156] Sometimes the less-than-fair procedures of the millers would be brought fully to light. Thus, the *cahier* of

Angoulême is quite specific on the subject: "To collect the corn well-dried, to remeasure it after swelling, to take the surplus that will thus accrue to every *boisseau* for themselves, and then to take a surcharge on the remainder to boot—that is how they go about it."[157] And the parish of Gandrange demands: "The monopoly of the lord's mill must be abolished, for the reason that the inhabitants here are thereby exposed to danger whenever they are obliged to take their corn to the aforesaid mill; for there is no bridge across the Orne, and they have to ford it to reach the mill, which is located over in Lorraine and in which there are neither scales nor weights."[158] The interesting point of this passage is that the seigneur's privilege is not being questioned in principle, but on technical grounds. It is not out of the question, however, that we should read between the lines of the argument: if the seigneur wishes to avail himself of his privilege, then he should see to the proper maintenance of the road and should build a bridge over the Orne. What is, at any rate, plainly pointed out is that the lack of scales and weights at the mill gives grounds for questioning the privilege. In some instances, the case was put more clearly: "Let the vassals be free to choose whichever mill they prefer when the miller fails to provide scales and weights."[159] Naturally, the matter of the *banalité* mill crops up time and time again in the *cahiers de doléances*, but rather infrequently in connection with the problem of measures (as when the millers had false measures or failed to provide weights and scales). The latter were, here, a side issue, unlike the matter of seigneurial rents in kind.

Finally, the third area of friction to which we wish to draw attention is the relationship between the peasant, selling his produce in the market, and his lord (represented, of course, by the tenant), who was entitled to levy a charge in the given town upon the compulsory remeasurement of the corn for sale—the so-called *droit de minage*. Let us cite two unequivocal examples.[160] In one, the source informs us: "In many towns of the kingdom, there is still a very onerous tax that goes by the name of *minage*. In some towns it amounts to 1/28, in others as much as 1/20 of the *boisseau*, no small burden upon the farmer. Only the nobles and clergy are exempt from it."[161] In the other case, the concise phrasing is most striking: "Let all customs charges and others deriving from the rights of *minage*, stallholding, measure, shopkeeping, bakehouse, press [for the wine grapes or olives,—W.K.], and the monopoly of the mill be abolished; for since the nobles and the clergy are not subject to them, they signify serfdom."[162] Nothing could be more emphatic; the entire reform program of the night of 4 August is adumbrated here.

Let us next consider the second set of motives, namely those of commercial derivation. The difficulties encountered in commercial relations because of metrological confusion were mentioned very frequently. To pick one example out of hundreds: "The diversity of weights and measures within each province, let alone the kingdom, poses the greatest inconvenience for trade, both by affecting accounts and by the unforeseeable occurrences it brings about in the market."[163] Except for specialists, everybody feels insecure, "for the variety of weights and measures opens the door to countless abuses which threaten all, except the professional traders; there should be but a single standard in the whole territory of our kingdom."[164] The proposition that weights and measures be standardized was an integral feature of the doctrine of free trade; for it postulated, in one breath, that internal customs and excise duties be abolished, allowing a free movement of grain between the provinces, and yet stipulated the erection of tariff barriers at the frontiers, both steps aiming to ensure the prosperity of the country's economy. So, at any rate, thought the goldsmiths of Orléans[165] in what was an ideological manifesto rather than a statement narrowly concerned with the well-being of their corporation. However, the context in which the document was issued helps us to appreciate the importance of the metrological postulate within the wider ideological program.

Commercial anxieties were at times expressed without any theoretical overtones, but directly, in a way that reflected the difficulties experienced by the people. So, for example, standardization of measures was sought "so that everyone will know just how much he is paying for."[166] "We want uniform weights and measures in all markets of the kingdom so as to avoid the need for numberless calculations, which most countrymen cannot perform and which daily confuse people."[167] Sometimes the standardization of measures would be demanded only for essential articles,[168] but these would be of genuine concern to the peasant as the small buyer in the market.

To return to the free-trade ideology of commerce-oriented arguments, we may cite an example from Brittany, where a certain parish demanded the standardization of measures because "a single measure, being universally used, would make things easier for many citizens who do not comprehend the diversity of measures and therefore cannot grasp the differences in prices between regions. However, a single measure, binding upon all, would enable them to apprehend the prices of all goods and wares, as well as the profits to be obtained by conveying them from one end of the kingdom to the other." Surely, this is no pronouncement of Breton peasants; it was not they who conveyed commodities "from one end of the kingdom to the other."[169] More likely, we should descry here the influence of the Physiocrats, filtered probably through some local Notable's acquaintance with their writings. Possibly also among their readers was the editor of the comprehensive *cahier* of the judicial district of Amiens, who saw in the standardization of measures "an effective means towards abundance."[170]

The tenor of the complaint from the shoemakers of Alençon was different: they wanted standardization "so that men of no learning might not be cheated."[171] On the other hand, there were complaints of peasants-as-sellers, such as: "In conclusion, the aforesaid inhabitants beg for the introduction of one and the same measure for all as far as the sale of corn and other produce is concerned, for the reasons that the present diversity of measures exposes people daily to swindlers and is of help to speculators taking advantage of the peasants' ignorance."[172] Or again: "Let all weights and measures be equal in our province, and thus all manner of things are to be bought and sold in different markets by the same measure, in order to prevent the frequent confiscations by the police, who will make no allowances for those who know little of different weights and measures."[173] The last complaint is voiced on behalf of the peasant wishing to sell his goods in the market: he may, indeed, have enough surplus for sale to motivate him to reach some markets outside his locality, but encounters obstacles from the police. The poorer peasant, on the other hand, favors standardization of measures so that "the consumer will not be cheated," one plea going on: "It is worth pointing out that in this parish three different measures coexist . . . the *pinte* of Orléans, that of Jargeau, and also that of the reverend Fathers of the chapter of Sainte-Croix."[174] Thus, different social groups for different reasons concur in seeking the standardization of measures for the sake of "increasing the scope of commerce, by simplifying its processes and arresting the proliferation of difficulties that discourage the merchant."[175] Standardization, it was felt, would "free the merchant from detailed studies . . . while the citizen will no longer suffer from cheating by those whose calling causes them to be experts in accounting."[176]

It is, of course, outside the scope of the present study to attempt to verify the extent to which the phraseology of the various demands corresponded to the social structure of particular parishes: for example, the high degree of proletarianization of some with the predominance of small holders, who acted mainly as buyers in the local market; or the dominance of others by larger farmers, whose role was more that of sellers. However, the present author is convinced that there was such a correspondence. Moreover, there is no doubt of a correspondence as far as regions were concerned. In some, there was much anti-seigneurial feeling and little commercial motivation, while in others the converse was true. Where the peasants' dependence upon the lords and the use of taxes in kind continued unabated, farmers had less chance of switching to production for the market.

The third set of arguments put forward in the *cahiers de doléances* for standardization of weights and measures may be termed "national-monarchist." They constitute an assortment as interesting as it is difficult to interpret in a good many instances, although readily comprehensible as a whole.

On occasion, it would appear that the relevant passages voice the emergent consciousness of modern national unity, and of the unity of the institutions destined to be its expression and bond; at other times, the arguments seem to be addressed primarily to the Dauphin—or, rather, the King. All in all, we are confronted by a mixture of genuine views of the authors themselves and of the views they assumed to be held by the addressee, and the two are far from easy to separate. For, if the arguments are often permeated by the ideology and spirit of the ancien régime, how can we determine how far the writers were merely seeking to convince the authorities, and how far they were themselves heavily influenced by them?

The terminology of the *cahiers* very often shows signs of the ideology of national unity. Naturally, matters other than metrological are also affected by it. Nonetheless, references to metrological issues are well to the fore because the uniformity of weights and measures was prominent among the desiderata of the time. For, in keeping with the value system of the Enlightenment, law was seen as the maker and not merely the mirror of the distinctive character of a people. Uniformity of customs and laws, of matrimonial law and laws of inheritance, and of many other institutions, not least of weights and measures, were seen as preconditions, both necessary and sufficient, of national unity. Thus the unity of weights and measures plays here a dual role—literal and symbolic.[177] And we should not sneer at the fact that the firmest ideological statement emanated from the *épiciers* of Rouen: "The hallmark of the perfection of our Constitution will be, above all, the integration of all *coutumes* into one whole. The uniformity of customs, viewpoints, and principles of action will, inevitably, lead to a greater community of habits and predispositions; the relevance and shrewdness, paragraph by paragraph, of the universal *coutume* will prevent numerous lawsuits arising from the ambiguity and diversity of the earlier obscure and barbaric *coutumes*. Hence, too, will follow a conversion of all measures in the kingdom to the same dimensions and names."[178] Admittedly, rather than written by the shopkeepers themselves, this passage was most likely penned for them by some ideologically committed intellectual of Rouen. But it is legitimate to surmise that they were indeed exercised by the diversity of measures and *coutumes* and wished this feeling to be plain in the *cahier*. More significantly still, we have here an interesting case of a highly literate statement of feelings that were often encountered elsewhere in less polished phrases.

Thus, we are told in one instance that "the same judicial order, equality of weights, measures, coinage, taxes, indeed more uniformity in everything, would form a firmer bond for all the inhabitants. The excessive diversity of customs affects people's characters";[179] in another: "Finally, let the same customs, weights, and measures be made binding in all provinces in order to bring about the harmony that all men of this state so much desire";[180] or again: "Let there be but one measure and one weight all over the kingdom; this will facilitate trade, as well as contacts among all its people."[181] A "model" *cahier* postulates:[182] "Let there no longer be in France provinces alien to one another," while another suggests that "to this end, it is necessary for each province to sacrifice some of the customs and institutions peculiar to itself, so that one law may obtain in the whole country."[183] Was integration of measures, then, to spearhead national integration? Paradoxical as it may sound, this was precisely the view—indeed, a slogan—presented by many *cahiers*.[184] And "the new measure" often, although by no means always, thus assumes a symbolic character.

There are many vigorous assertions of the close relation between the unification of measures and national unity. For example, "However can a nation, sharing one ruler and one government, have yet different *coutumes*, laws, weights and measures?"[185]—a view that would seem to accord well with the unifying efforts of the monarchy. Or again: "Let the different *boisseaux* of the seigneurs be reduced to the standard of our ruler"[186]—a wording reminiscent of Henry IV's decrees. Elsewhere, a parish statement noted that "it is astonishing that despite our country having long since been unified under one ruler, France's provinces are foreign to one another."[187]

An argument *sui generis* that is at first glance surprising runs as follows: "Let there be throughout the kingdom, one *coutume*, the same weights, and the same measures, for after all we have only one language";[188] yet there is sound logic here if we accept that the statement expresses at bottom a desire for national unity and a conviction that since laws and customs do mold people, lack of unity in them prevents the rise of a united nation. The idea is more fully developed in the following passage: "Since all Frenchmen constitute one people and are subjects of one ruler, and should, therefore, form one family, then following the same principles, it is absolutely essential that the States General should promulgate a single code of laws for the entire kingdom, whereby only one *coutume*, one weight, and one measure shall be instituted."[189] Similarly, an Orléans *cahier* states: "Common *coutumes*, laws, and measures for the kingdom are sadly lacking to aid its governance. . . . Posterity will find it hard to believe that one state, with one ruler, observed a thousand laws and diverse *coutumes*, often contradicting one another. It would be most beneficial for the States General to bring about a long-desired unity in these areas."[190]

The sources we have discussed afford a strange mixture of the polished and the uncouth, of the cultivated and the instinctive. The contention of the goldsmiths of Orléans, which we have cited,[191] was put forward by a group hardly lacking in self-confidence, faithful to the monarchy, and smacking of Physiocratic doctrines. Often, however, the documents reveal the time-hallowed appeal of the peasants to the "good King"—over the heads of their bad lords. "His Majesty alone has the authority to order for all France the same laws and the same weights and measures";[192] "Let there be in our kingdom one measure and one weight, just as there is only one King";[193] or more firmly: "May the King, considering his whole people as a father does his family, graciously declare one *coutume* and one measure for the entire kingdom";[194] and again: "It is desirable for France to have one law, one *coutume*, one weight, and one measure, uniform throughout the kingdom, for his Majesty [rules] a people that is uniformly loyal to him."[195] The last excerpt implies nothing less than a full-blown sociological theory, one that was basically the same for the peasants and for their kings, even if more emotively expressed by the former and based on intellectual argument for the latter. It is the theory that postulates the character-building nature of the law, the uniformity of laws leading to social uniformity of character, with the nation that manifests it being able to enter into an immediate *rapprochement* with its ruler. *Sans intermediaires*—this was the dream of absolute monarchs and emperors, as well as of other autocratic régimes, and it is a dream capable of being fulfilled when the subjects enjoy the bliss of living "under the sceptre of a king whose sole concern is their happiness."[196] What is at stake is precisely the attempt to turn "his peoples" into "his people," and this process is furthered by the unification of institutions, including weights and measures; or so the *épiciers* of Rouen were firmly convinced when they wrote the statement we cited above[197] [chap. 22, n. 178—TRANS.], and they saw it all as a straightforward assignment.

Many *cahiers*, while their arguments in support of the standardization of measures are couched in terms reflecting the spirit of the absolute monarchy and are likely to meet with its approval, refer to an older tradition, too. The King is being called upon to realize the ambitions of his forebears, to fulfil the desires they had merely formulated. The mixture of viewpoints here is intriguing. To achieve standardized measures throughout the kingdom would mean an almost revolutionary innovation. Nobody pretended otherwise, and the existing profusion of measures was denounced as a legacy of centuries of "feudal tyranny" or "barbarism," and thus as "the old," "the traditional."[198] Yet arguments from the past, from tradition, did not lack potency in late eighteenth-century society, particularly as far as the King himself was concerned, for his power ultimately derived from tradition, and it was to him that the *cahiers* were addressed. The problem was resolved nicely, in keeping with the ideology of enlightened absolutism, by extolling the tradition of the centralizing monarchy in opposition to the tradition of feudal particularism. This was the deeper meaning of the recollection by the Rouen *épiciers* of the attempt by Philip the Long in 1321,[199] which "the subsequent wars prevented from being effective." And our *épiciers* go on: "The introduction of uniform weights and measures to safeguard the honest citizen from a variety of abuses, frequently perpetrated and excused on the ground of diverse weights and measures, falls within the competence of the King, who sees the upholding of the rule of law as his prime duty."[200] Harking back even farther, namely to Charlemagne, was not uncommon. Apparently, local Notables knew of his reforms, while some sources reveal attempts to play on monarchic ambitions:

"For shame that Charlemagne's idea of bringing about uniformity of all measures in the kingdom has yet to find fulfillment."[201]

Frequently, demands for unification were introduced in the guise of a "return to the origins." Thus, the *cahier* of Troyes holds that "initially, measures in France were uniform, and became diverse with the diversity of customs. Taking advantage of inadequate law and order, each seigneur would establish what customs he pleased upon his estates."[202] The Crown was now, accordingly, called upon to reestablish its rights, which had been usurped by the feudal lords in the times of its weakness, especially seeing that the times were not orderly. There were many pronouncements like the following: "Louis XI greatly desired only one *coutume*, one weight, and one measure to prevail in France, but death stopped the execution of his great plan. But what is there today to stop our King in his overriding concern for the happiness of his subjects?"[203] Or, "There were equal measures and uniform *coutumes* under our earliest kings; it was the later turmoils alone that enabled seigneurs to introduce different arrangements in their lands."[204] The geographical extent of standardization postulated varied from—most frequently—nationwide, to *dans notre province*, and to the very modest, if rare, *dans notre paroisse* (see tabulation at the end of this chapter).

Should we attach any significance to their variety? In the view of the present writer, a fair amount. In support of this assessment, we should first note the goodly number of uncertain, hesitant suggestions; for example: "in the province, or even in France";[205] "in the whole kingdom, and particularly in the whole province";[206] "at least, in every province"; "for the entire province, or even the kingdom";[207] "in the province of Artois, and likewise in the whole kingdom";[208] and so on, *ad infinitum*. The indecision and the hesitation are of some interest. In some cases, they may reflect the vying with one another of the authors of the demands: one would suggest "the province," another "outbid" him with "even the kingdom." Yet they may equally well bear witness to overadvanced proposals being withdrawn, discretion overcoming the initial boldness ("in the kingdom, or at least in our province"). Again, the wording may point to the true and immediate concerns of the authors ("especially and above all in our province"). What is clear, however, is that the various postulates were not imitative copies of one another.

The second and perhaps weightier consideration supporting the above view is the presence of a certain regional pattern. In the more "feudal" western regions,—for example, at Rennes, Quimper, or Angoulême—there were proportionately fewer demands for countrywide standardization, relatively more that referred to the province, and still more that focussed on purely local practices. Surely, metrological privileges enjoyed by "the haves" of this world must have considerably vexed the populace if their boldest demand was the standardization of weights and measures within their own parish. Yet, despite these arguments, we are not inclined to accept the words of the *cahiers* literally in each and every case. Presumably, in some instances, even if nationwide standardization of measures was repeatedly mentioned, the true concern was with local measures only. There were, apparently, some converse cases, too. The parish of Mahalon-et-Cuiler was in favor of "the King's measure being the sole one in the entire province,"[209] thus specifying a province-wide standardization, but relating it to the central authority of the state. It seems legitimate to feel that the authors presupposed that each and every province would follow suit.

Third and finally, we should bear in mind the almost ever-present insistence, its context varying but the order never, that there should be one God, one King, one law, one custom, one measure, and one weight. The significance of this call becomes manifest if we heed two points. First, the wording is variable. If one parish urges, "Let there be in the entire kingdom only one God, one King, and one law, one weight, and only one measure,"[210] while another demands "one God, one King, one law, one weight, and one measure,"[211] then not only is the slight difference in phrasing interesting, as indicative of independent origin, but so is the identical assortment and order of the substantive terms, as indicative of the wide currency of what became a slogan. In fact, the "list" is not always the same: God sometimes figures and sometimes is absent; King, likewise. On occasion—and then always between *loi* and *poids*—*la coutume* is added, never affecting the rank order. Secondly, although we have come across identically worded postulates, they often appeared in *cahiers* that must have been independently prepared, emerging in different contexts and in parishes separated from one another by hundreds of kilometers. The critics of the *cahiers* as a historical source tend to see their occasionally identical texts as casting doubt on their independence and authenticity. But here the reverse is the case. The presence of identical phrases in unconnected *cahiers* confirms our conviction that they bore an ideological character that, reduced to a slogan, was current in virtually the whole country.

Among the millions of individual difficulties, anxieties, and injustices in daily life, we can discern certain situations typical of

215

larger groups and social classes, and the remedy proposed, even when the interests of different parties (such as buyers and sellers) seem to clash, is often the same one: standardization of weights and measures. The entire third estate was in favor of that, for a variety of reasons, and surely had been so long before 1789, although in deafening silence for lack of a platform. Only the privileged had been able to speak out then, and they were either hostile to, or skeptical of, the reform. Hence the astounding suddenness for the reader and the astonishing social and geographical compass of the requests for standardization in the *cahiers* of the third estate. Yet a closer reading all but enables us to discern the antecedents. Between the reality of those millions of individual concerns and injustices and the content of the *cahiers*, there had been an intermediate stage, whose elements included the formation of local opinion, the ripple effect of it, the emergence—with some help from the *philosophes*—of common catchphrases, which would then assume a life of their own apart from particular interests and would also become slogans, circulating all over the country and becoming an integral part of its ideological existence—a small but significant component of the mighty ideology of equality before the law and of national unity. Only a handful of the privileged, clinging to their privileges, remained outside that unity, yet although inimical to the reform, they obstructed but dared not openly oppose it or swim against the current of the nation's will. Are we exaggerating the importance of the processes under consideration? If so, then it is a common foible of historians. Yet it does seem to us that these processes are symptomatic of developments greater than the history of weights and measures—of the birth of an ideology.

The context of the metrological postulates figuring in the *cahiers de doléances* is itself deserving of some attention. Here, the term "context" means something quite apart from the essence of the argumentation: it means a certain coalescent mingling of matters spoken of "in one breath," or, in other words, the unspoken assumptions. To argue is to act consciously; the "context" in our sense unintentionally reveals the associations, hinted at by the authors of the sources, between the issue of direct interest to the student and other matters. Considered thus, the "context" tells us more of the mentality of a given group than does the conscious argument, mixing as the latter usually does the authors' own views and the views they feel will serve them best in influencing whoever they seek to influence. To conclude, such "contexts" of metrological postulates are typically either legal (judicial) or commercial or anti-seigneurial.

The legal "context" of the proposals for standardizing measures was that of pressure for the unification of laws and *coutumes*. Here, measures were seen as custom-derived institutions, as were the laws of inheritance, marriage, etc. Since it was widely held at the time that the distinctiveness of nations was above all expressed in and through their institutions and most particularly their laws, the demands for a unified system of laws—including statutory weights and measures—was tantamount to a demand for national unity. Strictly speaking, a legal "context" of this kind is met with frequently in the *cahiers*, but most often generically, rather than particularized; thus, "Let there be, in the entire kingdom, only one God, one king, one law, one weight, and one measure"[212]—virtually always in this order. Far be it from the present writer to play down the significance of this phraseology and to make light of its widespread prevalence; yet, on the other hand, we must not lose sight of the fact that the phrase "weight and measure" sometimes bore symbolic meaning and should not always be interpreted literally.

The commercial "context" promoting demands for the standardization of measures subsisted most frequently in the quest for free internal trade, that is, within the kingdom, and in whatever might aid the emergence of an internal market. The wording of the demands varied a great deal, ranging from the succinct "It is necessary to take steps to assist trade" to proposals that were highly specific and detailed. Requests for the abolition of internal customs duties were frequent—for example, "Let there be customs chambers only at the frontiers, and let there be only one *coutume*, one weight, one measure or container (*contenance*), and one ell in the whole kingdom."[213] Sometimes the demands stressed free trade between the provinces,[214] sometimes they would be linked with prohibitions of corn export,[215] and sometimes they just spoke of "freedom of trade and industry" in general.[216]

The influence of the leading ideas of the day is clearly discernible in such sources. However, this should not lead us to belittle their value, and to read into them only the Notables' and not the peasants' wishes. The preference for free trade in corn and, in particular, the ban on corn exports were deeply rooted in the popular psyche, and disturbances in famine years, both in earlier and later periods, bear witness to that. Whether the masses had absorbed ideas of the *philosophes*, or whether, conversely, the latter expressed popular feelings, must remain an open question.

The commercial "context" is most commonly discernible in more "negative" proposals such as those for the removal of obstacles to trade, and they were often very persuasive. "The deputies are expected to press for the abolition of the apprenticeship system, which is a brake upon trade and industry, and to demand freedom of arts and crafts as well as equality of ells, measures, and weights in this province"[217]—such phrases permit us to see the authors' economic liberalism. Yet it is interesting that the neighboring parish, whose demand is basically the same, puts it quite differently, in a manner clearly hinting at some earlier debate and compromise: "Let all monopolistic commercial privileges be abolished, and none such be instituted in the future, and let there be established one weight and one measure in the whole kingdom, and let freedom of arts and crafts be allowed (*qu'on hautorise—sic!*), if it is proper, in order to encourage industry. However, there are a few trades, of especial importance to public order, that should be exempted from the above."[218] Here, we have a postulate that was thought out, considered and agreed upon, and which reflects in concrete detail the local inhabitants' needs and concerns. But in the same area— as we shall shortly note—there were cases of metrological postulates showing contrary, anti-industrial assumptions. Finally, we should note some straightforwardly popular "contexts" among clusters of more "cerebral" ones. Sometimes they seemed to savor of the interests of the Notables ("Let there be in the entire kingdom one weight, one measure, one *coutume*, and one and the same term for repayment of the bills and promissory notes")[219] and, at other times, of the concerns of the masses, for example, the demand for stricter controls of bread prices.[220]

We find very different situations in the third group, in which metrological demands implied anti-seigneurial assumptions. We have a great many cases of this type, and the range of diverse associations that they reveal is considerable; their authenticity is quite clear, and catchword, banal phrases are virtually absent among them. The following examples bring together the various elements of the anti-seigneurial position: "The atmosphere in towns and villages would grow far calmer if a reform could be effected of the tithes and seigneurial monopolies (*banalités*), which cause so much confusion. Similarly, if there were in France but one measure to be applied to all things."[221] "We make bold to ask for a single *coutume*, equal weights and measures and the abolition of seigneurial monopolies, customs, and labor dues, and all seigneurial privileges derived from oldtime serfdom.[222] There are numerous associations here, and they appear, time and again, in a variety of combinations: measures and labor dues, customs duties and tithes, manorial monopolies, and all seigneurial rights at large.

Thus, some demanded "the abolition of seigneurial rents, improvement of roads and bridges, and equality of weights and measures."[223] Others coupled their metrological demands with the "abolition of unpaid labor of gathering (*cueillette*) rendered to the manor,"[224] and frequently the "context" was that of seigneurial monopoly rights in general, and of grinding in particular.[225] On occasion, the monopoly rights would be enumerated: "Let there be throughout the kingdom a complete uniformity of weights and measures; let the seigneurial ovens and mills, of which the use is compulsory, be abolished, and also the lords' exclusive hunting rights."[226] "We request that the lords' monopolies of mills, ovens, or presses [for wine grapes or olives—W.K.] be abolished, and equality of weights and measures in the whole kingdom be established."[227] And in an interesting contention, "Let the monopoly of grinding be prohibited, a ban imposed on the export of grain, and let there be just one measure for the corn in our parish,"[228] the several restrictions stipulated testify to its authenticity. At times, metrological requests would be coupled with demands "to abolish all judicial functions of the lords,"[229] or with a demand that "the keeping of pigeons for sport be abolished."[230] More often, the association would be with the problem of the tithes: "We petition His Highness to proclaim for the entire kingdom the same weights and the same measures, and also that the standing tithes be abolished,"[231] and elsewhere: "The deputies shall request that one weight and one measure be established for the whole province, and also uniformity of tithes."[232]

The three groups of "associations" do not, of course, exhaust the entire range of diversity. The demand for standardization of measures is sometimes coupled with tax issues, as in the following: "Let there be only one weight, one measure, one tax register (*rôle d'imposition*),"[233] on occasion specifying the salt tax (*gabelle*) or the allied questions of measures in salt stores, methods of weighing salt by the sellers,[234] and so forth; and tax issues were often linked with land measures. Sometimes the associations were nothing less than quaint, sometimes curiously "intellectual," touching on the freedom of the press[235] or the freedom of industry,[236] and, in one instance, on the legalizing of loans at interest.[237] At other times the apparently bizarre associations carried populist overtones, for example, the following request expressing the worries of poverty-stricken, largely proletarianized peasants, eking out a living by weaving and

219

yet buying cloth for clothes in the nearby country town: "We demand the end of tax farming (*fermiers généreaux*), and also that cotton-weaving machines be done away with, because they deprive of their earnings poor workers, for whom all too often no other work is available; and may it also be decreed that there be one measure and one ell in the whole province."[238] Plainly—for our few examples make that much clear—the "contexts" often overlap, just as their underlying concerns did in real life. For in the mighty process of the rise of a modern nation the cultural, legal, institutional, and economic factors were closely interwoven into one complete fabric.

The analysis of what we termed "contexts" has so far enabled us to illuminate the processes of "unconscious associations." Unintentionally, the authors of the *cahiers* have revealed to us their assumptions underlying metrological questions, and so have also revealed something of their attitudes towards the social institutions of their day. The question we have to consider in turn is that of the attitude of the two privileged estates towards the demands for the standardization of weights and measures. That is, we shall try to appraise their reactions in the light of the evidence of the *cahiers de doléances*, limited as it is, because the number of published *cahiers* emanating from the clergy was small, and from the nobles even smaller, while the *cahiers* we do possess are collective ones of the *bailliages* and *sénéchaussées*. In fairness, we should note that while the collective *cahiers* of the third estate are rather suspect as a source, their counterparts produced by the two privileged estates inspire in us more confidence, and we should, therefore, examine them all the more carefully.

It is the *argumentum ex silentio*, that is, what the *cahiers* fail to spell out, that is most informative for us on the present question. In the majority of the *bailliages* and *sénéchaussées*, even in those where the *cahiers* of the third estate mention standardization very frequently, the collective *cahiers* of the clergy and nobility contain no references to it.[239] Where such references appear, they tend to be hesitant, as if yielding to pressures from below. For example, we find the nobles of Troyes writing: "Let the States General be asked to consider whether or not it might be advantageous for trade, and promote the prosperity of the nation, to have the weights and measures standardized throughout the kingdom."[240] The same applies to Paris.[241] Sometimes the nobles actually proposed standardization, but only within the compass of a province;[242] and whereas one can by no means always take at their face value the demands from the third estate—here relating to "our province," and there to "the whole of France"—it seems legitimate to suppose that the phrasing in the *cahiers* of the nobles had been carefully considered. Only in one instance has this writer uncovered an unequivocally phrased proposal produced by nobles for the standardization of measures.[243] A few proposals of this type have been remarked in clergymen's *cahiers*[244] but perhaps more significant here is the fact that in the unique, large volume of *cahiers* of clergy from the judicial district of Orléans, there is only one that contains such a proposal.[245]

There is therefore much to justify the view that the *argumentum ex silentio* is of great importance for us. In order to apprehend more fully the meaning of that "silence," as well as the few-and-far-between instances we have cited of speaking out, it will be instructive to consider the collective *cahier* of the nobility of the area around Sens, for this is the only case where we are familiar with the antecedents of the matter. The *cahier* goes into numerous facets of the standardization of measures, although hardly without ambiguity; witness the following passage: "Since the question of gains to be derived from establishing equality of weights and measures throughout the kingdom is one of a kind, where it is not possible to come down on one side or the other without a very great risk of detrimental effects to property, to prerogatives bestowed by venerable capitularies, and to the interests of trade, it will therefore behove our Deputy to pay close attention to the deliberations of this most important matter. And since the States General [the authors of the passage were no fools!—W.K.] undoubtedly have had presented to them many submissions seeking to throw light upon it, accordingly the nobility of this district prefer to suspend judgment so as not to rue later any overhasty proposals." So they saw things in the spring of 1789—but let the reader recollect the same nobles' pronouncement of January 1788 quoted above[246] [p. 178, above—TRANS.]. For then, a year earlier, they were firmly in favor of leaving metrological privileges alone[247] and were not at all prepared to look into the question of standardizing rural measures, least of all those used in assessing dues in kind (in order to avoid "*un nombre infini d'inconvénients*"), simultaneously opposing the only reform worth considering at all, namely the standardization of the measures of corn in market transactions. Now, just over a year later, the nobles of Sens produced a document whose content had changed very little—but its form a great deal.

The peasants knew that there would be opposition to the reform from the nobles.[248] Sometimes they engaged in polemics with the nobles or, at any rate, anticipated the arguments they would voice openly without, of course, pleading their own material interest. A relevant *cahier* states:

> Our demand for standardized measures will, perhaps, be opposed on the grounds that it is proper for each region to hold in esteem its own customs, the privileges peculiar to it, the ways of its inhabitants, and the love of homeland, and, hence, that it would cause great inconvenience to bring in innovations against people's natural inclinations and the local customs they had been brought up with. Against such views we would maintain . . . that it is far more convenient . . . for the state to have one law applicable to all its subjects. . . . The inconveniences, and the objections levelled against the enterprise that we favor, will prove insubstantial and short-lived, and widespread profound happiness (*bonheur profond et universel*) will ensue.[249]

The *cahier* of 1789 in particular provides evidence that the nobles knew very well how widespread the desire for uniformity of measures was and realized that demands for it would be pressed hard by many parties in the States General, so hard indeed that they would not dare oppose them openly. That the authors were against the reform is made quite clear in the pages of the *cahier*, all the more so when we compare the text of 1789 with that of 1788. But by 1789 the nobles did not dare openly to oppose the call for standardization, and this is quite a revelation, for it shows that, even before the summoning of the States General, the idea of uniform measures had put down roots widely and deeply. In this situation, the nobles confined themselves to a reminder of their rights relating to their property and the august privileges upon which their metrological prerogatives rested. They concluded by menacingly pointing out the ultimate danger to the monarchy—food shortages. In the light of the documents cited, the silence on metrological matters in the overwhelming majority of the nobles' *cahiers* becomes quite deafening.[250]

The analysis of the *cahiers de doléances* has thus established that on the eve of the French Revolution, the problem of the standardization of weights and measures was an urgent one. Various strata of the third estate in the country had come to appreciate its importance and felt strongly about it, and the differences in the thinking behind the demands from the third estate faithfully reflected the internal differentiation of the estate itself.

In the backward provinces, like Brittany or Angoumois, where dues in kind were still very important, the peasants often complained of surreptitious increases by their lords of the size of the bushel used in collections. Such complaints were more numerous in Brittany than in Angoumois, but the former province was very large, while the marked increase in the frequency of such complaints in the latter area immediately after the outbreak of the Revolution inclines us to think that it was not the lack of similar views but the limited ability of the peasants in the preparation of the *cahiers*, or even their fear of manorial reprisals, that was responsible. The few explicit written sources at the historian's disposal amply compensate by their exceptionally vivid flavor and expressive realism.

In the economically more advanced provinces complaints about metrological chaos arose chiefly from market difficulties. For it was all very well for the nobles of Sens to aver that Savary's manual was all that was needed to understand the differences in weights and measures, but wherever were there any peasants setting off for the market with a copy of Savary in hand? So in some *cahiers*, we find traces of anxiety on the part of the peasant-producer reluctant to sell in any market but the nearest one for fear of an "unfair" measure. Other *cahiers* tell us of "proletarianized" peasants who either regularly or seasonally had to buy additional food and invariably felt cheated, even if they actually were not. And in the *cahiers* of the bourgeoisie related, if different, worries were mentioned, the road to whose elimination was ever posted "standardization of measures!"

Thus, these millions of individual worries and concerns produced situations that were typical of larger groups and social classes and that went into the making of broad, large-scale social trends. The seemingly narrow metrological perspective illustrates the process of the rise of rational unity in its social and spatial dimensions at the expense of the privileged class; it enables us to hear the prelude to the declaration by the third estate that it is representative of the nation, the prelude to the bourgeois, anti-feudal revolution. We can hear it in the ringing phrase of a Niort *cahier*: "Let there be, throughout the kingdom, one weight and one measure for *all the estates*"[251] [italics added—W.K.].

Unknowingly, the skeptical thinkers of the ancien régime were proved right. Metrological standardization was quite impracticable so long as the institutional framework of enlightened absolutism held; there were too many sectional interests to be resolved. Metrological particularism was part and parcel of a larger situation in

A TABULATION OF THE REFERENCES TO WEIGHTS AND MEASURES IN THE *cahiers de doléances*

Bailliages and Sénéchaussées	Number of References to Weights and Measures	Demands for Standardization			Motivation ("Context")		
		Nationwide	One Province	Local	Commercial	Legal	Anti-feudal
Amont	13	11			1	3	5
Angers	75	60	6		20	16	5
Angoulême	5	3		1		1	5
Arques	18	12	3		5		
Autun	7	7			1		
Auxerre	2	1					
Beaujolais	13	12	5	1	4		1
Bigorre	14	5	1				4
Blois	21	9	2	1	5	1	2
Bourges	27	12	2		5		1
Bretagne	6	2	2		7		3
Châlons-sur-Marne	31	18	2		4	1	
Châtillon-sur-Marne	15	9			2	1	1
Cotentin	12	7		1	2		1
Epernay	12	9		1	1		1
Flandre-Maritime	14	7	2	1	6		1
Nîmes	31	25	3			2	
Niort	13	10	1		9	2	2
Orléans	40	33	3			2	4
Pas-de-Calais	59	25	10	2	1		4
Quimper	26	14	1		6		13
Reims	55	39	4	2		5	1

Rennes	70	19	9	3	5		31
Rouen	57	37	9		9	4	2
Sarrebourg, Phalsbourg	4	3			1	1	
Sens	8	6			3	4	
Sézanne	11	5			3		1
Troyes	22	13	1		3	1	3
Varia	46	23	5		7		7
Total	727	440	69	13	110	44	98

Addenda: Demands for nationwide standardization include phrases like "dans la France," "dans l'Empire," "dans les terres soumises au meme roi," etc.

Demands for standardization within a single province include phrases like "dans notre province," "en Artois," "en Normandie," "dans notre election," etc.

Demands whose scope seemed uncertain or "intermediate," e.g. "dans tout le royaume ou au moins dans notre province," or "dans toute la France et surtout dans la province," have been categorized according to the "maximum" demand, since it was this that indicated the wish of the authors, and their qualifications would rest on their uncertainty as to whether such a scope was realistic.

The total in the first column is, of course, markedly less than the total resulting from the addition of numbers in columns 2 through 7; this is so because many demands in the *cahiers* do not specify the geographical scope of standardization, while many others complain about the metrological situation but do not explicitly state a demand for standardization.

Varia refers to *cahiers* found in dispersed publications and in the following *bailliages* and *sénéchaussées* which yielded very few metrological references: Nevers, Nogaro, Châteaux-du-Loire, Boulay and Bouzonville, Cosne, Varzy and Lignorelle, Autun, Ham, L'Isle Jourdan and Vigean, Caen, Rivière-Verdun Marseille, Civray, Havre, Cahors, Alençon, Toulouse and Commingues, Mirrecourt, Sarrebourg and Phalsbourg, Pont-à-Mousson.

Finally, the first six volumes of the *Archives Parlementaires* include many *cahiers;* of them, 113 contain metrological demands: 12 by the nobles, 7 by clergy, and the rest by the third estate.

the structure of privileges and vested interests upon which the monarchy's very existence rested. But for the Declaration of the Rights of Man and Citizen, but for the events of the night of 4 August, the standardization of weights and measures would never have come about. It could only become a social reality—being in the first instance an abstract, rationalistic creation, seemingly of armchair scientists—once the feudal privileges had been toppled, and equality before the law triumphed.

POSTSCRIPT.
A COMPARATIVE VIEW FROM POLAND

Writing a hundred years ago of the ancien régime, de Tocqueville was amazed to note that "virtually all Europe had the same institutions."[252] For the historian today, this is still more amazing because more recent studies of institutions, especially legal institutions, paint a false picture of their infinite variety; behind that picture, however, the forms of human relations in the different countries are fundamentally identical.

For the Polish metrological historian engaged over many years in the closest study of disputes between lords and peasants, the corresponding French picture is amazingly familiar. It is indeed amazing that, beginning with the Bible, the authorization of measures had everywhere been an attribute of sovereignty, and changes in it over time had corresponded to the changing concept of sovereignty. Amazing, too, that among all imaginable ways for the noble to exploit the peasant, and the town to exploit the country, metrological privileges were never absent; and no less amazing that in various countries, quite independently of one another, the same issues bound up with measures and measuring played the same part in social relations.

Nevertheless, there are also significant differences to note:

1. In Poland, clashes over dry measures of capacity formed the commonest expression of the antagonism between nobles and peasants; so it was in France, but less frequent and mostly in the less advanced areas, where taxes in kind were still important. Further researches should examine the geographical reach of this phenomenon and the question of its possible earlier occurrence in areas where it had disappeared by the eighteenth century.

2. The second most common object of metrological dispute between the Polish lords and peasants were the measures of length. In the sixteenth century, the lords were accused of secretly de-

creasing the measures in order to enlarge their demesnes at the expense of the peasants' holdings, and in the eighteenth century of increasing them so as to increase labor dues, which were, by then, based on the amount of work. This practice was not found in France because the labor dues—the corvée—had not survived well, and, moreover, the rights of peasants as landholders were far more secure. It remains, of course, for historians to investigate whether France too had known this practice in earlier times.

3. The *vox populi* demanding the stabilization and standardization of measures is in evidence in both Polish and French sources, the stabilization being more stressed in Poland, and both demands about equally in France.

4. In France, standardization was usually demanded nationwide and rarely for a single province. In Poland the compass was usually just the lord's demesne, which the peasants saw, with some justification, as a "state." Therefore, the demands were geographically limited, reaching at the most to the nearest country town. There were two reasons for the difference: (a) far more tenuous contacts with the market enjoyed by the peasants in Poland; and (b) the idea of national unity far more advanced among the peasants in France, where, even in out-of-the-way hamlets, they wanted a nationwide standardization of measures.

5. The conduct of the Crown was astonishingly similar in France and in Poland, the much stronger state authority in France being scarcely more active or effective in the field of metrological standardization. Indeed, at the time when the French monarchy had ceased to attempt such changes, an ambitious reform was carried out in Poland in 1764 with some degree of success—perhaps because it was not disfavored by the *szlachta*, whose commercial interest the reform might actually have advanced.

There were, then, considerable differences between France and Poland, but in the final reckoning, in both countries, and in others, too, in the great social game that was being played over the centuries between the Crown, the nobles, the bourgeoisie, and the peasants, they were all using in various combinations the same few simple cards that had always been available to them.

· 23 ·

"ONE OF THE BLESSINGS
OF THE REVOLUTION"

The seigneurial monopoly of weights and measures was toppled, along with the feudal system itself—that is, the feudal system viewed narrowly, as in the enactments of the night of 4 August 1789. Those enactments were followed by the ones of 15 and 28 May 1790, which completed the abolition of feudalism.[1] The void thus created had somehow to be filled, and the task was a gigantic one, the new powers seeking to fulfill the dreams of progressive thinkers and simultaneously to satisfy the aspirations of the masses, powerfully expressed in the *cahiers de doléances*. Not all the concomitant difficulties were at first appreciated, particularly since attempts to safeguard numerous vested interests could scarcely be anticipated.

Talleyrand, the bishop of Autun, who will be mentioned more than once below, was the first to prepare a submission on metrological matters[2] when he spoke out against the idea (frequently voiced under the ancien régime and favored by the monarchy) of making the Paris measures binding throughout France. What he proposed was the adoption of a brand new basic standard, "derived from nature" (*pris dans le nature*) and therefore acceptable to all nations; and he further suggested that preliminary work towards this objective should be undertaken jointly by the French National Assembly and the English Parliament and Royal Society. The National Assembly accepted Talleyrand's program on 8 May 1790, entrusted the Academy of Sciences with the management of the reform, and simultaneously decreed that all measures in use throughout the provinces of France (!) should be sent to the Academy. It would appear that the complexity and magnitude of this request were not appreciated, since the Assembly anticipated that once the new standard was adopted and its copies were distributed to all the parishes of France, the old measures would be dispensed with, and replaced by the new within six months.[3] In a sense, the reform seemed to realize the long-standing aspirations of the monarchy, but the King was in no hurry to bestow his approval on the plan. Eventually, he did so on 22 August 1790,[4] and then still took until 11 November to pass on the matter for action to Condorcet,[5] who was then Secretary of the Academy. The royal proclamation[6] repeated, word for word, the decree of the Assembly; and in it there were some interesting clauses and resolutions of several problems to which we shall return later. It was not until 26 March 1791 that the Constituent Assembly accepted the principle that the base of the new system of weights and measures was to be the length of the meridian, as measured between Dunkirk and Barcelona.[7] Altogether, therefore, a year had lapsed before the surveyors could commence their labors.

Meanwhile, however, important developments had been taking place in France. In the light of this study, the hypothesis seems tenable that the Revolution had released a great surge of complaints about metrological duplicities. The situation had indeed grown worse in many provinces during the period of the "seigneurial reaction" preceding the Revolution. Even in the areas where the peasants had so far been tolerating the injuries they suffered, it appears that they were no longer prepared to hold back once the existing institutions faltered. Because the majority of peasants' complaints were presented directly in the courts of law, our knowledge of them is scant and indirect. But some relatively important cases reached the files of the Committee of Feudal Rights in the successive revolutionary Assemblies, and a handful of them deserve mention.

The parish of Yvignac in Brittany had for long contested the enlargement of measures used in the collection of rents, and at the very time the *cahiers* were being compiled, the Parlement of Brittany had the case on its agenda, despite an earlier verdict, of 27 October 1787, which the parish refused to accept. The issue at stake was "that the measure pertaining to Plumaudan near Dinan, which was subject to the abbey of Beaulieu, had been considerably enlarged." The case grew more bitter after the outbreak of the Revolution, and a new verdict was pronounced on 5 September 1790.[8] Also in Brittany, the peasants were determinedly opposed to the measure used by the abbey of Bégard, and on 18 December 1790 the parishioners of Trézélan, Pédernac, Guénézan, Prat, and Louargat brought in a complaint that their rents were assessed by that measure. The verdict set the plaintiffs a condition that they could not, in practice, satisfy: that they produce documents with respect to their obligations in which it was explicitly stated that their rents were not to be assessed by the Bégard measure. But the peasants would not yield, and eventually, on 30 August 1791, the Directory of the *département* wrote to the National Assembly accepting their view of the case.[9]

In Normandy, in the barony of Hommet, there was a prolonged

dispute over the demand by some lords that their dues in kind be rendered by a measure containing as many as 29 *pots* instead of the 16 in the customary measure of the region. As late as 20 April 1792, a complaint in writing on the matter was lodged by the local inhabitants with the Committee of Feudal Rights.[10] The same committee received, on 7 August 1790, a letter in which no fewer than fourteen parishes of the canton of Moyaux (Calvados) protested the excessively large "heap" demanded from them with their grain taxes. The parishes sought "a decision that the former seigneurial grain dues be paid by the striked *boisseau* of the old measure of Arques, which has long been almost universally used, unless the creditor be able to present a different long-established standard on which he and the debtor have originally agreed, its legality being there for all to see."[11] As we can see, in this instance, it was the other side that insisted on the evidence of "old title deeds"—a wholly unrealistic demand for either party.

The Committee of Feudal Rights also received another complaint, of the same substance but set out at great length, on 28 November 1790 from the parish of Limousis (Aude). The parishioners here, "as our oldest fellow parishioners are ready to state under oath," had a tradition of paying the rent by a measure called *censuelle*. Now, the demesne had been acquired in 1774 by Sieur Rolland, "citizen Carcassonne," who, taking advantage of the "absence, in the written inventory, of a clear statement regarding the kind of measure to be used in collecting the due, demanded payment by the measure of Carcassonne," which was larger by a third than the customary one. Negotiations and remonstrations by the local inhabitants accomplished nothing, whereupon the parish prepared in writing the following illuminating statement: "Everyone, anxious not to lose what little he possessed and reluctant to fall foul of his seigneur, yielded to the oppression and paid up in a way that he was coerced to; so it has been until today, but now that the National Assembly, by means of the most just of all constitutions, has restored to man his original dignity and has done away with the last vestiges of serfdom (*servitude*), the parishioners of Limousis have refused to pay rent to Sieur Rolland, their erstwhile lord, by the measure of Carcassonne, and have suggested that . . . they will pay him what they owe him by the *censuelle* measure." All this was happening sixteen years after the purchase of the demesne and the attempt to foist the new measure on the peasants. "Sieur Rolland, disgusted by the proposal, and doubtless intending to avail himself of his still [*sic!*] influential position in the old Tribunal, brought in an accusation against the mayor by name," expecting to cow him into submission, and the others to follow suit. The parish, "submitting to the wish and just decree of the National Assembly, and far from following the insubordinate conduct of many other parishes [*sic!*] . . . does not refuse to render to Sieur Rolland what is lawfully due to him." The statement concludes by revealing that there was yet another bone of contention between the parish and the seigneur: "The parish wished faithfully to adhere to the decree of the National Assembly, which proclaimed the abolition of honorific privileges, particularly the right to separate pews in the church." Truth to tell, the parishioners of Limousis were correct in their thinking: the link between the lord's rights to occupy a special seat in the church and that of imposing his own measures on the tenants was a direct one.[12]

The most vivid and eloquent description of a dispute about measures between the peasants and their lord in the wake of the outbreak of the Revolution comes from another document bequeathed by the Committee of Feudal Rights and written by the inhabitants of the barony of Marthon (Charente).[13] The dispute was an old one, although it was only at the time of the compiling of the *cahiers de doléances* that numerous parishes came out into the open with their grievances about the seigneurs' metrological abuses.[14] The *cahier* of the town of Angoulême informs us that "there are almost as many local measures to collect dues by as there are estates." True, the municipal Notables write circumspectly that "it is commonly believed that those seigneurs who personally collect their corn-rents from their tenants do not employ measures other than the lawful ones." They, nevertheless, appreciate that the seigneurs "avail themselves also of the crafty dealings of their underlings, who recover from the tenants the large amount they had themselves been forced to "invest," as the price of their position, by means of employing two different measures: one of them is larger than the proper one and is used in the collection at the granary. The payers are not blind to this injustice, but are too afraid to protest."[15] Again therefore, the peasants were disgruntedly submitting to the practice—until 1789. Thereafter, however, things did change, and the vigorously written submission from Marthon affords us a detailed insight into the continuance of the dispute.

The municipal authority of Marthon in Angoumois, not long after constituting itself, was confronted with demands for verification of the *boisseaux* that were used in the levying of rents. The members, armed with an official extract from the judicial book of

Angoulême, went along personally to attend the rent collection. The measurement of the seigneur's *boisseaux* showed them to be "considerably larger than they ought to be," and the *boisseaux*, at the request of the parish prosecutor, were thereupon sealed up and deposited with the notary. At the same time, the decision was taken not to proceed with the matter until "all the municipalities of the barony of Marthon, to whom it was of concern, had been notified of what the verification had brought to light." Subsequently they all assembled at Marthon, and a delegation, to which each contributed one representative, was dispatched to see the lord. They had returned—the Assembly continuing its deliberations meanwhile—to suggest that negotiations should be commenced in quest of an agreement. All the parishes were informed accordingly, their representatives reassembled at Marthon and called upon the lord to attend their meeting. He complied and was then informed of the outcome of the verification of his *boisseaux*, and—his antagonists referring to the aforesaid relevant judicial document—he was requested not only to rectify his *boisseaux* but to restore to the parishes the excess amount they had paid during the previous sixteen years! The seigneur flatly refused to accept the validity of the document he was shown, which he called a scrap of paper, and since no more valid document was extant, he deemed it to be within his rights to employ the same *boisseau* as his predecessors, even though it differed from the *boisseau* of Angoulême, and consequently, to put it briefly, "it was not proper to demand that he should accept any restrictions or agree to repayments." But the deputies proved to be more than barrack-room lawyers. Their rejoinder was, first, that the document they quoted, although unsigned, was valid evidence, for "it had been kept in a public place with other archives of high importance, in care of the appropriate official"; secondly, that the register in question "has always been accepted as the foundation for judicial decisions . . . and in the law of the province, and has been scrupulously observed as the exemplar"; third, that the seigneur was not at all entitled to question the said document, since all the decisions he had taken against the tenants were in keeping with the said document, and, moreover, "his own declaration of income for the last six months of 1789, submitted to the tax authorities, stated, over his signature, that he had so many *boisseaux* of rent by the Marthon measure"; fourth, the *boisseau*, insofar as it has been used to measure essential goods, should be unalterable; fifth, that it is unreasonable to insist unreservedly on the antiquity of the standard, since every standard presupposes the existence of an earlier one; sixth, at the very site of rent collection there had once been a *boisseau*-fastened with a chain, and there had been a stone *boisseau* as well, but both had vanished without trace, and the seigneur was "quite unconcerned at this," although it was his business to look into the matter; and, finally, if the lord was not prepared to accept the validity of the document presented, then the onus was on him to present another, more reliable one.

The lord "submitted, but it did not appear that he was convinced by the above arguments." Since the Assembly was seeking a compromise he was then asked to put forward his proposal. His answer was remarkable, for he contended that "as many as eighty parishes in this province were involved in similar disputes, and he did not wish to preempt any collective decision by the lords through acting on his own." His antagonists replied "that only a moment earlier his brother-in-law had been haggling with his tenants in a similar situation."[16] And so the matter stood as of 26 April 1790, and we do not know how it was eventually resolved. However, the statement of Sieur Rolland's tells us a great deal: first, of the extent of the trouble, with no fewer than eighty parishes being involved as early as the spring of 1790; and second, of his (surely not unusual) readiness to assume solidarity among the seigneurs.

This solidarity was matched by the peasants' solidarity: we possess a document (also from Angoumois, although concerned with other parishes) that testifies to the peasants' resistance to the payment of any rent at all, justified by their complaint that they had for long been compelled, through the enlargement of the seigneurial measure, to pay more than was due. It is, in fact, stated that "the better-off peasants would prefer to pay the seigneurs what was owing to them," as they were anxious about a possible accumulation of arrears, but "they were intimidated by their poorer fellows."[17] Thus, the solidarity of the peasants was threatened by the economic inequalities among them; but the poorer were, apparently, able to force the relatively prosperous minority to toe the line.

To return to the document in question, its authors were the mayors and municipal officials of Blanzac, Cressac, Pérignac, Rouffian, and other townships in the *département* of Charente, and they despatched it on 15 March 1791 to the Committee of Feudal Rights in Paris for the attention of Merlin de Donai. The authors feared that the moment was approaching "when the authority will have no option but to use force to compel the people in our Cantons to pay up what was owing to the seigneurs. . . . The abuse of power wielded by the seigneurs was reaching unprecedented heights; the people groaned but had to bear it . . . groaned under the overwhelming burden of enormous taxation. . . . The people repeat, over and over again, that the seigneurs have been gradually increasing their *boisseaux*, until, in effect, the payments of dues they exacted were quite arbitrary. Are the seigneurs not obliged to present to us some documents in support of the size of the measures they employ? What is the use to us of the abolition of the feudal system, if the seigneurs remain at liberty arbitrarily to increase or decrease their measures?" (*A quoi nous servirait la destruction de régime féodale si les seigneurs resteraient les maîtres d'agrandir ou de diminuer leurs mesures à volonté.*)[18]

The peasants, therefore, maintained that, far from continuing to pay anything at all, they ought to have restored to them by the seigneurs the amount they had overpaid over many years because of the increased measure. And that the *boisseau* had in fact, been, increased "there is no doubt, because the seigneurs have destroyed their former *boisseaux*." The officials who prepared the document viewed the situation as critical and asked for forbearance: as the law was still new, they might not be able to interpret it aright, and even the lawyers whom they have consulted have offered different counsel; some say that the seigneurs ought to present documents fixing the magnitude of the measure employed, but others, on the contrary, have maintained that it is not feasible "to demand from the seigneurs documentary evidence in support of the correct size of the *boisseaux* by which they assess rent; to furnish such evidence would be a rank impossibility, and the law is too wise to ask for the impossible."[19] Accordingly, the authors were asking the Committee of Feudal Rights to interpret the law. However, the phrasing of their questions seems to anticipate the answers:

1. Are the former seigneurs entitled to retain arbitrary, self-determined measures, or must they produce documents in support of the size of their measures?

2. In cases where the seigneurs' measures are found to be in excess of what they should be, are they not obliged to restore to their rent payers the resultant excess amounts?

3. Over how many years, retrospectively, may the payers demand such "restitutions"; does the usual time-limit applicable to private prosecutions, thirty years, hold here, or should the metrological abuses be deemed criminal fraud and be subject to no rule of lapse at all [clearly, an issue of major significance—W.K.]?

4. What is to be done in instances when, the seigneur having failed to produce documents in support of his right to use the measure he has adopted, a proper agreed-upon measure has to be determined?[20]

The whole situation was transparently clear: the abolition of the feudal system had not been worth the effort if the seigneurs remained in control of measures. Talleyrand asserted in his program of May 1790 that under the ancien régime, feudal dues in kind had been the principal obstacle holding up the reform of measures and was then convinced that since the night of 4 August 1789, "happily the obstacle has been removed." In fact, precisely because of the intractability of the problem of measures, the entire system envisaged by the decree of 4 August—that is, allowing the retention of rent—was being questioned by the peasants, who were flatly refusing to pay up at all and affirming that they had paid more than was due over a great many years in the past when the seigneurs kept increasing their measures.

It was hardly surprising that a compromise was difficult to reach in situations like the one in Charente, for both sides were acutely aware of how much was at stake; the seigneurs were unwilling to negotiate individually with their tenants, preferring to act in group solidarity, while the peasants went so far as to terrorize those of their fellows who breached *their* solidarity. Nothing but open conflict lay ahead. A case in point occurred halfway between the aforementioned district of Charente and Aude (Lot); the following résumé of it focusses only on aspects relevant to our theme. It happened that in the *département* of Lot a rebellion broke out in the autumn of 1790,[21] when some 5,000 peasants armed with shotguns, axes, and scythes, led by Joseph Dinars, began an open fight against feudal rents. Troops sent against them proved ineffective, if only because the officers could place little reliance on their rank and file. It was known that the "unrest" arose from trouble over rent: "In many parishes the people were firmly under the impression that they were now wholly exempt from it, and in others payments would be made only after a rigorous examination of the relevant documents," at any rate those for 1790, rents having already been paid up for 1789. The call that the "original documents" be produced went up everywhere, coupled with the demand for repayment of large excess sums which the seigneurs had unlawfully collected in the past. Among the variety of complaints, there were some about "the falsified and oversized measures."

The commissioners sent out to deal with the revolt went beyond setting down the details of the strife in progress and endeavored to explain its causes. In their opinion, nowhere was feudalism as oppressive as in Quercy; the dues were huge, and kept increasing—not, of course, that this was the seigneurs' doing, but it was the fault of their stewards, agents, and "most of all of their *feudistes*, busily digging up old 'title deeds', both genuine and spurious. Receipts in respect of the dues rendered were never issued, and taking advantage of the peasants' illiteracy, the measures employed in collection were being increased." In a stunning phrase, the commissioners noted that "the measures horrified and terrified the people" (*un object d'horreur et d'èffroi*). The verification of "entitlements," which they carried out in some places after 1789, convinced them that the peasants' complaints were often justified, for example, that the payments had been so excessive that they had paid off the principal and, in truth, had bought out their dues. The commissioners felt certain that peace would not be restored as long as legitimate grievances were not satisfied, including the demand that "falsified measures" be reduced "to their original capacity"—a proposal by no means absurd in the eyes of the law. Another demand sought municipal control over measures in the future. Altogether, the demands were not exorbitant. However, the commissioners, appreciating the nonexistence of "the original documents" in most cases, realized also that the whole conflict could not be resolved other than by the sovereign decision of the National Assembly. The examples that we have cited originate from only two parts of France: the northwest (Normandy and Brittany) and the southwest (Charente, Lot, and Aude). This was no mere coincidence. Nevertheless, the attacks upon the seigneurial monopoly of measures, as part of the question mark the people had placed over the level of rents, were universal. Throughout the length and breadth of the country, the cry arose "Show us your old title deeds!"

A year and a half had by now elapsed since the decrees of 4 August had accepted the new base of the length of the meridian for a new system of measures. The *philosophes* were intoxicated with the boldness of the idea. Countless communications on the subject were arriving in Paris from the provinces, some penned by the administrative officials, others by spontaneous associations or by private individuals[22]—from the Société des Amis de la Révolution à Carcassonne, the Société Républicaine de Livry (Calvados), the Conseil Départmental de Dordogne, and the Justice of the Peace of Burrat (Ardeche), to the professor of philosophy at Annecy, the Member of the Academy at Arras,[23] and assorted savants, inventors, and cranks. As late as 1793, a letter from the parish of Castelnau-de-Médoc (a cantonal capital in the *département* of Gironde) was read out at a meeting of the Convention, in which "the inhabitants, unanimously accepting the new constitutions, requested speedier action from the Academy in preparing its long-awaited plan of standardizing weights and measures throughout the Republic."[24] In numberless petitions there were abundant references to the chaotic market conditions responsible for the injustices afflicting the small-scale buyer; however, no references can be found to the attitudes of the peasants towards the lords. A gulf was opening up between, on the one side, the ineffectual decree of the National Assembly, the procrastinating labors of the Academy, the enthusiasms of the *philosophes*, and, on the other, the urgent needs of the peasants; it was a gulf that was deeper than it appeared, for, after all, we must not forget that the initial intention of the blueprint for metrological reform was to respect fully the rights of the rent-earning landowners.

In November 1790, it was stated that "some of the municipal standards are incorrect, notably those of corn measures, and others have been deformed through the passage of time and no longer agree with what was stated in their original validating documents," and furthermore, that to regularize their mutual relations "would be to sanction errors or bad faith."[25] There were virtually identical expressions used in Condorcet's letter of 11 November 1790 to the Chairman of the National Assembly, and in the Assembly's decree dated 8 December 1790.[26] This standpoint was in accordance with the spirit of the proposals contained in the *cahiers de doléances* and with the widespread insistence on "the original documents." Yet, at that very juncture, as Condorcet was writing his letter and the National Assembly was poised to pass its decree, there were 5,000 peasants, in the *département* of Lot, offering armed resistance to any payment of rent at all. The peasants of Lot were fighting against privilege as such; while the National Assembly merely condemned the abuse of privileges. Nonetheless, all that the *philosophes* had thought up, and all that the National Assembly approved, aimed at the good of the people, for were they not all "friends" of the people?

Metrological reform at its core was militant in its ideology. It was such in its objective nature, it was presented as such to the nation, and the nation perceived it as such. The metric measures were officially termed "republican" by the resolution of the Convention dated 18 germinal of year III (7 April 1795), by analogy to the new "republican" calendar. The ideological association of the two reforms—although the metric system was actually pre-Republic—was generally perceived; this became even clearer when, in later years, the republican calendar was gradually abandoned. Additionally, metrological reform was closely linked with the reform of the administrative division of the country. Speaking at the meeting of the Convention on 11 August 1793, Fourcroy extolled the reform as "one of the greatest blessings of the Revolution"; in speaking of the work in progress, entrusted to the *ci-dévante* Academy of Science and in need of being continued, despite its dissolution, he particularly referred to the cause of metric reform: "The National Convention . . . [is desirous] of gratifying the French nation as speedily as possible by that blessing of the Revolution, thereby wiping out the very last traces of the territorial and feudal divisions, which had given rise to the diversity of former measures."[27]

Official documents, legislative texts, minutes of revolutionary meetings and assemblies, letters by the hundred in the archives of the Temporary Agency of Weights and Measures—all of them are movingly replete with revolutionary grandiloquence. To those familiar with the centuries-long story of "man's inhumanity to man," there is nothing ludicrous about that grandiloquence, for it was born of deeply felt needs and dreams of the masses. Indeed, the public and private phraseologies of the reform merit some consideration. The royal proclamation of 22 August 1790, repeating the words of the decree passed by the National Assembly on 8 May 1790, hopes that "all France will, as soon as possible, enjoy the advantages that ought to come in the wake of the standardization of weights and measures." This is a modest ambition, well within the framework of thought under the ancien régime—an impression confirmed by its later emphasis on the necessity of the most accurate conversion of all the old measures into the new.[28] As time went on, however, the phraseology was enriched and injected with more heat, both in Paris and in the provinces.

Thus, for the members of the Société des Amis de la Republique à Carcassonne, writing to Paris on 11 December 1792, the reform was essential in order "to buttress the edifice of liberty and equality and to bestow further splendor and prosperity upon the Republic."[29] At the meeting of the Convention on 31 July 1793, Danton presiding, Arbogast employed novel turns of phrase: he spoke of the fulfillment of "the old dreams of the philanthropists," of a "new bond of Republican unity," of the deceptions that the craftsmen and urban plebs had suffered (the peasants being at that point nobody's concern); the new measures would be "a symbol of equality and a guarantee of fraternity that should unite the people." The measures ought to be such that the legislators could rest assured that "no revolution would overturn the existing order of the world and bring new uncertainty into that field" and, furthermore, they "would form a system that belonged to no single nation and could therefore be universally adopted." Philosophers in centuries to come would undoubtedly "admire the scientific wonder and humanity marching on, through the storms of wars and revolutions, but enriched by the fruits of peaceable labors and profound meditations of men as modest as they were renowned"—so Arbogast concluded.[30]

Again, for the Minister of Finance in year II of the Republic, "the introduction of new weights and measures was extremely important on account of its association with the Revolution, and for the enlightenment and in the interest of the people."[31] Three months later, the Ministry of the Interior was even more explicit when, in a circular letter to the administrative heads of the *départements*, it referred to "the reform of the monstrous diversity of measures that had come into existence in many parts of France in times of despotism and feudalism . . . to help the people of the Republic now in their endeavor to erase the last traces of serfdom, from which they have freed themselves."[32] Those concerned with the reform had to be "men of integrity, zealous for the public good,"[33] so a purge was duly carried out by the Committee of Public Safety in the Agency of Weights and Measures. The purge was defended in highly characteristic terms: "The Committee of Public Safety, being persuaded how very important it is to the raising of the public spirit that those entrusted with the function of governing should delegate their mission and tasks only to men tried and trusted for their republican virtues and their detestation of kings," decided to dismiss from the Agency the very luminaries of learning of the day—Bord, Lavoisier, Laplace, Coulombe, Brisson, and Delambre(!)—while the remaining members were being called upon to devise with all possible speed the means of "enabling all citizens to use the new measures, on the wave of their revolutionary enthusiasms."[34] Since the "devising of means" was not easy without Lavoisier's participation, he later attended the Agency's meetings, being allowed out of prison for the purpose, but under an escort of gendarmes. Perhaps we should not judge too severely, but just confine ourselves to noting that in the light of what we know of the history of class struggle insofar as it impinged upon metrological issues, such expressions as "republican zeal," "revolutionary enthusiasm," or "detestation of kings" were not mere stylistic embellishments. In year III, the authorities called upon booksellers and editors to "furnish proof of their loyalty to the one and indivisible Republic"[35] by applying the metric system. On another occasion, the Convention wrote that "we are convinced that uniform weights and measures are one of the greatest benefits we can offer all the citizens of France" (*Un de plus grands, bien faits qu'elle puisse offrir*)[36]—a strongly expressed sentiment indeed.

Later on, the phrases continued to be purple but the admixture of red heat disappeared. An excellent example of this tendency is provided by the report presented to the Convention on behalf of the Committee of Public Enlightenment by Deputy Prieur of Côte-d'Or on II *ventôse*, year III. Admittedly, the Convention was being requested by him to carry out the reform of measures, "the old ones to go, as the last vestiges of tyranny have gone," especially since this was what "public opinion demanded, and had done since the outset of the Revolution"—this attitude of the people testifying to their virtue "and love of justice and good faith." The report, however, was primarily concerned with "the need to encourage trade because a flourishing trade ensures for everyone a fair share of necessities." Second place in the report was given to the cause of national unity: "The unity of the Republic requires that there be a unity of weights and measures, just as there is a single coinage, one language, one law, one government, and a single-minded readiness to defend our country from foes without and to march together towards universal prosperity." Third, the report advanced anti-feudal sentiments such as the following: "However could friends of equality suffer the most inconvenient ragbag of measures, which preserves vestiges of the shame of feudal serfdom; however distasteful must it be for true republicans to have to measure their fields by the *arpent royal*, or to use the *toise* or *pied de roi*, for all that they are resolute to do away with all traces of any tyranny whatsoever." Only then, having duly paid this tribute to the dominant values of the day, the report considered the reform at length in terms of solidarity and mass education. The reform should bring it about "that one and all will wish to learn arithmetic"; from this, there will ensue "an augmentation of universal education, whereby the arts will gain," and also, "a flourishing of Man's reason, so indispensable to national happiness and prosperity." The report goes on to assert that "education, learning and the arts, trade and the police—all these are bound up with the renewal of weights and measures. . . . The labors of men of talent . . . promoted by the decrees of your members [i.e. of the Convention—W.K.] have resulted in an excellent system of new measures, and, accordingly, the National Convention can now offer the Republic the firm promise of the enjoyment of this benefit of progressive legislation in the near future. Some further effort is still required, but the labors will soon be crowned with success" (in fact it turned out that "soon" was 45 years). However—and here the tenor of the report undergoes a change—"it is a matter of crucial importance to make sure that all possible efforts have been made to smooth the path of transition to the new order, to assuage the inevitable frictions, and to make the new practice simple." Former attempts at reform proved unsuccessful "because of the influence of people of consequence, whose concern was to preserve their feudal dominance." Only the Republic will now prove equal to this mighty task and will carry it through alone, for the hopes of the National Assembly to secure the cooperation of England have turned out to have been naive. Prieur continues: "This goes to show that, at the outset of the Revolution, the French greatly deluded themselves regarding the prospects of alliance with a people [the English—W.K.], whose character had been vitiated by an unjust government and who were themselves deluded into thinking they were free, while their crimes and plunders justly aroused the implacable hatred of all true Republicans and friends of humanity among all nations." He concludes by saying that "the interest and the dignity of the Republic demand on every score that the reform of weights and measures be effected."[37]

In year V of the Republic, the Minister of the Interior put this postulate more concisely: "Uniformity of measures has always been desired by the people," and therefore the new system will be "a splendid instrument for the molding of public opinion (*la raison publique*) and for binding all Frenchmen more closely, and also for the saving, through the simplification of accounts, of valuable time, thus facilitating trade and raising the level of commercial honesty."[38] And so we witness a return to the pre-1789 categories, since the Minister argues along the same lines as the *cahiers de doléances* had done.

The ideological justification of the reform of measures during the revolutionary period passed through a very curious evolution, from universalism to patriotism. Thus, Condorcet's report of the work on the reform by the Academy, submitted to the National Assembly in March 1791, admiringly states that "the Academy has done its best to exclude all arbitrary considerations—indeed, all that might have aroused the suspicion of its having advanced the particular interests of France; in a word, it sought to prepare such a plan that, were its principles alone to come down to posterity, no one could guess the country of its origin."[39] Less elegantly, the same idea was expressed in the decree of the National Assembly of 26 March 1791 adopting a length equal to a quadrant of the meridian as the base of the new system of measures: "The only hope of extending the standardization of measures to foreign nations and persuading them to accept it, lies in selecting a unit that is in no way arbitrary, nor particularly suited to the circumstances of any one nation."[40] Not very long afterwards, however, on 1 *brumaire* of year II, the Committee of Public Enlightenment, in its report to the National Convention, could not conceal its pride in the results of the relevant surveys and enthused: "Surely this shows . . . that in this field, as in many others, the French Republic is superior to all other nations."[41]

There is no point in attempting to choose between the thrall of Condorcet's universalism embracing all mankind, and the splendor of the Jacobin upsurge of patriotic pride. We can, nevertheless, discern humane undertones in both the stances. If, however, we recollect the idea of national unity, its red thread running through the hundreds of the *cahiers de doléances*, then it is easy to see that Jacobin patriotism was nearer to the sentiments of the masses than was the rationalistic universalism of Condorcet.[42]

Condorcet's position had, nonetheless, some important practical consequences. The reader will remember that, as early as in Talleyrand's first program, it was urged that the preparation of a new system of weights and measures should be a collaborative venture, with the French inviting the participation of the English Parliament and Royal Society. Efforts to establish contacts with other countries included negotiations with Thomas Jefferson. The report (above) of Condorcet of 11 November 1790 referred to these negotiations, mentioning that initially Jefferson's studies focussed on the latitude of 38°N, this being close to the center of the territory of the United States, but eventually he accepted 45°, proposed by the French National Assembly as being better suited to an international destiny.[43] The Spanish authorities, Condorcet added, dispatched to Paris a scientific observer, whose task was to watch the progress of the reform, and they were prepared to verify the measurements by replicating French surveys at the 45th parallel in the southern hemisphere, that is, in the New World. However, by year VI of the

Republic the selfsame Talleyrand was busy trying to turn the joint Franco-English project into an anti-English venture, inviting the collaboration of all the allies of France and of the neutral countries.[44]

To pass a reform act and to extol innovation is one thing, but to put it into practice is quite another. In their "republican zeal," the legislators, for all that they represented the nation, did not begin to foresee the extent of the difficulties. First of all, there was a major problem for the surveyors; those concerned, like Archimedes in antiquity, desperately searched for a fulcrum that in the ever-changing world would furnish a fixed basis for the novel system. Three possibilities came under consideration: the length of a pendulum beating seconds; a length equal to a quadrant of the equator; and a length equal to a quadrant of the meridian. To adopt the first would have certainly been the easiest solution, but it appeared to provide an insufficient degree of exactitude. The decision was eventually taken, despite many doubts as to the feasibility of precise measurement of the earth,[45] to adopt a length equal to a quadrant of the meridian. Scientific literature provides us with a good deal of information relating to the work involved. The geodetic measurements, which were carried out between Dunkirk and Barcelona, gave rise to many descriptions which are still extant. In truth, the participants themselves, in order to secure immortality for their labors, marked the more significant spots with commemorative obelisks. The fellowship of interested astronomers (e.g. Arago), geographers, and physicists of later times led them—and justifiably, too—to commemorate their predecessors' achievements.

However, it turned out that it was far easier to measure the terrestrial globe than to cope with some major and unforeseen social complications. For while the much admired and ambitious scientific work proceeded, a great variety of difficulties arose. Given the metrological confusion prevailing in France before 1789, the National Assembly's resolutions of 1790 created a void, an intolerable situation of vacuum and lawlessness. The decision to adopt the meridian as the basic unit of length was not taken until March 1791, and the decision to dissolve the Academy of Sciences certainly did not advance the preparatory work. Small wonder, then, that many voices were raised asking for some sort of unit, say, a provisional meter,[46] and this was at long last decreed by the Convention on 1 August 1793.[47] "At long last" is no exaggeration because the metrological situation in France had been extra-legal for four years, since the night of 4 August 1789! We could, however, also say "as early as" August 1793, since four years was no unreasonably long period for carrying out an operation as complex as metrological reform. Whether those pressing for some speedy practical arrangement like the "provisional meter" had any conception of the endless difficulties the administration and, still more, the people in general would have to undergo again when a "definitive" meter was approved, only the future would reveal. Let us consider some of these difficulties.

The whole country—every parish and every shop—had to be supplied with sets of weights and measures of identical shape and dimensions. The accumulation of the necessary raw material alone for such an undertaking presented a major problem. Military authorities were asked for help, but measures could not compete with guns during the Revolution, although, admittedly, they won out easily enough against church bells, the army lending a hand with the transport of the metal.[48] A remarkable brainwave of the Agency of Weights and Measures is worth mentioning here as an indication of the extent of the difficulties. New small-change coins, they suggested, should be minted with their weights corresponding to the new scale of weights, that is, the 5-centime piece would carry on its obverse side the legend *1 sous ou 5 centimes R.F.*, and on the reverse, *Poids de 1 gramme. Une et indivisible*; and there would be an inscription, *mutatis mutandis*, on all other denominations.[49] The simplicity of the project amounted to genius. However, the bulk of the raw material needed to produce the new measures was to be provided by scrapping the old measures. Merchants and craftsmen, too, if they had the right to keep standard measures, had to deliver up their *aunes* in exchange for the meter, or, otherwise, to purchase the latter.[50] This practice came about partly precisely because of the shortage of raw materials, but the danger now was that the private production of new and unfamiliar measures, with little control and supervision, might result in much confusion. Moreover, many overzealous officials tried to verify the old measures that were being yielded up; the Agency, naturally, sharply reprimanded them for this practice,[51] since the purpose of the exercise was that they should be withdrawn from use, and the anxiety caused by the simultaneous checkup could not but slow down the delivery. At any rate, the securing of sufficient raw material was only the first step. It was more difficult to organize the production of the new measures in vast quantities. There were many offers from would-be entrepreneurs to take on the task, but the Agency fully realized that they were far from disinterested parties. Technological considerations, moreover, dictated that for the sake of precision in the new measures and of quality control, the process should be centralized. The dilemma was how to launch mass production of a product that had to be finished with a precision attainable at that time only through individual craft workmanship? The archives of the Agency reveal that there were many inventors sending in proposals for machinery that, they claimed, would solve the problem, but no satisfactory solution was found.

Furthermore, another urgent issue was to determine the authorized ratios between all measures formerly used in various parts of France and the meter. Old obligations of all kinds, including taxes, had to continue being honored, and this demanded that the amounts involved be converted into metric quantities. At first, the legislators saw little difficulty here. The decree of 5 May 1790 simply ordered all parishes in France to send in to Paris, through their district administrations, the standards of all measures used in their areas, as well as documents confirming their authenticity. The Academy would then carefully verify their dimensions, relate them to the metric standard, and publish comparative tables. But the idea was insane! In the first place, those responsible for it clearly failed to perceive how complicated the existing national metrological reality was—still less to foresee the problems to which their own enactments would give rise. As far as rural parishes were concerned, the peasants simply possessed no standards of measures. The seigneurs had gone and with them, the seigneurial standards of measures, which generally had been the only reliable ones; or rather, they had been reliable at the outbreak of the Revolution, for we have already discussed the frequent tampering with them, and the seigneurs' continual attempts to effect on the sly the disappearance of old standards, including even some that had been on public display. Now, when the Academy circularized the parishes, virtually none of them possessed their own standards; at best, they would verify their measures by the standards available in the nearest country town. When—as often—the only available standard had been kept by the *ci-devant seigneur*, it would, in most cases, have disappeared with him. Thus, the parish of Mathay, in the *département* of Doubs, stated: "The standards of our measures are the property of the *ci-devant duchesse de Corge* and should have been deposited with the clerk of the court at Daubelin, while we ourselves are unable to present any documents." The parish of Grand-Cheint stated: "We have no standards whereby to verify our weights and measures, for the practice was for the functionaries of the seigneur

245

alone to see to this." Again, the parish of Rigot was exceptional in possessing a measure of its own, but the measure applied by the *ci-devant seigneur de Reaumont* in collecting his dues was different and larger. The parishioners of Curcey inform us that "in days of yore, the seigneur's men used to check our measures"; the parish of Maiche writes: "the municipality holds no standards of measures or weights because they are still [*sic!*] in the hands of the seigneur Guyot, who was our seigneur"; finally, this from the parish of Mont-de-Laval: "The standard is not in our possession, but in that of the men of the former seigneur of Reaumont."[52]

The examples from Franche-Comté that we have cited were not peculiar to that region alone:[53] From Touraine we have evidence of the refusal by many seigneurs—in the teeth of unequivocal legislation—to yield up their standards to the parishes. In Angoumois, we note a similar complaint dated 2 *ventôse* from the parish of Chalair in reply to the questionnaire of year VI: "We hasten to forewarn you that we possess neither standards nor models of old measures. All we know is that there had once been in existence a copper *boisseau* kept at the *ci-devant château*, and that it is still in the hands of the men . . . of citizen Talleyrand Périgord."[54] The standards, then, had gone together with the seigneurs, despite the unmistakable wording of clause 18, section II, of the decree of 15-28 March 1790. The true state of affairs that ensued is very well reflected in the replies to the questionnaire on metrological matters that the Academy of Sciences, prior to its dissolution, sent out to all districts. The districts, in their turn, sought information from the parishes; it is the latter's answers that are of real value to the historian, since the summary reports of the *départements* generally oversimplified things, omitting issues, which in their opinion were not of prime importance or wide application, and thereby lightening the burden of their own labors. Unfortunately, however, the replies of parishes inevitably came to rest in the archives of the *départements*.[55] Although, to repeat, most rural parishes did not possess the standards that the Academy so optimistically had asked them to send in, much effort was expended in trying to satisfy Paris. The authorities at the *département* level would tell the parishes that it was not enough to state that they did not possess the standards in question, but it was their duty "to carry out the most thorough searches in the archives of their former lords, and also in the files of the law courts."[56] Exhortations from the Paris bureaucrats were repeated month after month, and year after year, continuing—irrelevantly—as late as year VI.[57]

There were other difficulties, too. Provincial organs initially received with composure the demands for specimens of all measures, but, on second thought, they were seized with horror. For example, the council of the *département* of Nièvere reacted, by December 1790, with this statement: "On closer consideration the task has horrified us, since all estates have their own corn measures, and the expense of sending off specimens of all of them would be exorbitant."[58] This reaction is easy to sympathize with: to prepare accurate models of all measures in use throughout the areas, and then to dispatch them to Paris, well packed against any damage, by cart-and-horses hired for the purpose—this would have cost plenty! Furthermore, the specimens arriving in Paris would have uselessly cluttered up the allocated space in the buildings of the secularized churches and convents,[59] while the Academy recruited and organized the huge personnel required for the task of verifying them! And while the administrative machinery went on grinding, albeit slowly, the process of the Revolution itself continued to pass through several stages, with concomitant changes in the program of metrological reform.

The task of authorized conversion of all existing measures into their metric equivalents was one of prime importance, an essential precondition of the reform itself. For, on the one hand, numberless transactions and obligations, hitherto calculated in traditional measures, had now to be reassessed in metric units while, on the other hand, the people at large had to be taught and habituated to them and to resultant price changes. An important proviso here, in the early stage of the Revolution, was that no one should lose by the reform. For example, the measures employed in the collection of corn tax had to be converted into metric ones in a way that ensured that exactly the same amount was received as previously. The conversion, then, had to take into account both former seigneurial measures and the method of measuring: for example, if "heaped" then the heap was now to be expressed metrically. At most, some known "abuses" committed by the lords were corrected through resorting to old documents or to well-authenticated old standards.

As time went on, however, things changed. The Revolution moved on from fighting the abuse of privileges to attacking privilege as such. To succeed in the latter task was not possible without mobilizing local forces and thus aiding decentralization. But the Revolution was inseparably associated with centralizing tendencies, and in an enterprise that called for much effort, tact, and technical expertise and equipment, centralization was eminently justified, and it was only against its better judgment that the Agency embarked upon limited decentralization. Thus, the conversion tables were to be prepared at the level of the *département*,[60] with the assistance of local "friends of arts and sciences who cared deeply for the success of the new metric system."[61] Nonetheless, the results of the work had to be approved by the Agency, which was not slow to praise or chastise, often to correct and sometimes to stop publication,[62] while all maverick publications were ruthlessly suppressed.[63] More importantly still, the very scope of local enterprise was restricted, if only because the task proved impossible to accomplish. The headquarters in Paris may at first have forcefully insisted on the verification of all measures in use, without any exceptions whatever, but as time went on, it began to restrain the local producers of conversion tables, sometimes actually forbidding them to delve into details or, in particular, to bother with measures used only locally and with seigneurial measures;[64] the idea was to condemn them to oblivion.

The end product of the work, however, was a comprehensive set of conversion tables for all France by *départements*. Today, they constitute a priceless historical source, and despite some errors in them that the Agency in Paris had not managed to remove, their level of accuracy is generally high. In some cases, where the original *étalons* have survived until today, the 200-year-old conversions have recently been verified and found accurate.[65] It should, nevertheless, be borne in mind (and it often is not!) that the tables intentionally did not include a wide variety of purely local measures, notably seigneurial ones. Still, as time went on and the victorious French armies overran other countries, "the meter followed the flag," and conversion tables were duly constructed—either by the French or by the local administrative apparatuses—in many other lands.[66]

To return to the homeland of the metric system: during the years it took to establish the metric system, daily life in France had to go on. Millions of small commercial transactions of buying and selling, countless court trials of litigious cases, and the paying, day by day, of rents and taxes continued as usual. Yet the disapproval "from above" of former measures without a prompt promulgation of the new system created a state of affairs that was extra-legal, as it were, and, moreover, one that favored speculators. Inflation and the consequent imposition of maximum prices contributed to further discontent. Three and a half years after the portentous night of 4 August 1789, *les citoyens composant la Société des Amis de la République à Carcassonne* were still calling upon the representatives of the nation to hasten the arrival of the new system of measures, adding: "Be mindful, citizen representatives, that once that is done, it should check—partially, at least—the speculation of the greedy profiteers, constantly sucking the blood of the people."[67] Speculation, in fact, was rampant; every day was a field day for its practitioners. Psychologically, this was a state of affairs scarcely conducive to a favorable reception of the new measures by the masses.

As prices soared, we may suppose that in order to attract customers, many a seller would conceal part of the increase by reducing the measure sold. The many pertinent complaints reaching Paris mentioned waxing prices and waning measures in one breath.[68] Demands for imposing maximum prices on essential goods went hand in hand with requests to speed up the standardization of measures. The Committee of Public Safety of the *département* of Paris, on numerous occasions, had to look into dishonest practices in the weighing of the sacks of grain arriving in Les Halles;[69] while in the vicinity of Paris, at Rambouillet and Saint-Germain-en-Lay, there were popular disturbances brought about by the practices of market swindlers.[70] Though the terror had gone and the formal maximum prices no longer prevailed, there was as yet no cessation of dishonesty in retail trade. Depending on the circumstances, traders alternated the use of old measures under new names[71] with the use of new measures under old names.[72] Shops equipped with two (and from 1812, even three) sets of measures proliferated. Letters from the provinces told of "how cruelly the poor classes were cheated by all petty retailers among whom you would not find a single one adhering to only one weight and one measure."[73] And Paris, meanwhile, kept reissuing orders to destroy and break up former measures, and the prefects did so with alacrity,[74] encouraged by a firm directive they received from Chaptal, dated 2 *frimaire*, year XI: "Old measures must be broken up and sold for scrap."[75]

It does not require much imagination to appreciate the situation of the ordinary citizen, uneducated as he was, when faced with some daily transaction in the local market; he was not very likely to know of, let alone be able to read, the conversion tables—duly hanging on the wall since the law had ordered the shopkeeper to display them—and more often than not, he felt cheated. The feeling, after all, had been bred into him over the many generations that had experienced measures as an instrument of extortion. And often, indeed, he did not just feel cheated, he actually was cheated. As other reforms of the kind, so the metric reform, too, afforded egregious opportunities for deceiving the common man to whoever was more experienced in commercial matters, whoever held the whip-hand in any given transaction, whoever could calculate more efficiently than the other party. The relevant complaints are countless, and their substance often confirmed by the ensuing official enquiry. Fraudulent practices included deliberate miscalculations, taking advantage of the confusion of terms in verbal deals, and the dishonest application of new names to old measures or *vice versa*. Millers, taking advantage of the general uncertainty, quite commonly increased their charges for grinding; bakers would diminish the weight of loaves; throughout the length and breadth of France, shops and markets reverberated with quarrels and contentions.

The circular by Chaptal we have already cited concedes that "the general public suffers from dishonest practices . . . and loses faith in a system that seeks, above all, to benefit it," and goes on: "It is most important to enlighten the citizens of all classes as to where their true interest lies."[76] And so the poor citizens were being enlightened by all manner of means. In Paris, public meters were provided in places where people often gathered,[77] and these served to settle disputes; schools were instructed to teach the metric system;[78] there was encouragement and speedy evaluation for writers of appropriate manuals; and many towns organized public courses of tuition in the metric system,[79] the course in Paris taking place in the courtyard of the Louvre. As far as we can judge, however, these educational endeavors achieved little, for the subject matter was difficult, in particular the decimal system and the new nomenclature. We may, instinctively, feel surprised at the difficulties in mastering the decimal system, but we should not, for, on closer consideration, this system with its much vaunted perfection is by no means as simple in use as it appears to us today. It is certainly far from simple for people who, for all that they are quite familiar with dividing into halves and then again into quarters and quite fluent in multiplying by two and then doubling yet again, are nevertheless quite inexperienced in the use of decimal fractions and the decimal point. In the next chapter, we shall cite a document prepared by the Commission of Weights and Measures of the Cisalpine Republic, which makes it crystal clear that while every ignoramus of a tailor knew full well the meaning of "half a quarter of the ell," even trained accountants did not know that "half a quarter" is the same as 0.125. Consequently, the decimal system certainly appealed to the protagonists of the Revolution and to reformers who imagined that it would "release, for the good of the state, a great many additional hands [for work]" by simplifying the reckoning in millions of daily calculations.[80] But, in reality, it turned out to be a major obstacle to the introduction of the metric system.[81]

The nomenclature of the new system, too, gave rise to many difficulties. Contemporaries, including some supporters, were unhappy about it, and as the preparations for the reform proceeded, numerous modifications were introduced. Paris generally did not trouble to answer the odd letters asking that former nomenclature be left alone,[82] but as the requests multiplied, firm rejection became the rule.[83] Critical voices were also raised in revolutionary assemblies and even within the Convention, and the Press joined in, too. Thus, *La Feuille du Cultivateur* of 7 *messidor*, year III, writes: "Some consideration is surely due to old habits," but the Agency's rejoinder did not share this sentiment.[84] Nonetheless, the new terms were hard for the people to master, not merely because they broke away from the familiar but because they *were* utterly novel. They were composed of sounds that were alien to the French language, and this resulted in frequent misunderstandings. Small vocal differences would in some cases signify considerable differences in magnitudes; prefixes such as *hecto, centi,* or *kilo* were far from easy to master. The creators of the metric nomenclature were extremely proud of it, and it was officially called "methodical," but its methodical structure did not make it easier for the people.

Moreover, traditional metrological systems had fairly often been—as we have noted elsewhere—functional; they stood for some social reality: they were bound up with man, his work and the fruits of it. To take one, admittedly pre-revolutionary example, textile merchants, in buying cloth by the piece as it came off the peasants' looms, had been accustomed to pieces of unvarying dimensions, since these were dictated by technical factors such as the width of the loom, or by rigid social custom (the length). The raw cloth would then be subjected to the processes of fulling, drying, clipping, dyeing, and so forth, in the course of which it would stretch or shrink, but this presented the merchant with no difficulties, for he knew that he had bought by the traditional "length" and he would sell by the half-length, the quarter-length, or the half-quarter; his accounting was simple. But now that he had to calculate in metric terms of length, his accounting was thrown into indescribable confusion.[85] Again, as far as the measures of grain were concerned, the entire pattern of commercial thinking was attuned to selling inferior grain by the "better" measure, or using the "better" measure with poorer-quality or impure grain. The change from thinking

routinely in such terms, to thinking in categories of invariable measures and variable prices (the latter being adjusted to the quality,
purity, dampness, etc., of the grain) amounted to a complete mental
revolution: it was not enough simply to learn the measures, or even
to master the decimal way of counting.

The concept of measure, the manner of apprehending it, and
even its true magnitude in practice—all these were deep-rooted
categories of habitual thought, certainly too deep-rooted to be dis-
missed with facile references to "traditionalism," "conservatism," or
"obscurantism." This is attested by the tardiness with which the new
system was received not only by the masses of common people, but
by the most highly educated groups, too. There is much evidence
that this was so: the Agency of Weights and Measures repeatedly
admonished a wide variety of governmental agencies, reminding
them of the obligation to use the new measures; it complained
about the failure to observe this obligation; it chastised, appealed
and thundered "in block capitals." Extant sources include the Agency's epistles to the Commission of Public Works,[86] to the Agency
of Mineworks,[87] to booksellers and editors of journals.[88] It complained to the Council of Ancients about the conduct of the postal
services,[89] as well as to the fiscal and general administrative organs.
Furthermore, the Agency was aggrieved at the failure to stipulate
metric measures by the Council of Ancients, the Council of Five
Hundred, and by the Directory in their very legislative enactments.[90] Paris was exercising pressure on the administrative officials
in the *départements*, and they in turn—repeatedly, but to little effect—urged the municipal authorities.[91] As late as year VII, François de Neufchâteau admitted in a letter to the prefects that he had
deluded himself, thinking that the officials of public administration
would set an example in employing the metric system. He goes on,
"But it pains me to see that these considerations have not so far
been heeded, and many central administrators write to me daily,
using old measures and, worse still, not bothering to mention the
new ones in order to aid comparison or conversion," and the Minister concludes by threatening the dilatory officials with disregard-
ing their communications.[92] But six years later the Ministry was to
send out identical admonitions. It clearly took a very long time
truly to internalize the metric measures, even for highly qualified
civil servants, well-educated booksellers and journalists, or Treasury officials, to whom reckoning was bread and butter—and all of
them, surely, were anxious to curry favor with the higher authorities.

Perhaps this procrastination should not baffle us. For the uneducated folk, the new system was hard to grasp cognitively; but
it seems legitimate to hypothesize that for educated people it was
scarcely less inaccessible. They had all been educated and profes-
sionally trained—those mining engineers, landtax assessors, and
accountants—within a system of pre-metric measures and former
ways of reckoning. Their minds and imaginations had been set to
react in certain ways when faced with magnitudes and numbers,
and it was automatically in terms of the old dimensions that their
faculties operated in coping with their professional tasks and problems. So among the great many difficulties attendant upon the
introduction of the metric system into daily life, the forces of social
psychology, contrary to what the innovators had anticipated,
proved the most intractable; routine reactions, habits of thought,
specific technical difficulties of the new super-rational system, and
the practical advantages of the traditional measures all contributed
to the slowness of the adoption of the reform, even though it had
been wanted for centuries.

The promulgation of the new system was now followed by decades of strife between the administration and the people. Closest
to the line of fire, it would seem, were the prefects on one side,
and the mayors on the other. The former, invariably keen to press
the government's cause and by virtue of their office representing
the forces of centralization, did their utmost to impose the reform
upon the community;[93] the latter, contrariwise, intimately associated with the local citizenry, stood for their interests, doubts, vac-
illations, or recalcitrance.[94]

Meanwhile, Paris saw a succession of governments and each successive régime entailed changes in the structure of the central authority, in the competences of various offices, and consequently,
there was a succession of different bodies passing from one to
another the task of overseeing the progress of metrication.[95] The
Academy of Sciences was disbanded on 8 August 1793, and the
Interim Commission of Weights and Measures then came into existence, only to give way, from 18 *germinal* of year III (7 April 1795)
to the Interim Agency of Weights and Measures. The latter was
finally replaced, on 24 *pluviôse* of year IV (13 February 1796), by
the special section of the Ministry of the Interior. The successive
changes in administration were accompanied by successive purges
of "politically unsound" elements, all of which hindered the realization of the great and difficult reform. And yet, despite all the

upheavals, the successive agencies carried on with the most impressive steadfastness and perseverance.

Every time a government in Paris fell and another rose, rumors would circulate all over France that "the metric system would not last much longer";[96] yet after every change, the new rulers would issue assurances of their firm adherence to the reform. The Girondins had scarcely been in power for ten days, when on 3 April 1792, speaking from the tribune of the Legislative Assembly, Roland, in a statement characteristic of his party, insisted that the reform be accelerated and the provisional meter be adopted because "the diversity of weights and measures is the chief obstacle to free circulation of goods throughout the kingdom."[97] The stance adopted in the matter by the Jacobins is still easier to guess at; we have already quoted some typically Jacobin pronouncements, their proposal to erase the old feudal divisions and regional distinctions of customs being particularly characteristic.

It seems that Champagny, Napoleon's Minister of Internal Affairs, was sound in his judgment when, much later at the opening of the 1806 session of the Legislative Chamber, he said:

> There is yet another institution left by the Revolution, namely the new system of weights and measures, whose usefulness is indeed sensed by those who find it hardest to get used to it. It is an achievement of science, presaging rational rule over an enlightened people; let me say that this institution is here to stay, and the government will increasingly take it upon itself to propagate the use of the new measures, and it shall support them against old usages and superstitions with steadfastness and firmness of purpose, rather than by the violent but short-lived exertions of the innovators. Time is on our side and the government shall overcome all obstacles, nor shall its activity cease until that triumph has been secured.[98]

To return to the 1790s, the Directory, on learning of rumors of the likely demise of the new system, proclaimed its faith in it in the announcement of the Minister of the Interior on 5 *germinal* of year IV, and his statement was repeated annually while the Directory remained in power. Thus, on 23 *fructidor*, year V (9 September 1797), a similar circular went out to the prefects: "We must not, citizens, lose sight of the fact that the standardization of measures has always been wanted by nations. . . . I put it to you that it will offer us an excellent means of forming the public spirit (*pour former la raison publique*), of strengthening by a uniformity of customs the bonds uniting the French . . ., of facilitating trade by loyal merchants, of bringing more glory upon the name of France among all nations."[99] That his egregious statement mingles, in a peaceful coexistence, the views of the absolutist monarchy, of the Girondins, and of the Jacobins is quite an achievement! And a year later, less eloquently but no less firmly, the Minister was putting pressure on the prefects again.[100]

Following the coup d'état on the 18 *brumaire*, rumors revived of plans to do away with the metric system. Again, however, these were vehemently denied by the Minister of the Interior, who charged the rumormongers with groundlessly attributing to the Consuls ideas more proper to the ancien régime.[101] And the story was repeated in the early days of the Empire, when on 18 *pluviôse* of year XIII (7 February 1805), yet another circular letter to the prefects from the Minister of the day informed them: "His Imperial Majesty has instructed me to inform you, gentlemen, that it is most definitely His unalterable wish to maintain the new system of weights and measures in its entirety, and to accelerate its extension throughout the Empire."[102] However, the matter did not end there, for in the following year the prefects received another similar circular from the same source, repeating the same assurances regarding the Emperor's intentions and concluding: "His Majesty has ordered me to demand from you, gentlemen, that you double your efforts and exertions in order to bring to a successful conclusion this undertaking, which the government fervently desires to see completed and will not suffer to stagnate, for stagnation would deal a blow to its hopes for the future and would contribute to a growth of disorder and abuses."[103] Nevertheless, no sooner had Napoleon been deposed, and scarcely had the restored Bourbon monarch resumed the throne, than still another affirmation of the metric system was issued on 4 July 1814. On that day the new Minister of the Interior presented to Louis XVIII a report on the matter, justifying his promptness quite outrageously: "Sire! The uniformity of weights and measures has long been desired in France, and your royal predecessors sought, in several reigns, to establish it. I therefore presume that your Royal Highness will wish to uphold an institution that accords so excellently with his great ideas of public utility and whose effective spread, in any case, is by now very much advanced."[104] The final legislative step in the story of the metric system's beginning, however, was to be taken by the July Monarchy, exactly fifty years after the outbreak of the Revolution. To sum up, governments came and went, rulers too, but successive Ministers, following instructions from their successive masters, never ceased pressing their civil servants to impose the metric system upon the community.

This consistency in several different régimes was nothing short of astounding, although there was one exception to it that may be termed "the Napoleonic compromise." The passive resistance to metrication by the people at large, almost from the outset exercised the authorities. As early as year III, the Commission of Public Works issued a circular to the chief engineers of the *départements*, requesting information about the people's attitudes to the reform.[105] It has not, most unfortunately for our researches, proved possible to find any answers to the circular. The matter was raised again on a few occasions, most forcefully in the last year of the Empire. Napoleon, like all rulers confident of their strength, preferred to ignore the passive resistance of the citizenry and busily searched for a compromise, but it proved difficult to find. On the one hand, the authorities had to appease surviving scientists from among the founders of the meter, as they desperately defended the system in its rational perfection (its Jacobin purity, one is tempted to say), thereby indirectly defending their own collective standing and reputation; on the other hand, there was the civil service, headed by the prefects, who had for a dozen years or more invested much labor in their attempts to impose the system upon the nation, and who, from the very nature of their work, favored centralization and standardization wherever possible, and who now set themselves up against compromise in any shape or form, both in order not to "lose face," and in order to prevent any revival of metrological confusion.

The whole question was debated two or three times by the Council of State, and much resistance was in evidence. Again, we cannot avail ourselves of primary source material, for the protocols of the meetings, as well as the evidence presented to the Council, were burnt with the rest of the archives in the conflagration that destroyed the Tuileries. Eventually, at any rate, on 12 February 1812, the Emperor issued a decree setting out the new compromise. Even though we do not have the materials that went into the making of the plan, we do possess two relevant, extensive circular letters from the Minister of Internal Affairs, Montalivet, to the prefects, dated 28 March and 10 July 1812. The first of these explains the intentions of the government and the principles of the reform, and the second rebuts the prefects' criticisms of new system, which apparently had recently begun to reach Paris.[106]

Let us pause to explain the principles of the Napoleonic reform.

Basically, it reaffirmed the binding power of the metric system (a "cult of perfection" had by now come to surround it) and permitted the use of measures amounting to fractions or multiples of the metric units. The radical change that was now introduced, however, allowed those fractions and multiples to be other than decimal. Thus, it became possible to use, for example, as a unit of measures of capacity, one-eighth of the hectoliter, or to weigh things by the half-kilogram. A further step in the same direction was to permit the application of the name *boisseau* (bushel) to the eighth part of the hectoliter; indeed, the differences between the old and the new bushel, or the old and the new pound of weight, etc., were minimal.

The reform represented a major concession to popular practice and to tradition. The metric system was modified in three ways: first, the banned traditional names were restored to use since, after all, the new hybrid Greco-Latin terms had not won much popularity; second, some units very close to former ones were restored, since it was appreciated that they had many functional associations in daily life that the new units had failed to acquire (e.g. in his circular Montalivet admitted that the horse's daily ration of oats was still referred to as a quarter of the *boisseau*); and, third, successive divisions by two, as well as multiplication by doubling, were now allowed—Montalivet accepting that the difficulties of dividing and multiplying by ten were insurmountable for most people. The continued dominance of the metric system, however, would now be assured: first, the new units were fixed as fractions or multiples of metric units, and consequently the new *boissaux*, *perches*, or *livres* would be that little more or less than their pre-metric namesakes; second, the newly legitimated units were to serve retail transactions only, whereas wholesalers were assumed to be able to cope with the decimal system, and the wholesale trade had to apply the metric system as heretofore; third, the newly legitimated units were permitted but not mandatory; and, fourth, they had to have their magnitudes marked upon them both in traditional and in metric nomenclature.

A couple of excerpts from Montalivet's circulars will not be out of place here. His point of departure is that the Emperor has been taking interest in the reasons for the slow rate of progress in the universal adoption of the metric system; he has been informed that the major obstacle is the fact that "metric units for practical daily use are not suited to practical daily needs. The exclusive employment of the decimal system may suit accountants but is by no means well adapted to the daily dealings of the common people, who have much difficulty in understanding and applying decimal divisions."

The Emperor, therefore, has proposed "to find out whether the objective in question might be more readily attained by permitting the use of certain instruments of measuring and weighing that seem better suited to the needs of ordinary people, but which could readily be integrated with the legally binding units [i.e., the metric ones—W.K.]." The newly permitted measures closely approximated the traditional Paris ones, a useful coincidence not to be thrown away. True, the Paris measures were not used everywhere in France, but at least they were known virtually everywhere. Montalivet duly notes:

> The *boisseau*, being one-eighth of the hectoliter differs from the former *boisseau* of Paris by only four-hundredths [it is smaller by $4/100$—W.K.], and may be used with ease on all occasions by the common people. They may not easily grasp the sense of the double deciliter, or the ratio of the deciliter to the hectoliter, but should have little difficulty in grasping the relation of either to the *boisseau* and will no longer fear that they may pay for a quarter when buying a fifth, or for an eighth when buying a tenth, etc.

The passage is very important, for it confirms our interpretation above of the popular system of reckoning in the times prior to universal schooling—that is, that the normal practice was to divide or multiply by two, and then by two again, and so on. People's minds were attuned to quarters, eighths, etc., but not at all to fifths or tenths.

As far as weights were concerned, the decree of 1812 brought in a new pound, equal to one half of the kilogram, but "no different from the old pound *poids de marc*, save by two percent upwards." Characteristically, the new pound divided into sixteen ounces, and one ounce into eight *groschen*. This, in fact, was a clear departure from principle, since neither the ounce nor the *groschen* was even a nondecimal multiple of the gram: the ounce equalled 31.25 grams, and the *groschen*, 3.9. The Minister considered the question of weights as supremely relevant to the daily business of the people, appreciating that merchants "still employ in retail trade the new weights for no other purpose than to be able to sell by the old weights and measures, to convert them dishonestly into the new ones, and so to arrive at equivalents that are not true." This elucidation is very valuable for us, and it refers to a situation that is not difficult to imagine. While ordinary consumers put their orders to shopkeepers or market stallholders in terms of old measures, the police were compelling the sellers to use only the new ones; in seeking both to satisfy the customers and to obey the police, the seller got involved in conversions, in searching for equivalents of the old measures in terms of the new, and, naturally, his conversions would be generally "rounded off" in his own favor.

The Minister had no doubts that "the majority of consumers, either through habit or through negligence, continue to phrase their orders in terms of the former weights and measures; but the merchants must not take advantage of the slips or ignorance of members of the public. To prevent such abuses, article 12 stipulates that all orders phrased in former weights and measures are to be understood as phrased in the newly permitted measures." Montalivet, moreover, expresses the hope that the slight increments of weight, which accrued to most new units when compared with the old (the official adjective was "analogous"), would of itself incline the public to ask for the use of the new ones. The employment of the "analogous" measures was permitted only in the retail trade, and even there this was an optional practice. However, to enable customers at least to request that they be used, all merchants were required to possess them, and the public were to see to it that they did, and that the metric and the "analogous" measures were kept separate. The decree was optimistic about the Emperor's future: he stipulated in it that a report of the effectiveness of the new reform be prepared in ten years' time, in 1822.*

How the people of France received the decree of 1812, we do not know. What we do know is that it was not well received at the vital administrative level, namely by prefects, usually so well drilled and disciplined. Many of them, indeed, did not hesitate to communicate their doubts to the Minister—for example, that the principal hindrance was neither the nomenclature nor the decimal system, but the continual need for conversion, a nuisance that was unavoidable whatever type of new system was tried. Montalivet begged to differ, and his prompt circular letter of response contends:

The names given to the metric measures do not harmonize with the spirit of our tongue because the names are long and too similar to one another. These drawbacks, however, are not great deterrents, provided the measures in question do discharge, in their divisions, their purposes in daily life. Here lies

* Napoleon fell from power in 1814, and again, finally, in 1815, and died in exile in 1821—Trans.

the real rub. It was precisely because the divisions were ill-adapted to practical use that the authorities' efforts on behalf of the metric system were foiled. The consumer was forced constantly to calculate the ratios of the new measures to the old, which retained their hold on his mind . . . all because of the decimal divisions. They helped the bookkeeper but not the ordinary man, unaccustomed as he was to endless calculations. He should not be forced into this corner, and once measures that are easy and simple to divide are available to him for coping with his needs, he will no longer bother his head about the ratios of the new measures to the old, and, before long, forget the latter.

The prefects were also concerned that given the diversity of traditional measures in pre-revolutionary France, the re-legalization of the old names would revive in people's minds different magnitudes in different regions, and there were even some few regions quite unfamiliar with the Paris measures. This objection was accepted by the Minister, but he countered by referring yet again to the structure of divisions. True, there had been a variety of measures in France, but still the *toise* was everywhere equal to six feet, the foot to twelve *pouces*, the *aune* would always be divided into halves and thirds, and the pound was invariably equal to sixteen ounces.

Next—an issue we have already exemplified from the parts of the pound—it was pointed out that while it was easy to relate the basic units of the so-called "analogous" measures to metric measures, this was not so as far as the divisions of the former were concerned. The Minister evaded this criticism, for, in fact, there was no answer to it. Insofar as the central objective of the new reform was to return to the traditional ways of division, even though it proved possible (with marginal differences) to assimilate the pound into the metric system, there was absolutely no way to ease in, as it were, one-sixteenth or one hundred-twenty-eighth of the pound. Some of the prefects were anxious lest even minor discrepancies between old and new measures bearing the same name should interfere with commercial dealings or the practicing of crafts. To this, the Minister's reply was that surely the "analogous" measures were intended for use in retail trade only. Next, the prefects anticipated that the latest reform might be sabotaged by merchants taking their time to equip themselves with the new measures, "both because of the expense involved and [because of] the problems of storage of the new pieces separate from either the legal metric ones or the pre-metric ones that the majority of them were in no hurry to get rid of." The Minister's answer was straightforward and severe, namely that "the merchants were not at all acting properly in retaining the old measures or weights; the police were answerable for any omissions in their performance of the duty, which devolved upon them, of seeing to the elimination of such practices where they still occurred." Again, the prefects felt that the simplicity and integrity of the system was becoming flawed; commerce, administration, and the arts would now all be afflicted with the ceaseless conversion of "analogous" measures into metric ones. On this score, the Minister tried to mollify them: "The labors that may arise thus will fall to the lot of educated people and not simple folk, and for the former this will be no great burden." This sounds like an assurance to the prefects of the arithmetical prowess of their own functionaries, which the prefects were perhaps inclined to doubt.

Finally, some prefects unhappily remarked that "the latest changes go against the very grain of the central principles of the decimal division and uniformity; the former would be violated by the substitution of a different system of division; and the latter, because retail trade will no longer employ the same methods as trade of higher order." This criticism was answered by Montalivet at length, for the very heart of the system was at stake and he had to reassure the critics that the government had no designs on the integrity of the metric system. His counter-argument ran:

> The decimal system is genuinely useful for accounts only, whereas ordinary retail trade calls for no accounts save those simple enough to be carried out by means of dichotomous divisions. ... The essence of uniformity has been misconceived, for it subsists in the sameness of weights and measures, which is not threatened at all; however could this be effected by the popular use of the *toises, pouces,* livres, and *onces*? It is the meter and the kilogram that are the basic units, and the others will have to bear reference to them. Uniformity, therefore, is unimpaired, and shall continue so, since even in retail trade the newly permitted measures will be the same everywhere.

Montalivet's undoubtedly disingenuous argument can scarcely have satisfied the prefects, but it was too late for them to query the decree: it was, a *fait accompli* and it only remained for them to implement it.

Nevertheless, the prefects—those of Montalivet's day and their successors, too—kept faith with the metric system, for it was, arguably, bound up with their function and the part they played. That part, to repeat, whatever the central régime, inclined them to favor centralization and standardization.[107] And so the reform was launched yet again, the "analogous" measures now playing the leading part, very like the provisional meter and the definitive meter on previous occasions. Yet again conversion tables had to be produced, sent to shops, and their dutiful display had to be checked on.[108] The parishes had to be furnished with standards of measures.[109] Paris demanded six-month reports of the progress of the reform,[110] and new opportunities for new abuses and new ways of arresting them came about. We have evidence in some regions[111] of a very disgruntled popular reception of the reform that was intended to help the common people. And the merchants, who were often equipped with three sets of measures rather than two as hitherto, would manipulate them to their own advantage.[112] Meanwhile, Napoleon, in his famous bulletin describing the disastrous crossing of the Beresina by his Grande Armée in Russia the following February, saw nothing wrong in referring to the river as 40 *toises* wide and the bridge as 300 *toises* long,[113] for what emotive reaction could a metric expression possibly have kindled in the breast of the French reader?

As already mentioned, on 4 July 1814 Louis XVIII gave his approval to the modified Napoleonic version of the metric system. The 1812 compromise arrangement was accepted as the best way of "combatting effectively the reluctance of the people, since it related mainly to the 'methodical' nomenclature and the decimal system."[114] Within a year, however, concessions were in the air. The juggling with three systems would create situations that were not to be tolerated, and complaints were flooding the competent Paris offices. The Minister of the day, Vaublanc, now writes to the prefects as follows: "Complaints have become so widespread and have made it clear to me that there has been so much confusion and so many serious abuses, that I considered it my duty to inform the King; and His Majesty, in His wisdom, has determined that the government must no longer delay the application of preventive measures."[115] What this turned out to mean was the prohibition of the use of weights and measures with decimal divisions—naturally, in retail trade only, as indeed the 1812 decree had provided. The metric system was to continue unaltered otherwise. Obviously, this was a surrender to expediency even more abject than the Napoleonic compromise. We have noted that the *philosophes* had endeavored to stop the reform of 1812, calling it "a retrograde decree and a bastard system";[116] and twentieth-century students of metrology were to adjudge it as a "dastardly blow against the integrity of the metric system."[117] But it was that "retrograde and bastard" system that was to hold sway in France for a quarter of a century.

Not until 1837, did the matter come again, at the government's instigation, before the Parlement. The deputy from the *département* of Nord, Martin, presented the case: "We must admit that the legislators of 1812 did not fully appreciate that it would be the habits rather than the needs of the people that would sustain the resistance against the metric system. People's needs are sometimes basic and unalterable, but that is not the case with their habits, which arise by happenstance [*sic!*—W.K.] and which can always be combatted and—sooner or later, easily or arduously—overcome by the law-giver."[118] Martin's simplistic argument is permeated by a right-wing social philosophy. It was, apparently, the likes of him and his colleagues who were entitled to decide what ought to be respected as a popular "need" and what might be disregarded and changed by the government, being merely a "habit." On this occasion, his colleagues shared his point of view. The post-1830 Bourbon monarchy restored the metric system in its "purity" by the decree of 4 July 1837 (binding from 1 January 1840). The struggle between the state and society was now concluded; it had lasted half a century.[119]

Postscript.
The Irony of History

The soldiers were marching and the drums were beating on the day of 2 *nivôse* of year IV of the Republic (22 December 1795), when, following the law promulgated on 18 *germinal* of year III (7 April 1795), the people of the 48 districts of Paris heard the glad tidings: "The centuries-old dream of the masses of only one just measure has come true! The Revolution has given the people the meter!"[120]

Was it indeed a centuries-old dream come true? Was it indeed the metric system that had been awaited by the peasants rebelling against seigneurial abuses, or by the urban plebs deceived by millers and bakers, or even by the rural assemblies that the "good King"

Louis XVI had so precipitately invited to set down their *doléances*? Certainly not. Hopes for the "just" measure had related to particular, local measures serving a certain parish or two; often, this would not be the currently binding measure, "altered of late" by the say-so of the seigneur or abbot, but the earlier "immemorial," "true" one, still visible as a stone standard in the marketplace, or at least living in the memories of the oldest inhabitants. What was wanted was restoration of old measures, not the invention of new ones. No one questioned the names of existing measures; other appellations were beyond imagining, as was the abandonment of the perfectly convenient dichotomous system of division. Instead, they were given the meter, a strange, new-fangled measure allegedly relating to the very land people walked, but in a manner that no one understood; a measure allied to the decimal system of division that imposed intolerable strain on the minds accustomed only to the operations of mental arithmetic, and that constantly demanded the use of excessively difficult fractions; a measure that entailed a confusing and indistinct nomenclature;[121] a measure, finally and least comprehensibly, that would be equally applicable to textiles, wooden planks, field strips, and even to the road to Paris.

"So you wanted a new measure? Well, you've got it now." The nation had desired a standardized, invariable measure. Men of learning, the *philosophes*, came up with the meter, and their new device was seized upon by the rulers and the bureaucrats, who imposed it upon the people. They satisfied the longings of the nation, and in so doing, they subordinated the nation to their own control. The nation received what it wanted—only it was not what it wanted. Or rather, it both was and was not what it wanted: immediately, it was not, but in the long run, perhaps when some two generations had passed, it became just that. The reform that standardized weights and measures, which had been so ardently desired for centuries and so widely demanded by the common people on the eve of the Revolution, extolled by so many of the truest revolutionaries and conceived by the finest scientific minds of the day, had, ultimately, to be imposed upon the people. But if we pause to reflect, we cannot but note that metric reform was not, in this respect, alone among the achievements of the Revolution.

PART FOUR

· 24 ·

"FOR ALL PEOPLE,
FOR ALL TIME!"

The reader will recollect that Talleyrand's initial plan for the standardization of measures envisaged collaboration between the French Academy and the English Royal Society. He hoped, moreover, that "perhaps this scientific collaboration for an important purpose will pave the way for political collaboration between the two nations."[1] Indeed, he was not the only writer in the field to whom, apparently, the main appeal of the metric reform lay in being a means of *rapprochement* with England. For example, La Rochefoucault, speaking in the National Assembly on 8 May 1790, felt that "we cannot make enough haste over promulgating this decree, which should bring about fraternal relations between France and England."[2] And even when this hope had come to nought, Condorcet still emphasized the universal intent of the reform, the absence of any peculiarly national (i.e. French) features about it, which should facilitate its adoption by other nations, and the fact that its standards were "derived from nature," which, particularly to the thinkers of the Enlightenment, is the true bond of mankind. And even though the Jacobins did later come to regard the invention of the metric system in all its perfection as a triumph for revolutionary France, this was precisely because it was France's superb gift to all mankind. As the phrase of the day had it, the metric system was "for all peoples, for all times." It was quite obvious that, once tyrannical rulers had been done away with, all peoples, in their new-found brotherhood, would adopt the metric system for the sake of better mutual understanding and cooperation.[3]

Universality was thus the central principle of the metric system. Before long, it was to begin its all-conquering march, and at first, like freedom, it marched in the wake of French bayonets. Let us not, however, forget the "perfection" of the system: its perfect logic and internal consistency. This "perfection," this wholeness, meant that there was no way for partial adoption: it had either to be accepted or rejected in its entirety. The system could not be adapted to local circumstances; like the French Republic, it was one and indivisible. And yet, as it proceeded on what was to be its worldwide progress, the local circumstances it encountered did vary. An inflexible system grafted on to diverse situations presents, of course, a classic case of cultural borrowing and of the wanderings of isolated elements of cultures.[4] The metric system was created in the context of a particular society, at a particular stage of its historical development, and even there the realization of metrological reform was a centuries-long process. Attempts at reform were bound to fail as long as other concomitant reforms were neglected. As long, that is, as feudal rights remained in force, equality before the law was not accepted, and provincial particularisms remained in force. Thus, metrological reform was not practicable unless and until it was an integral part of a wide complex of reforms. The meter was carried by the armies of the Revolution, whose bayonets also served to remove crowns from regal heads; it all fitted together. Later on, when the crowns were restored, albeit on new heads, the matter became complicated.

When the barefooted troops of revolutionary France invaded Lombardy, they planted trees of liberty and they brought the meter with them. Before long, the soldiers were well shod, but the trees of liberty grew poorly and soon withered, while the meter never took root at all: local circumstances could not stop the French *sabreurs*, but they proved resistant to the meter. For Lombardy had already experienced a long history of resistance to metrological standardization. It began with the failure of a Renaissance attempt: in 1597, Governor Juan Fernandez de Velasco, having first consulted a variety of magistrates and syndics, ordered the standardization of all local measures according to the Milanese standard. Attempts to carry out his reform, however, failed so badly that his successor, de Fuentes, called it off as early as 1605;[5] before long, the matter was forgotten. However, some twenty years before the arrival of the French revolutionary troops, the Austrians launched between 1771 and 1782 a great metrological reform in the "state of Milan," and their attempt was supervised personally by the Chancellor Kaunitz.[6] There was some resistance, and the authorities in Vienna did not press the matter too hard for fear of upsetting their subjects excessively. Thus, they took a long time deliberating over such questions as: Should the new measures for Lombardy conform to those in their hereditary German lands? Should they be introduced only in the "Stato di Milano," or throughout the Austrian possessions in Italy, including Mantua, which had, not long since, experienced a costly reform of weights and measures and, besides, had a system of law peculiar to itself and its own coinage? Should all weights and measures be reformed forthwith, or initially only the measures of length, it being assumed that this part of the task was the easiest? The local authorities feared *una rivoluzione troppo sensibile . . . in una materia delicata e irritabile*, and Vienna's response was to advise them to proceed *successivamente, e pocco a pocco*, while the greatest Italian scientist, Cesare Beccaria, was instructed to check most thoroughly the length of the Milanese ell.[7]

In the end, on 30 January 1781, Joseph II and Ferdinand signed in Vienna a decree that laid down a "minimalist" reform program: measures of length alone were to be standardized for the time being, the ell of Milan providing the standard, and the reform was restricted to the state of Milan. Justification for the reform was highly characteristic of the period: "The dimensions and diversity of the measures in use in different regions of our state, although it has one law and obeys one ruler, and the absence or inexactitude of the standards employed by the agencies of control—these are matters that have always been considered as sources of errors . . . and a hindrance to the flow of secure internal and foreign trade." An identical document might have had its provenance, at that time, in any one of the European countries ruled by the absolute monarchs of the Enlightenment. It took a long time for this limited reform to become accepted, and we do not know whether or not the task had been completed when the plain of Lombardy was flooded by regiments under the French tricolor.

The new Cisalpine Republic was then involved in numerous major reforms, but it would seem that it was chary of introducing the metric system. And no wonder—for how could the authorities have asked the population to put up with yet another change of measures so soon after the Austrian reform? On 6 *ventôse* of year VI (24 February 1798), the Great Council declared that in future, weights, measures, and coinage would follow the decimal principles;[8] but then, it did not cost much to make declarations. Within three years, a state of utter chaos prevailed. On 9 *nivôse* of year IX (30 December 1800), the decision was published to convert weights and measures in all parishes of the Cisalpine Republic to the Milanese standard, and this step revived the spirit of the reform of Joseph II. Presumably, the aim was now to complete the latter, eliminating some survivals of earlier units, and to extend it to those areas of the Republic that had not previously formed part of the Austrian "state of Milan." And then, just 36 days later, on 15 *puluviôse* of year IX (4 February 1801), the meter was introduced in the Cisalpine Republic.[9]

The Commission for Weights and Measures, established in 1794, had long since intended, at a distant point in time but quite definitely, to assert the use of the meter: tables of conversion, assimilating the Milanese measures to the metric system and covering not only measures of length but also the weights and measures of grain and of wine, had been prepared and distributed in all parishes; the production of standards for distribution in the communes was under way; moreover, attempts were made to spread the knowledge of the metric system and of decimal arithmetic—the latter, in the opinion of the Commission, was likely to cause the people the most headaches. The Commission for Weights and Measures of the Cisalpine Republic was right and knew its business, but it was too late to argue; it was time to curry favor with Paris. Only the *philosophes* and the prefects were delighted with the new turn of events—as doubtless Beccaria would have been had he lived to witness it. A certain Paolo Ricchini de Voghera, formerly a vice-prefect, sent off to Champagny, the Minister for the Internal Affairs of the Empire (*sic*!), a screed extolling the virtues of the metric system, bemoaning the variety of measures in Italy, and remarking that "this Gothic (*sic*!) diversity was the ineluctable outcome of the partition and fragmentation of Italy into small seigneuries, as well as a direct consequence of the feudal law." He went on to demand tough measures anticipating strong resistance from conservatives and traditionalists. The outcome of the change in 1801 could easily have been foreseen: the meter did not flourish any better in the soil of Lombardy than had the trees of liberty planted by the French troops.

Let us next consider the course of events in Piedmont.[10] That old kingdom, still proudly clinging to its ancient title of "Kingdom of Savoy, Cyprus and Jerusalem," had experienced more than one metrological reform. As in many other European states, so here, too, albeit with some delay, the Renaissance ushered in a great reform.[11] It had been carefully thought out, as we know from surviving seventeenth-century plans for the establishment of an apparatus of control and from sources telling of numerous metrological disputes. Nonetheless, the reform appears before long to have been given up, our evidence of this being eighteenth-century descriptions of metrological confusion, looking to the reforms of the Enlightenment.

By the 1750s, the administration embarked upon attempts to tidy up the disorder in weights and measures. The 1750 instance of prescribing the manner of measuring grain—namely, that it be poured into the container *de poca altezza* is relevant here. In 1777, a careful verification of all measures in use in Piedmont was made. Initially, it was of course the scientists whose interests prompted this activity, the authorities gladly accommodating them for their own purposes. Beccaria was enjoined by the Duke to measure the meridian passing through Turin; on a distant day, after the Vienna Congress of 1815, Turin's Royal Academy of Sciences would have to explain to the authorities, who wished to possess their very own "sovereign" meridian, that the question of just which meridian was used in the measurements was of no consequence.[12] In the wake of the metric reform's success in France, the *philosophes* went about their task with a renewed will, their strictly scientific interests tending to go hand in hand with their favorable view of the revolutionary Republic. The main propagandist for the metric system was A. M. Vasalli-Eandi, Professor of Physics at the University of Turin, acting in conjunction with French scientists. As early as 1797, he published a book on the metric system, and its second edition appeared in 1802, by which time Piedmont was obeying French laws.[13] The way in which matters were beginning to move is indicated in a gentle but grandiloquently persuasive letter from the ambassador of a friendly power, namely the French representative in Turin, Ginguené, writing on 5 July 1798 thus:

> The Directory does not wish that an enterprise of such great import should proceed without the participation of scientists abroad . . . the commission established in Paris has at heart not only the progress of the arts and sciences but also the glory of the nations prepared to collaborate in this undertaking, while French scientists would be happy to share the achievement with all, keeping for themselves only the preparatory deliberations and researches. The unflagging enthusiasm with which they dedicated themselves to this task, in the midst of political storms, has been a phenomenon proper to the French Revolution.[14]

However, further paragraphs of the letter imply that the timetable of the reform prepared in Paris envisaged its completion as early as 6 October. One cannot resist the inference that the very late invitation to participate, issued to the scientists of allied and neutral countries, was but a piece of propaganda emanating from the Directory. Nevertheless, an appreciation of the offer had to be shown, and the King of Piedmont passed the matter on to the Academy in Turin on 27 July 1798, while Balbe, his ambassador in Paris, who happened to be a member of the Academy, was to join the international commission.

Meanwhile, Turin was the scene of work and contention: on the one hand, the Academy ordered—and proposed itself to supervise—production of the most accurate standards for the Turin measures, but on the other, it formed a working party to consider the principles of the reform. Clearly, the opinions of its members were far from unanimous, for it was decided that they should all submit individual reports, which were duly dispatched, with the standards, to Ambassador Balbe in Paris. But before long Piedmont was occupied by French troops; the kingdom was no more, nor the King, nor the ambassador. And, on quitting his post, Balbe bequeathed all his materials to Vassalli-Eandi, who at the time happened to be sojourning in Paris.[15] The reform, nevertheless, did proceed, for on 2 *thermidor* in year XI, the Commissione dei Pesi e delle Misure dell Accademia Imperiale delle Scienze, Lettere ed Arti di Torino was founded, with the abbé Valperga di Caluso as its chairman and Vassalli-Eandi as secretary.[16] The authorities, meanwhile, saw to it that the standards from the provinces were collected and verified, and conversion tables constructed. At last, on 17 August 1809, the Mayor of Turin—to be followed by other municipal leaders—declared that from 1 October the metric system would be obligatory. However, its adoption for practical purposes was to drag on. Local archives are replete with relevant correspondence, bearing dates both before and after August 1809,[17] between the authorities in Turin and the communes. It would appear that by 1814 the reform was well-nigh complete, all the communes having by then been supplied with standards of the metric measures, and the conversion tables published. Moreover, the network of local inspectors was now strong enough to deal with abuses, their verdicts on the transgressors receiving wide publicity to impress the population.[18] Although as late as 1810 the merchants of the department of Stura put in an application for postponement,[19] the work of completing the reform went ahead.

However, when the restored king's decree of 21 May 1814 repealed all French laws in an attempt to restore the legal *status quo* of 1800, the government turned to the Academy[20] with a request—veiled, but all too easily seen through—to give up the meter and return to the former Piedmontese measures. The *philosophes*, the majority of whom were enamored of the metric system, now tried to salvage as much of it as possible. In the event, a contentious compromise was arrived at by the Academy,[21] and in due course the government pronounced it as binding in law. The solution proposed by the Academy was inventive, based on the calculation of the Piedmont foot as equal to one second of the Turin meridian of long. 2°E, which Beccaria had calculated. Thus the old unit of the Piedmont foot was retained for use, and the metric system's vital principle of reference to nature, namely, to the dimensions of the earth, was preserved, too. Indeed, the new arrangement appealed to patriotic sentiment: *Di modo che per singolarissima combinazione possiamo veramente dire che la nostra misura piemontese ha in qualche senso una relazione certissima alla nostra posizione geografica.*[22] The new Piedmontese measures and the metric measures were now commensurable, and the "epoch of happiness with the return of our rulers to their possessions"[23] dawned upon the metrological scene.

Strangely enough, and perhaps bearing out the considerable degree of progress to that date in metric reform in Piedmont, although the metric system was "rescinded," it yet continued to be employed by the civil service in the administration of taxes, highways and bridges, water supplies and forestry, official statistics, and so forth. In 1814, the metric system in Piedmont had not yet been universally accepted, although it would appear to have been widely applied. The resulting confusion and the scope for abuses in daily life must have been endless in a situation where a less-than-fully accepted system had now been formally "revoked," and yet remained in use—by the state agencies, at that! Accordingly, in 1816, the Academy was called upon to consider the situation and in the following year the administration concocted a new half-hearted reform project—fortunately soon to be dropped—whereby the decimal system was to have been combined with pre-metric measures and their traditional nomenclature.[24] Local administrators kept complaining to the central authority about disorder and abuses, which they were unable to curtail, and kept pressing for a return to the metric system. Local petty merchants, for all that they had dragged their feet before 1814, like those in Stura, now favored the metric system, too (e.g. the merchants of Casale). After hesitating for over ten years, the royal decree of 7 July 1826 was issued, which in the words of the Regia Camera di Conti was "the best possible means of tackling the urgent task of combatting abuses, which had been proliferating in the production and control of weights and measures—of both the customary local and the metric variety."[25] The decree (for the nth time) stated that new standards were to be produced in all communes and sent in to the Regia Camera di Conti, where they would be verified and defined in terms of the meter. The Chamber would then destroy standards of incorrect shape and produce new *archetipi*. Every standard would bear its own name and its metric equivalent, and the approved standards would be deposited in the Chamber's archives; and, of course, new conversion tables were to be published.[26]

The decree apparently was intended to legalize, for the time being, the coexistence of the pre-metric and metric systems of weights and measures. The longer-term aim, however, was a "painless" completion of the rule of the metric system in Piedmont, witness the direction of several administrative bylaws during the following two decades. Two routes were followed: one, of the gradual diminutions of the number of old-style measures that it was permissible to use; and the other, of the conversion of such measures to the simplest possible commensurability with the units of the metric system. The relevant regulations proliferated in particular from the late 1830s, culminating in the decree dated 1 July 1844.[27] By this decree, the number of permitted measures was severely reduced, and all of them were, in fact, simple multiples of the units of the metric system. To exemplify:

la canne = 3 meters
la palme = 0.25 meter
10 starello = 40 ares (measure of area)
10 starello = 50 liters (measure of grain)
la botte = 50 liters (measure of fluids)
la quertara = 5 liters
la cantara = 100 livres = 40 kilograms
la livre = 400 grams

Clearly, this arrangement closely approximated the full metric system. Another royal decree, dated 11 September 1845, proclaimed that the metric system would be binding as from 1 January 1850. Since, however, the Parliament did not legislate the necessary directives, a law dated 6 January 1850[28] delayed the effect of the decree by three months (1 April 1850). At last, after being tolerated for a quarter of a century and quietly "eased into" popular acceptance, the meter was now finally supreme.

The five years between 1845 and 1850 witnessed a great many intensive labors preliminary to the adoption of the metric system. Standards were commissioned both from Piedmontese manufacturers and in Paris. Of the former, a number returned their contracts, finding themselves less than equal to the task, while some others supplied measures that the commission had to reject as failing to comply with the norms. A year after the formal completion of the reform, a third of the communes still had not been supplied with standards! The costs rose, the standards were more expensive to produce than had been expected, and moreover, payments charged directly for their work proved inadequate to pay the corps of inspectors, and they had to be paid state salaries.[29] Also, while formally the number of valid measures was being gradually reduced, the traditional ones were going out of use so slowly that to take all of them into account in the conversion tables would have resulted in a publication of several volumes.[30] As in France in the period of the Revolution, some measures whose currency was restricted to very small regions were simply consigned to oblivion (and, as in the case of France, this practice hinders today's historian, reliant as he is on the tables).[31] Meanwhile, courses for teaching the new system were organized in the parishes, and countless denunciations—generally anonymous—of dishonest practices had to be dealt with, while the continued coexistence of the two systems aided their perpetrators.

Now, when the King issued his decree in 1845, the various territories under his rule were unequally ready for the metrological reform. Savoy and Piedmont were quite ready for it, but what of, for instance, Sardinia, where French rule had never reached? Small wonder that the Sardinian peasants did not wish or were not able to change over to the new measures, but stubbornly adhered to the old ones, either out of traditional attachment or because they could not afford—in the wake of a poor harvest—to acquire new ones.[32] Indeed, peasants actually rioted in Sardinia against the introduction of the metric system. Yet the government in Turin did appreciate the importance of the matter. What was at stake was not only the introduction of the reform throughout the lands of Piedment (if need be, with some delay, as in the case of Sardinia where the decree of 1845 was not published until 8 April 1850!), but the winning over of other Italian states—namely, Parma, Piacenza, Modena, and Reggio—and therefore nothing less than paving the way to the unification of Italy.[33] In fact, ultimately the introduction of the metric system throughout Italy was not accomplished until after the unification of the whole country. As in the case of many other institutions, the system was finally—let us not mince words—imposed by the North upon the South. So much for Piedmont.

Let us now consider how the meter fared in another region where it had been brought by the troops of the tricolor—the Republic of Geneva, otherwise known as the *département* of Leman.[34] Its incorporation into the highly centralized French state naturally brought in its train administrative integration, and with it the metric system, too. The cantonal authority prepared appropriate instructions[35] and organized classes to teach the metric system to the citizens,[36] while in the marketplace of Geneva stone standards of metric measures were set up. By a decree dated 16 *floréal* of year X, the prefecture of the *département* of Leman made the metric measures obligatory as of 1 *prairial*, and the municipal masters of Geneva fell into line on 16 *fructidor* of the same year. Thereafter, the authorities and the community fought each other on the issue for ten years, with the pressure from Paris never ceasing. Thus, Chaptal, in his letter of 8 *thermidor*, year XIII, to the prefect of Leman, notes that the Geneva merchants do not apply the new measure. Again, a letter from the Ministry, dated 20 November 1806, concedes that there has been some progress made in Geneva itself in the matter of introducing metric measures, but no similar headway in other districts of the *département*. And on 30 May 1807 a directive from Paris orders that recalcitrant merchants shall be dealt with by courts of law. The exchange of letters between Paris and Geneva went on for years.

However, the situation within the territory of the former Republic of Geneva differed diametrically from the situation in the Cisalpine Republic. Geneva's area was small and firmly under the control of the capitalist régime in its large urban center,[37] while its villages had long since parted company with feudalism and been commercialized to a high degree. All this seemed to suggest that the introduction of the metric system here would be relatively easy. And so it proved. Over the years, the hortatory letters from Paris grew fewer, those praising local efforts multiplied, and the great many reports that the administrative officials of the *département* received from the superintendents of the markets and shops clearly indicated that metrication was gaining ground in a very healthy fashion. Then, however, very late in the day, disaster struck in the form of the Napolenoic compromise.

The reader will recollect that the objective of that compromise was to assimilate the new system to the traditional usages. Consequently, divisions and multiples other than decimal ones were revived. So were old terms, as names for the resultant, "rounded-up" units. Thus "the pound" was now once again the appellation applied to the half-kilogram, two meters became the former *perche*, and so forth. However, like the Republic itself, the metric system was one and indivisible, even if local usage varied, and a practice that constituted concession to tradition in one region would be quite unknown in another. Even in the older *départements* some prefects—as we have already noted—were aware of this inconsistency, and it was certainly in evidence in the former Republic of Geneva.

On 25 September 1812, the Geneva Chamber of Commerce contacted the prefect to express their regret that the objective of the reform, having been all but attained, was now given up. The prefect, in his turn, wrote on 2 October 1812 to the Minister for Internal Affairs saying: "The latest new measures will, in this *département*, be almost as outlandish to the people as the metric ones . . . having little or nothing in common with our old local measures." It would have made sense to fight on for the introduction of the metric system, given that the manufacturers, wholesalers, and merchants were appreciative of its convenience in use, and the people in general were getting used to it. It would have made no sense, however, to change over to measures that were in use only in some regions of France. Thus, as far as the *département* of Leman was concerned, the Napoleonic metrological compromise, contrary to its intention, placed the metric system farther away from the understanding of the people.[38] That the days of the *département* of Leman were by then numbered was another matter. It would seem that it was precisely because of the compromise arrangement of 1812 that the metric system, which had been so close to being fully accepted, was dropped by Geneva almost as soon as Napoleon fell from power. A new law, dated 1 October 1816,[39] spelt out a revision of weights and measures: the *pied* and *setier*, as dry measures of capacity for grain; and the *livre* and *once*, as weights. All units in all categories were given with their equivalents in smaller units, resulting in the most bizarre ratios, for example, 1 *aune* = 3 *pieds*, or 7 *pouces*, or 10.5 *lignes*; none of these—except by use of fractions—was convertible into multiples or divisions of the meter. The resulting inconvenience in practice is easy to imagine: it was the price the merchants and manufacturers had to pay for the independence that Geneva now recovered! To complete the story, in 1851 a federal law, passed on 23 December, announced the unification of weights and measures for the whole Swiss Confederation;[40] much later still, in 1876, the Confederation adopted the metric system,[41] and so at last it came to Geneva, too.

Next, although our knowledge here is far less complete, let us consider yet another region that became acquainted with the metric system as a direct result of coming under French rule—the annexed *départements* of what was later to become Belgium, and the Batavian Republic.[42] Here, too, administrative unification brought in its train the metric system.[43] In the *département* of Dyle (Brussels) the change was carried through swiftly, and the metric system became legally binding as early as 19 *frimaire* of year VIII (10 December 1799). However, in the area of Antwerp (the *département* of Deux-Nethes) the changeover took much longer.

Now, in every *département* (although, for some unfathomable reason, with much delay in Deux-Nethes) a commission for weights and measures would be established as a matter of course. The commissions would then start collecting standards, at least from all larger towns, and this task went on for years. Proud of their centuries-long autonomy, the municipalities were by no means ready to accept the *diktats* of centralized administration. Yet, until the collection of standards was completed, the commissions were unable to tackle the task of constructing the conversion tables. Meanwhile, however, the law demanded the use of metric measures. The uncertain situation favored private enterprise in the production of numerous publications, which were incomplete and full of errors. In the *département* of Dyle the matter was largely settled by the relatively early appearance, in 1803, of an authorized conversion table,[44] but in other *départements* confusion prevailed. So it continued for many years. The fall of Napoleon, however, ushered in a new order with little delay. On 27 August 1816, William I promulgated for the united Netherlands a new system of measures that combined the old-style nomenclature and the decimal system of counting. Perhaps it was because of their anti-Dutch sentiments, or out of sympathy for the post-1830 bourgeois France, that the newly independent Belgians returned to the metric system in 1836. So much for the unhappy adventures of the metric system as it followed the tricolor into four different lands.

When the tricolor was carried farther east, and came to flutter over Prussia, Austria, and Poland, no attempt was made at all to propagate the metric system in its wake. This omission was of symbolic significance. The meter was not introduced in the basin of the Vistula, nor was there established on its banks a new republic that might well have been given the name *Sarmatian*, had it come into being a few years earlier.* All that Napoleon offered the Poles was the Duchy of Warsaw, reigned over by the Elector of Saxony, newly promoted to the dignity of King. Clearly, the era of export of republicanism by the French was over and done with; so was

* The Sarmatians were legendary forbears of the Polish people—TRANS.

the export of the metric system; and so, too, the export of the Revolution itself.

As we pointed out in an earlier chapter, Tallyrand, who had originally proposed collaboration with England's Royal Society in the production of a new system of measures, attempted by 1798 to turn the whole undertaking into an anti-English venture, shared with France's allies and with neutral countries—albeit still international in its scope.[45] In the days of Napoleon's Empire, the metric system assumed an exclusively French character; there was no longer anything international about it, and in the countries that the French had conquered or dominated, the metric system collapsed with the passing of their rule. Yet its idea lived on as a symbol of modernity, of democratic aspirations, of equality, and of progress.

Germany adopted the metric system as an integral part of its "revolution from above" in 1768,[46] with Austria and Hungary following suit before long, in 1871 and 1874, respectively. East European, and also the Baltic, countries accepted the meter as a concomitant of their independence: Serbia in 1863, Rumania in 1883, Poland in 1919, Yugoslavia in 1919 too, Latvia and Lithuania in 1920; and Czechoslovakia, in whose lands as part of the Austrian empire the meter had been binding anyhow since 1871, became a member of the Metric Convention in 1924. Except for Czechoslovakia, the association of former measures with hostile occupying powers and loss of independence smoothed the way for the metric system. Soviet Russia accepted it as early as 1918, symbolizing its quest for modernization, and in 1921, in a very different political context but also to express its drive for modernization, Japan accepted the metric system. China, too, opted for the meter on the victory of the revolution that looked to modernization, while India—not without some procrastination—adopted the metric system in the wake of winning independence in 1947.

The process of evolving an international organization to correspond to the international ethos of the metric system took a long time. In 1868, following the World Exhibition in Paris, amid the tawdry splendors of Napoleon III's foredoomed Second Empire, there met in the French capital a conference of representatives of thirty states, who determined to found an international metrological organization. The participants asked the French government to invite all interested countries to launch it, and the invitations went out in the following year. But dramatic events of 1870–1871— the Franco-Prussian War, the collapse of the Second Empire, and the Paris Commune—delayed the proposed conference. It did, however, take place in 1872, when the decision was taken that all participating countries should receive copies of the French metric prototypes. The next conference met three years later and, on 8 May 1875, passed the resolution to form the International Convention and to establish in Paris the International Bureau of Weights and Measures, its costs to be defrayed by the members. The Convention was initially signed by Russia,[47] Austria-Hungary, Germany, Belgium, Brazil, Argentina, Denmark, Spain, the United States, France, Italy, Peru, Sweden, Norway, Switzerland, Turkey, and Venezuela.

England was the country to hold out most stubbornly against the reform. Having vanquished Napoleon, having kept away from its shores the revolutions raging in continental Europe, England saw no reason whatever for changing over to the metric system with its revolutionary overtones—and indeed, there was no reason. England had for centuries been, arguably, Europe's most efficiently administered kingdom, and it would seem that at the turn of the eighteenth century, her measures had been relatively more uniform than those of other countries. True, the pound sterling divided by twenty into shillings; the shilling by twelve; and the pound of weight, by sixteen; nonetheless, everybody had for generations been accustomed to these divisions. Furthermore, as we have argued in an earlier chapter, the vigesimal, sexdecimal, and duodecimal systems all have considerable merits. As far as overseas trade was concerned, there were no problems between England and her colonies, where English settlers had taken their native measures,[48] while commerce with continental Europe had for centuries run in certain well-worn grooves, and if anything interfered with its smooth running, it was surely not the different measures.

As time went on, however, and as more and more countries adopted the metric system, while the volume of international trade was growing apace, attention began to be paid to the question in England. Compromises were suggested—but the metric system is exclusive and not susceptible to flexible arrangements. A project suggesting an amalgamation of the decimal system with the English units of measure was rejected by a Parliamentary commission in 1862 on the not unreasonable grounds of expense, which would have been no less than the costs of accepting the full metric system but without offering equal gains.[49] The date 1862 is worth bearing in mind; the matter was to be put before Parliament time and again over the following hundred years.

Before long, both the supporters and opponents of the metric reform organized themselves. The Decimal Association was formed and counted some 1,500 members in its heyday, and in 1897 the use of the meter was optionally accepted, but made virtually no headway in practice. Parliament gave the matter renewed consideration in 1904, aware of the support for the meter by such reputable bodies as the Council of the Chambers of Commerce and the British Medical Council, but again the outcome was negative. Meanwhile, the opposition to the metric system was active, too. Above all, they emphasized, if not exaggerated, the cost of the proposed reform. The estimate of Austin Chamberlain, the Chancellor of the Exchequer, was five percent of the national income, while A. A. Haworth, M.P. for Manchester South, suggested the figure of one hundred million pounds! All manner of difficulties to which reform would give rise, should Anglo-Saxon countries overseas adhere to the Imperial measures while England went metric, were being pointed out. F. A. Halsey's pamphlet of 1905, *The Metric Fallacy*, informed the readers how, in countries that had long since adopted the metric system, it continued to be disregarded in daily practice by the common people, who went on using traditional pre-metric measures; although his work was biased and tendentious, many of the examples he cited were genuine. During a debate in the Upper House of Parliament in 1904, Lord Lansdowne contended: "Considering this question with due attention to the units in actual use and to their relations, the British system of weights and measures is simpler and more uniform than any other system in the world. No traditional system of weights and measures has ever, in fact, been ousted by the metric system." In response, the Decimal Association produced a publication containing facts and figures on grain measures in use at the beginning of the twentieth century in various parts of England. The government, for its part, requested its diplomatic and commercial representatives abroad to report on the measures that were actually in use in the countries of their posting and to give their views on the likely consequences for British overseas trade of the adoption of the metric system.[50]

As the European state of armed peace turned to one of tension presaging war, and the great powers competed more uncompromisingly in world markets, the metric question was again revived in England in 1911. One of the leading antagonists of the reform, Sir Frederick Bramwell, wrote: "It is in the interest of British industry to retain British weights and measures because they are the best and most practical. Moreover, industrialists from the metric countries have been encountering difficulties in the Far Eastern market where the British weights and measures had gained earlier acceptance, and this constitutes an advantage that helps our merchants and industrialists in retaining their hold on those regions." His view was put more tersely by the Australian journal *The Surveyor* of Sydney: "Sir F. B. is wont to assert that we enjoy an advantage over foreigners because we can readily grasp their metric measures while they will never be able to understand ours" (31 January 1911). The South African magazine *Engineering* supported Bramwell: "Whatever the inconveniences of the British system, it does have the great merit of keeping Continental machinery out of the markets that had earlier adopted it. For this reason alone, we should do all we can to resist the adoption of the metric system in South Africa and wherever else machines are bought. If South Africa herself were a major producer of machinery and were a less important market for engineering imports, then the question of our system of measures would be less significant. As long, however, as we adhere to the British measures, an advantage accrues to us over our continental competitors who use the foot less readily than we can use the meter."

The 1897 facultative ruling on the use of metric measures did not advance the reform at all. However, metric measures would be occasionally used in medicine and pharmacology in some scientific enquiries, and in commerce in order to comply with the stipulations of foreign buyers; instances of the last type were at first extremely rare, but gradually grew more frequent, causing more and more headaches.

Eventually, in 1949, a committee for metrological legislation was established.[51] Its very knowledgeable report two years later quite unambiguously supported the introduction of the metric system into England. The report stated that the British or "Imperial" system did not have the standing of the metric system and was not truly international. The United States used British units but also related them to the meter as the standard of reference. The Imperial system was not based on an international convention, nor could it boast an international administrative center as the metric system did. The report conceded that the cost of metrication would be considerable, but the sooner the inevitable reform was introduced the less expensive it would be. Finally, three conditions of going over to the metric system were set out: first, an agreement with other members of the Commonwealth binding them to a similar reform simultaneously; second, decimalization of currency;

and, third, an intensive propaganda-*cum*-education campaign on behalf of the new system. Nevertheless, contrary to all expectations, the Minister of Trade in November 1952 decided against the reform.

It may be mentioned here that the decennial inspection of the London standards for the yard and the pound showed in 1947–1948 that the standards were wanting in the required degree of accuracy. This, moreover, implied another danger: since the British system of weights and measures binding in the United States bore reference to the Paris meter, it was not unlikely that the identically named American yard and pound would begin to diverge substantially from their British namesakes, causing further difficulties and an altogether paradoxical situation. Accordingly, the British Empire Scientific Conference insisted that the British measures should be redefined in terms of the standard meter in Paris, and the necessary investigations were concluded in 1951, establishing the ratios of one yard equal to 0.9143 meter, and one pound equal to 0.453592 kilogram.

Today, Britain has, in principle, decided in favor of the metric system, but much toil and labor have yet to be expended in preparation for the changeover. The cost will run into millions—immeasureably more than if the step had been taken in 1862 or 1904 or 1911. The price that Britain will pay for the unwillingness over a century and a half to accept on its soil an invention of revolutionary France will be high. And when the metric system is finally adopted in England, the great French Revolution will have scored its first victory over her.

It would be overambitious for us to try to consider the vicissitudes of the metric system in the course of its adoption by the great ancient civilizations (with their ancient measures) of Japan, China, and India. Having some knowledge of the European experience, however, we do not imagine that the reform there had a smooth passage without difficulties corresponding to those encountered in Europe—or greater. Nor does it surprise us to learn that as recently as 1964 an investigation in an Indian village established that of 176 adults questioned, not one was acquainted with the metric system.[52]

Wherever true equality before the law is lacking, we are bound to find cases like that in Brazil's Nordeste, described by the renowned agriculturalist, Dumont. There, agricultural laborers employed by the local landowner were paid with tokens redeemable in the shop owned by the same personage and bought flour by the "liter." In the presence of some other chance customers, who watched the scene with interest and did not miss its significance, Dumont made so bold as to reweigh the "liter"—and the weight was short by almost half of what it should have been.[53] However, when the international expert moved on, the wretched laborers remained at the mercy of their "feudal" employer and his men, just as before. From Brazil too, information comes to us of many instances of landowners taking unfair advantage, through dishonest use of scales, of their sharecroppers or of their tenants paying rent in kind.[54] In circumstances like that, we can safely assume that even educating ordinary people in the principles of the metric system and in the ways of weighing and measuring, as has been done in the Cameroons or in the Central African Republic, will not help them much.[55]

What we know of China is relevant here.[56] In China, the many successive attempts to strengthen the central government were accompanied by attempts to standardize measures, whereby the rulers hoped to gain popularity with the masses. Major unifying reforms were carried out by Emperor Shi Huang Ti in 221 B.C., and by Emperor Wang-Mang in A.D. 9 ("the good measures of Wang-Mang"). After every reform, however, with the decline of central authority over the decades, regional measures not only underwent growing differentiation but also increased in size—with the (capacity) measure of grain, naturally, increasing most markedly. By the period of the Ming dynasty, the basic measure of grain, the *chi* or foot, was 40 centimeters longer than under Wang-Mang, the standard unit of weight doubled, and the grain measure of capacity quadrupled! Official investigations carried out in all of China in 1936 uncovered 53 dimensions for the *chi* varying from 0.2 to 1.25 meter; 32 dimensions of the *cheng*, between 0.5 and 8 liters; and 36 different *tsin* ranging from 0.3 to 2.5 kilograms.

European powers sought to impose on China their own measures in foreign trade transactions by the Treaty of Tientsin in 1859. The Chinese eventually began to prepare for the metric system during the closing years of the Empire.[57] In 1908, the imperial ambassador in Paris took the trouble personally to visit the International Bureau of Weights and Measures at Sèvres, and the Chinese law of 29 July 1908 looked to the coming of the metric system: it proposed that traditional measures be retained but be defined in terms of the units of the metric system, the central frame of reference being the Paris meter.[58] Wherever possible, the decimal system would also be simultaneously introduced. The French were delighted, all the more so since English merchants had for long sought to bring in their own Imperial measures. However, scarcely had the Chinese Imperial reform been launched, when the Empire itself vanished overnight.

Before long, in 1912, the republican government decided to continue and even to accelerate the realization of the reform. C. S. Chen and L. M. Tseing were dispatched to Paris to look into the matter. However, the reform met with the greatest difficulties and proceeded slowly because the very terminology of the metric system was alien to the Chinese language and full of sounds that simply did not exist in it. At last, in 1929, the Kuomintang government proclaimed a compromise, whereby retail trade would continue to use the measures validated in 1908, while wholesale trade and official transactions would rely on the metric system. The years that followed, of the civil war and then of the war against Japan, made it all but impossible for the partial metrication to make any progress. In 1959, the government of the People's Republic of China again decreed metric reform. In order to facilitate the acceptance of the reform by the people, the tried and trusted route was to be followed—of retaining many of the former names while standardizing dimensions and integrating them with the metric system. Thus the *cheng* was made equal to the liter, the *tsin* to half a kilogram, the *mu* to one-sixteenth of the hectare, and the *chi* actually to one-third of the meter. The State Bureau of Weights and Measures had been established earlier in 1955, and the decision to go metric was taken in 1958. Old measures and weights were confiscated by the thousand, melted down and then used in the production of new ones with considerable saving of raw materials as well as the safeguarding of the people against potential swindles. Within five months, the operation apparently resulted in there being enough new weights and measures to equip all China; they were made obligatory from 1959, while the former measures of feudal lords were consigned to museums.

Thus, with a delay of some 150 years, in a faraway country with a far different civilization and following a very different revolution, the metrological achievement of the great French Revolution was reinvested with its revolutionary complexion. Let me explain: in the district Tegu (Szechuan province), in the house of the former landowner, the two different measures of capacity that he once used are exhibited to instruct the visitors. The measure is the *tou*, which should be approximately the equivalent of ten liters, but the two specimens differ by 3.6 liters. The large *tou* was used in collecting dues in kind from the peasants, and the small one when loans of grain were made to them. This is why the Kuomintang reform remained a dead letter; for it turned out, yet again, that to establish the metric system, with its standardized and invariable measures of convention, is quite impossible without the simultaneous destruction of feudalism itself, without doing away with great landowners, freeing the peasants from the dues they have to render them, and consigning to the fire the inventories of such burdens, and without, at the end of the day, erecting museums of false measures used by landowners for purposes of unfair exploitation, like the one in Szechuan province.

Postscript. In Praise of Prefects

"One King, one law, one measure, and one weight"—so rang the cry of the peasants throughout France on the eve of the Revolution, in a mighty surge of reformist and patriotic demands and dreams. And through the unity of institutions, they sought to attain the ultimate ends of liberty, equality, and fraternity. In 1813, a shocked Benjamin Constant wrote:

> One code of laws for all, one system of measures, one set of regulations . . . this is how we perceive today the perfection of social organization . . . uniformity is the great slogan. A pity, indeed, that it is not possible to raze all towns to the ground so as to be able to rebuild them on one and the same pattern, and to level mountains everywhere to a single preordained plain. Indeed, I am surprised at the absence, as yet, of a *ukase* ordering everybody to wear identical clothes, so that the sight of the Lord will not encounter lack of order and offensive diversity.[59]

It was through uniformity of institutions that the peasants of different regions of France hoped to become members of one great family. But that uniformity alarmed Constant, who naturally felt that it was he who was the mountain that would be levelled.

Now, when Constant wrote these lines, the Napoleonic metrological compromise had recently been decreed and it held up—even more effectively after the restoration of the Bourbons—the progress of the metric system in France. Yet there were still men in France who kept faith with the idea. Chateaubriand, writing in 1828, tells us who they were: "Whenever you meet a fellow, who instead of talking of *arpents*, *toises*, and *pieds*, refers to hectares, meters, and centimeters, rest assured, the man is a prefect"; and he had just a paragraph or two earlier defined the prefect as "a person characterized by garrulousness and petty tyranny, a bureaucrat concerned with conscription."[60] So, such were the prefects, whom we have come across time and time again in these pages as enthusiastic executors of the metric reform.

Let me confess that I have not read all three thousand pages of Chateaubriand's *Mémoires*, finding the passage quoted in the course of my labors, but my attention was directed to it by a friend of mine, a young French historian, in a letter that continues as follows: "You know that I have always been interested in the social history of the relations between the authority of the state and individuals and groups. In this area, it must be admitted that my own country's history abounds in cases of coercion and the exercise of undue authority. The important lesson France has taught the rest of the world regards the effectiveness of centralized administration. Our right-wing, as well as our left-wing, parties have inherited this legacy; it applies to De Gaulle and Mitterand as well as to Robespierre and Bonaparte. My kind of history is concerned with those who swam against the current. There is a long tradition here from the Fronde to the Paris Commune, but little political achievement, although as an ideological and intellectual tradition, Montaigne and Rousseau, Pascal and de Tocqueville, have a lot to offer. Hence, the Frenchman is most sensitive to whatever emanates from the state and the state apparatus, and this sensitivity is by no means coextensive with what one might casually term his reactionary reflex. In fact, once we start viewing matters from this standpoint, many phenomena assume a novel complexion.

Let me reply: My friend, as befits a historian of the metric system, I am its admirer. Therefore, I cannot but rise to the defense of the prefects whom you have so abused. They were men of great merit. It was their steady, enduring labors that enabled the ideas of the *philosophes*, so beautiful in their rationalist purity, to materialize in daily practice and to permeate the very thinking of the entire nation. And thanks to this, gone are the countless, daily opportunities for the strong to injure the weak, for the smart to cheat the simple, and for the rich to take advantage of the poor. Many centuries ago the Greek philosopher Architos was enthralled with the new arithmetical basis of weights and measures, and said: "This will prevent many discords and bring about more harmonious human relations; there will now be no scope for fraud, and equality shall reign."[61] Today we know that he was mistaken. Nevertheless, the hopes he attached to the invention of measures, as such, might well be restated with reference to the invention of the metric system.

Through this innovation, moreover, the whole nation was made to acquire common ways of thinking, to share the same perceptions of space, dimensions, and weights, and to grasp—albeit with the greatest difficulty—the principles of decimal division. When two peasants from two different regions met, they used to find it hard to understand each other when the conversation touched upon miles, *arpents*, or *toises*, for each would have different magnitudes in mind. But they understand each other without difficulty when talking of meters or kilograms. And to have imposed upon men common ways of perceiving, and thereby to have enabled them the better to understand one another, was surely an admirable achievement.

The prefects shall not desist from their labors. It is reasonable to expect that they shall seek, and achieve, in the areas of their administrative competence, further unification of ever new perceptions among men. Many further achievements will yet redound to their credit. They shall continue to bring about a higher level of mutual understanding among people. And in the end, a time will come when we shall all understand one another so well, so perfectly, that we shall have nothing further to say to one another.

NOTES

CHAPTER 1

1. Flavius Josephus, *Starożytności Żydowskie* [Jewish Antiquities], I, 2, 2, Warsaw, 1962, p. 105.
2. A. Spirkin, *Pochodzenie Świadomości* [The Origins of Consciousness], Warsaw, 1966, p. 416, fails to understand this point.
3 The pace was used, however, only when the distance was small; ancient Slavs do not appear to have paced out large distances like the Roman mile.
4. K. Moszyński, *Kultura ludowa Słowian* [The Folk Culture of the Slavs], II, pt. 1, Cracow, 1934, p. 118.
5. Niangoran-Bouah, "Weights for the Weighing of Gold. One of the Aspects of African Philosophical and Scientific Thought before Colonisation," *First International Congress of Africanists* (mimeo), Accra, 1962; see also D. Paulme, "Systèmes ponderaux et monetaires," *Revue Scientifique*, no. 5, 1942.
6. R. Mauny, *Tableau géographique de l'Ouest Africaine au Moyen Age d'après les sources écrites, la tradition et l'archéologie*, Dakar, IFAN, 1961, pp. 410-419. For distances measured by time taken to walk them, with and without a load, in sixteenth-century Congo, see F. Pigafetta and D. Lopez, *Description du Royaume de Congo et des contrées environnantes par . . .* , ed. W. Bal, Louvain-Paris, 1963, p. 121.
7. S. Czarnowski, "Kultura," w *Dzieła* [Culture, in *Collected Works*], I, Warsaw, 1956, p. 57.
8. *Ibid.*, p. 58.
9. Technical differences would demand that some measures should differ for different textiles. For cloth and linen in Poland, see S. Hoszowski, *Ceny we Lwowie w XVI i XVII wieku* [Prices in Lvov in the 16th and 17th Centuries], Lvov, 1928, p. 62.
10. K. Sochaniewicz, "Miary i ceny produktów rolnych na Podolu w XVI w. [Measures and Prices of Agricultural Produce in Podolia in the 16th Century], *Lud*, XXVIII, 1929, p. 149.
11. E. Tomaszewski, *Ceny w Krakowie w latach 1601–1795* [Prices in Cracow in the Period 1601–1795], Lvov, 1934, p. 18.
12. J. F. Bergier, *Genève et l'économie européenne de la Renaissance*, Paris, 1963, p. 194.
13. A donkey's or horse's load as a measure was known elsewhere, too; e.g. see *Cahiers de doléances*, (hereafter *Cdd.*) district Rennes, parish Laillé, II, p. 211.
14. R. Dumont, *Terres vivantes. Voyage d'un agronome autour du monde*, Paris, 1961, p. 31.
15. M. Luzzati, "Note di metrologia pisana," *Bolletino Storico Pisano*, XXXI–XXXII, 1962–1963, p. 201.

16. L. Musioł, *Dawne miary zboża na Górnym Śląsku. Przyczynek do metrologii śląskiej* [Old corn measures in Upper Silesia. A contribution to Silesian metrology] (mimeo), Opole, 1963, p. 35.
17. E. le Roy Ladurie, *Les paysans de Languedoc*, Paris, 1966, p. 210.
18. A. Keckowa, " 'Bałwany' wielickie. Z badań nad historią techniki górnictwa solnego w Polsce w XVII–XVIII wieku" [The salt-blocks of Wieliczka. An enquiry into the history of the techniques of salt-mining in Poland in the 17th and 18th centuries], *Kwartalnik Historii Kultury Materialnej*, VI, no. 4, 1958, p. 637.
19. E. le Roy Ladurie, *op. cit.*, p. 271.
20. S. Streltsin, "Contribution à l'histoire des poids et des mesures en Ethiopie," *Rocznik Orientalistyczny* XXVIII, no. 2 1965, p. 86.
21. For the use of the bowshot in Hungary and Slovakia in the 17th century, see A. Huščava, "K dejinam najstaršikh dlžkowykh mier na Slovensku," *Slovenský Narodopis*, V, nos. 3-4, 1957.
22. *Ibid.* The hachet's throw backwards equaled half the distance of an ordinarily executed cast.
23. *Ibid.*
24. J. K. Zamzaris, "Metrologiya Latvii v period feodalnoy rozdroblennosti i razvitogo feodalisma (XIII–XVI vv.)," *Problemy Istočnikovedeniya*, IV, Moscow, 1955, p. 189.
25. P. Burguburu, *Métrologie des Basses-Pyrénées*, Bayonne, 1924.
26. H. Navel, *Recherches sur les anciennes mesures agraires normandes. Acres, vergées, et perches*, Caen, 1932.
27. H. Rybarski, *Handel i polityka handlowa Polski w XIV stuleciu* [Trade and the Commercial Policy of Poland in the 16th century], II, Poznań, 1929, p. 332. The range [for "lengths" of cloth] was only slightly less in the Cracow market in the 16th century; see J. Pelc, *Ceny w Krakowie w latach 1369–1600* [Prices in Cracow, 1369–1600], Lvov, 1935, pp. 35-36.
28. R. Bazavalle, "Zur Geschichte des Grazer Masses," *Zeitschrift des historischen Verein für Steiermark*, XXV, 1929, pp. 47-48, and especially his "Zur Geschichte des Judenburger Masses," *ibid.*, XXVI, 1931, pp. 190-199.

CHAPTER 2

1. Leviticus 19.35-36. Hesiod, if taken literally, offers similar moral suggestions relating to measures in his *Works and Days*.
2. Deuteronomy 25.13-15.
3. Proverbs 16.11.
4. Amos 8.5.
5. Micah 6.11.
6. Mark 4.24; cf. Matthew 7.2.
7. Luke 6.38.

8. Mohammed, *Le Coran Traduction intégrale par E. Montet*, ed. Payot, II, Paris, 1958 [from the French].
9. In the Bourges cathedral the scales are real, made of metal.
10. A. Pigafetta, *Premier voyage autour du monde par Magellan, 1519–1522*, ed. Ed. P. Peillard, Paris, 1964, p. 139 (under 10 April 1521).
11. F. Pigafetta and D. Lopez, *op. cit.*, p. 142. Presumably, the authors were unable to recognize and understand local metrological systems.
12. Montaigne, *Essays*, Paris, 1950, p. 1018.
13. Niangoran-Bouah, *op.cit.* Relevant literature is extensive; e.g. D. Paulme, "Systèmes ponderaux et monetaires," *Revue Scientifique*, LXXX, no. 5 1942; see also H. Abel, "Poids à peser l'or de Côte d'Ivoire," *Note d'information*, 1961, p. 69.
14. I. Ehrenburg, *Ludzie, lata, życie* [Men, Years, Life], III, Warsaw, 1966, p. 157.
15. O. Kolberg, *Przysłowia* w *Dzieła wszystkie* [Proverbs, in *Collected Works*], 60, Wrocław, 1967, p. 178.

<h2 style="text-align:center">CHAPTER 3</h2>

1. For instances of popular hostility to censuses, see W. Kula, *Problemy i Metody Historii Gospodarczej* [Problems and Methods of Economic History], Warsaw, 1963, pp. 347-348. Note also the resistance of the peasants in western Bulgaria to the introduction of "sinful" birth certificates, which they felt would increase infant morality; D. Marinov, *Živa starina*, III, Russe, 1892, p. 187.
2. Č. Zibert, in *Česky lid*, VII, 1898, p. 204.
3. Private communication from Professor K. Koranyi. See also P. Sartori, "Zählen, messen, wägen," *Am-Urquell. Monatschrift für Volkskunde*, VI, 1895, pp. 9-12, 58-60, 87-88, 111-113. E. Hoffman-Krayer and H. Bächtold-Stäubli, "Mass, messen," in *Handwörterbuch des deutschen Aberglaubens*, V, Berlin, 1932–1933, p. 1852.
4. S. Ciszewski, "*Początki miernictwa i pierwotne miary*" [Early measurement and primitive measures] (unpubl.).
5. "Miara," *Słownik Staropolski* [Measures, *The Dictionary of Old Poland*].
6. This is accepted by B. Chmielowski, *Nowe Ateny* [New Athens], Warsaw, 1966, p. 221.
7. R. Bloch, *Etruskowie* [The Etruscans], Warsaw, 1966, p. 83.
8. B. Baranowski, "Pośmiertna Kara 'za złą miarę' w wierzeniach ludowych," *Łódzkie Studia Etnograficzne* [Posthumous punishment for 'bad measures' in folk beliefs, *Łódź Ethnographic Studies*], VII, 1965, pp. 73-85; idem., *Pożegnanie z diabłem i czarownicą* [Farewell to the devil and the witch], Łódź, 1965, pp. 52-56.
9. *Źródło cudów i łask* [The source of miracles and grace], Warsaw, 1729; cited after Baranowski.
10. P. Krevelakis, *Słońce śmierci* [*The Sun of Death*], Warsaw, 1967, p. 73.
11. C. Bobińska, "Pewne kwestie chłopskiego użytkowania gruntu i walka

o ziemię," in *Studia z dziejów wsi małopolskiej w drugiej połowie XVIII w.* [Some problems of the utilization of soil by the peasants and the struggle for land, in Historical Studies of villages in Lesser Poland in the second half of the 18th century], Warsaw, 1957, p. 353. S. Askenazy, "Trybun gminu," in *Dwa stulecia* [The tribune of the common people, in *Two Centuries*], Warsaw, 1910, p. 371. H. Grynwaser, "Przywódcy i 'burzyciele' włościan," in *Pisma*, [Leaders and ringleaders of the peasantry, in *Collected Works*] II, Wrocław, 1951, p. 213.

CHAPTER 4

1. E.g. Exodus 30.13, 38.24-27, Leviticus 27.25.
2. E.g. 2 Samuel 14.26. Cf. Firdausi [Persia, *ca.* 935 to *ca.* 1020—TRANS.], *Opowieść o miłości Zala i Rudabe* [The love of Zal and Rudabe], ed. F. Machalski, Wrocław, 1961, p. 80.
3. W. Ciężkowska-Marciniak, "O greckich i rzymskich wagach i miarach" [Of Greek and Roman weights and measures], *Meander*, nos. 1-2, 1956; pp. 40-56; *idem., Jak ważono w starożytnej Grecji?* [How did the ancient Greeks weigh?], National Museum, Warsaw, 1957.
4. Cf. D. Paulme, *op. cit.*
5. In ancient China, apparently in periods of weak government, the craft guilds effectively controlled measures; F. Tannenbaum, *Une philosophie du travail. Le syndicalisme*, Paris, 1957, p. 25.
6. H. van der Wee, *The Growth of the Antwerp Market and the European Economy*, I, The Hague, 1963, I, p. 56.
7. W. Beveridge, *Prices and Wages*, I, London, 1940, p. 16.
8. J. F. Bergier, "Commerce et politique du blé à Genève aux XVIe et XVIe siècles," *Revue Suisse d'Histoire*, XIV, no. 4, 1964, p. 528.
9. A. Machabey, "Les sources," *Revue de Métrologie Pratique et Légale*, no. 4, 1952.
10. G. Schmoller, "Die Verwaltung des Maas- und Gewichtswesens im Mittelalter," *Jahrbuch für Gesetzgebung, Verwaltung und Volkswirtschaft*, XVII, 1892 (*contra* G. von Belov, *Die Entstehung der deutschen Stadtgemeinde*, Düsseldorf, 1889, and *Der Ursprung der deutschen Stadtverwaltung*, Düsseldorf, 1892, and the later rejoinder to Schmoller, *Die Verwaltung des Maas- und Gewichtswesen im Mittelalter; Antwort an Herrn Schmoller*, Münster-Regensburg, 1893).
11. H. van der Wee, *op. cit.*, I, pp. 66, 75-76. For a case of a new mayor receiving at his installation, along with the city keys, "an iron standard of the ell . . . and metal scales and weights for bread," see J. A. Brutails, *Recherches sur l'équivalence des anciennes mesures de la Gironde*, Bordeaux, 1912.
12. J. W. Thompson, *Economic and Social History of the Middle Ages, 300–1300*, II, New York, 1959, pp. 596-597.
13. M. Luzzati, *op. cit.*, pp. 191, 192, 200.

14. E.g., H. van der Wee, *op. cit.*, p. 67.
15. *Ibid.*, pp. 68-69.
16. C. Biernat, "Stanowisko Rady Gdańskiej wobec nadużyć mierników zbożowych w XVII i XVIII w." *Roczniki Dziejów Społecznych i Gospodarczych* [The Attitude of the City Council of Gdańsk towards the Abuses by the Corn Measurers in the 17th and 18th centuries, *Yearbooks of Social and Economic History*], XV, 1953.
17. *Ibid.*, p. 14.
18. J. K. Zamzaris, *op. cit.*, p. 181.
19. Gregory the Great, *Listy* [*Letters*] trans. into Polish by J. Czuja, Warsaw, I, pp. 73-74.
20. J. Rutkowski, *Historia gospodarcza Polski do 1864 r.* [Economic History of Poland to 1864], Warsaw, 1953, pp. 40, 61, 243; G. J. Rolbiecki, *Prawo przemysłowe miasta Wschowy w XVIII w.* [The Industrial Law of the Town of Wschowa in the 18th century], Poznań, 1951, p. 442; W. Smoleński, *Komisja Boni Ordinis Warszawska, 1765–1789* [The Warsaw Commission *Boni Ordinis*], Warsaw, 1913, p. 11. In Poznań the municipal scales had gone out of use but were restored by the Commission *Boni Ordinis*; T. Ereciński, *Prawo przemysłowe miasta Poznania w XVIIIw.* [The Industrial Law of the City of Poznań in the 18th Century], Poznań, 1934, p. 723.
21. E.g. in Poznań the use of officially approved scales was compulsory for butchers; T. Ereciński, *op. cit.*, p. 723.
22. *Ibid.*, p. 723.
23. L. Musioł, *op. cit.*, p. 13.
24. Mensura Sancti Adalberti at Trzebnica; *ibid.*, p. 16. See also M. Gumowski, "Najstarsze systemy wag w Polsce," *Studia Wczesnośredniowieczne* [Earliest Systems of Weights in Poland, Studies in the Early Middle Ages], II, 1953, p. 35. For bishops' measures in Silesia, see Musioł, *op. cit.*, p. 22.
25. *Ibid.*, p. 95.
26. *Ibid.*, p. 97.
27. *Ibid.*, pp. 78-79.
28. *Ibid.*, p. 82.
29. *Ibid.*, *passim*.

CHAPTER 5

1. P. Krevelkis, *op. cit.*, p. 100.
2. See P. Lafargue, "Badania nad pochodzeniem pojęcia sprawiedliwości i pojęcia dobra," in *Pisma Wybrane* [Studies in the origin of the concept of justice and the concept of good, in *Selected Works*], I, Warsaw, 1961, pp. 74-75.
3. In Ethiopian medicine the measure of the "earhole" was applied; S. Streltsin, *op. cit.*, p. 77.
4. *Ibid.*, e.g. prescription no. 992. But see A. Machabey, *Histoire générale*

des techniques, II, *Les premières étapes du machinisme*, p. 315, for an instance of an anthropometric measure—to wit, the abstract *finger*—among medieval Mohammedans being defined in nonanthropometric terms: six grains of barley placed end on, each of them of a width equal to six hairs of a mule's tail.

5. Among the 16th- and 17th-century geographical discoverers, the *arquebus shot* was used as a large unit in the measurement of distances; see A. Pigafetta, *op. cit.*, p. 148; F. Pigafetta and D. Lopez, *op. cit.*, pp. 25, 44, 76, 93-94; Jean-François de Rome, *Brève relation de la fondation de la mission des frères mineurs capucins . . . au Royaume de Congo . . . par . . .*, Rome, 1648; F. Bontinck, ed., Publications de l'Université Lovanium de Léopoldville, Louvain-Paris, 1964, p. 19.

6. It has been empirically possible to establish usual correlations between parts of the human body, e.g. that the circumference of the head, measured vertically, is commonly one *ell* in length; K. Moszyński, *op. cit.*, I, p. 117.

7. S. Streltsin, *op. cit.*, p. 76.

8. J. K. Zamzaris, *op. cit.*, p. 180.

9. K. Moszyński, *op. cit.*, p. 118.

10. B. A. Rybakov, "Russkiye sistemy mer dliny XI–XV vekov. Iz istorii narodnykh znanii," *Sovyetskaya Etnografiya*, no. 1, 1949, pp. 67-91.

11. S. Streltsin, *op. cit.*, p. 78.

12. B. A. Rybakov, *op. cit.*, p. 70.

13. A. Keckova, *op. cit.*, p. 627.

14. E.g. see K. Dobrowolski, "Dzieje wsi Niedźwiedzia w powiecie limanowskim do schyłku dawnej Rzeczypospolitej," in *Studia z historii społecznej i gospodarczej poświęcone Franciszkowi Bujakowi* [The history of village Niedźwiedź in district Limanów until the decline of the Republic of Poland, in Studies in social and economic history in honor of Franciszek Bujak], Lvov, 1931.

15. B. A. Rybakov, *op. cit.*, pp. 82, 91.

16. M. Bataille, "Viollet-le-Duc, jardinier des pierres," *Archéologie*, no. 6, 1965, pp. 50-56.

CHAPTER 6

1. H. Hauser, *Recherches et documents sur l'histoire des prix en France de 1500 à 1800*, Paris, 1936, p. 28.

2. S. Orsini-Rosenberg, *Geneza i rozwój folwarku pańszczyźnianego w dobrach katedry gnieźnieńskiej w XVI w.* [The genesis and development of the corvée-based manorial demesnes in the estates of the Gniezno cathedral in the 16th century], Poznań, 1925, p. 93; the definition is Pliny's (8.3).

3. C. Estienne, *L'agriculture et Maison Rustique de MM . . . et Jean Liebault . . .*, Lyon, 1618, p. 486.

4. J. Rutkowski, "Studia nad organizacją własności ziemskiej w Bretanii

w XVII w." [Studies in the organization of landed estates in Brittany in the 17th century], *Przegląd Historyczny* 1913, p. 13.

5. J. W. Thompson, *op. cit.*, II, 736.
6. P. Vilar, *La Catalogne*, II, Paris, 1962, p. 278.
7. J. Richard, "Arpentage et chasse aux loups. Une définition de la lieu de Bourgogne au XV s., *Annales de Bourgogne*, XXXV, no. 4, 1963, pp. 239-250.
8. G. Prato, *La vita economica in Piemonte a mezzo il secolo XVIII*, Torino, 1908, p. 470.
9. J. K. Zamzaris, *op. cit.*, p. 191.
10. M. Confino, *Domaines et seigneurs en Russie vers la fin du XVIIIᵉ siècle*, Paris, 1963, p. 110.
11. G. W. Abramovič, "Neskolko izyskanii iz oblasti russkoi metrologii XV–XVI vv. (Korobia, kopna, obzha)," *Problemy Istočnikovedeniya*, XI, 1963, p. 371.
12. K. Moszyński, *op. cit.*, p. 124.
13. Outside Europe, the land measure of *kula* or *kulya* in India from the 4th to the 6th centuries A.D. referred to an area for two plows; S. K. Maity, "Land Measurement in Gupta India," *Journal of Economic and Social History of the Orient*, I, 1957/1958, pp. 98-107.
14. For obvious reasons (the labor theory of value) this was noticed by Karl Marx (*Kapital*, vol. I, p. 76).
15. J. Kolendo, *Postęp techniczny a problem siły roboczej w rolnictwie starożytnej Italii* [Technical progress and the problem of manpower ancient Italian agriculture], Wrocław-Warsaw, 1968, p. 61.
16. *Cdd.*, Bourges (Azy), p. 768; and similarly for vineyards, *Cdd.*, Bourges (Gron), p. 770. Gustav Laurent, who edited the *Cahiers de doléances* for Sézanne and Châtillon-sur-Marne, states the following ratio of the labor measures to geometric measures: the local unit was an *arpent* of 100 *perches*, and 75 *arpents* kept one plow at work for a year, preface, p. xxx.
17. The earliest Polish reference to the use of seed measure known to me is dated 1388: "pro agris duo coreti possunt seminare"; the first fuller definition is dated 1498. In Burgundy, too, seed measures came later than work measures; cf. J. Richard, "Arpentage."
18. *Cdd.*, Bourges (Aubigny), pp. 767-768.
19. E. Stamm, "Miary powierzchni w dawnej Polsce" [Area measures in old Poland], *PAU* (Dept. of Hist. and Phil.), XLV, no. 2, 1936, p. 1.
20. "Perché intra in terreni nel gittare il seme fa qualche cosa differenzia, secondo la qualità della terra peggiore o migliore, di piano o montuosa" ("Petri Mariae Calandri Compendium de agrorum corporumque dimensione," *ca* 1596, in G. V. Soderini, ed., *I due trattati della agricoltura*, Bologna, 1902, pp. 292ff.).
21. Poix de Frémonville, "La pratique universelle pour la rénovation des terriers et des droits seigneuriaux . . . par Edme . . . , bailly des villes

et marquisat de la Palisse, commissaire aux droits seigneuriaux," Paris, chez Morel ainé et Gissay, 1746, avec l'approbation et privilège du Roy, pp. 558, 562. Similarly, in Normandy, Rutkowski, *Studia*, p. 14.

22. M. Luzzati, *op. cit.*, pp. 208-209, 219-220.
23. J. A. Brutails, *op. cit.*, p. 13.
24. *Ibid.*
25. *Ibid.*
26. S. K. Maity, *op. cit.*, pp. 98-107.
27. Le Comte de Tournon, *Études statistiques sur Rome et la partie occidentale des États Romains, contenant une description topographique et les recherches sur la population, l'agriculture, les manufactures, le commerce, le gouvernement, les établissements publiques, et une notice sur les travaux exécutés par l'administration française par . . . , Pair de France, Grande-Officier de la légion d'Honneur, préfet de Rome de 1810 à 1814*, 2 vols., Paris, 1832. "Weights and measures," II, pp. 26-27.
28. J. Leclére and P. Cozette, "Les mesures anciennes en usage dans le canton de Noyon," *Bull. Historique et Philosophique*, Paris, 1903, p. 10.
29. R. Richard, "Arpentage."
30. S. K. Maity, *op. cit.*
31. The data hereafter have been taken from *World Weights and Measures. Handbook for Statisticians*, United Nations Statistical Papers, series M, no. 21, rev. 1, New York, 1966. All refer to measures still in use. For parallel names of dry measures for corn and agrarian area measures, see also Machabey, *Histoire générale*, II, p. 312.
32. As in Burgundy; J. Richard, "Arpentage," and "La gréneterie de Bourgogne . . . ," *Mémoirs de la Société d'Histoire*, X, 1944–1945, pp. 117-145.
33. For Russia, see medieval examples in Z. A. Ogrizko, "K voprosu o yedinitsakh izmerenya zemelnykh ploščadey v XIII v.," *Problemy Is-točnikovedeniya*, IX, 1957, pp. 258-261.
34. I. Rychlikowa, *Studia nad towarową produkcją wielkiej własności w Małopolsce w latach 1764–1805* [Studies in commodity production of large estates in Lesser Poland], Wrocław, 1966.
35. See works by H. Grossman, Z. Kirkor-Kiedroń, W. Grabski, *et al.*
36. Rutkowski, *Studia*, p. 13; J. Richard, "Arpentage."
37. S. G. Strumilin, "O merakh feodalnoy Rosii," in *Voprosy istorii narodnogo khozyaystva SSSR*, Moscow, 1957, pp. 7-32.
38. *Ibid.*, pp. 11-12.
39. A. R. Jarry-Gueroult, "Économie rurale: sur la signification de la métrologie rurale traditionelle," *Comptes rendus des séances de l'Académie des Sciences*, 234, 1962, pp. 1197-1199 (séance 10.3.1952).
40. Hence rent-based estates would until modern times be larger than corvée-based ones, averaging double the size of the latter, according to J. Rutkowski, "Studia nad położeniem włościan w Polsce w XVIII

w.," in *Studia z dziejów wsi polskiej XVI–XVIII w.* [Studies in the situation of Polish smallholders in the 18th century, in Studies in the history of the Polish countryside], Warsaw, 1956, pp. 159ff.

41. The *łan* was one size for wooded areas and another for cleared; E. Stamm, *op. cit.*, p. 35, and J. Szewczyk, "Włóka: pojęcie i termin na tle innych średniowiecznych jednostek pomiaru ziemi," *Prace Geograficzne* [The concept and term of *vloka* in comparison with other medieval units of land measurement, *Geographical papers*], no. 67, Warsaw, 1968, pp. 41ff.

42. E. Stamm, *op. cit.*, pp. 4, 7, 18; see also J. Szewczyk, *op. cit.*

43. This is the central idea of S. Śreniowski, "Uwagi o łanach w ustroju folwarczno-pańszczyźnianym wsi polskiej" [Notes on the *łans* in the corvée-based manorial demesnes in the Polish countryside], *Kwartalnik Historii Kultury Materialnej*, III, no. 2, 1955, pp. 301ff.

44. S. Śreniowski, *op. cit.*, pp. 321, 337; see also I. T. Baranowski, "Dobra puławskie między pierwszym a trzecim rozbiorem," in *Wieś i folwark* [The Puławy estates between the first and third partitions of Poland, in The village and the manorial demesne], Warsaw, 1914.

45. S. Śreniowski, *op. cit.*, pp. 317ff.

46. *Ibid.*, p. 316.

47. *Ibid.*, p. 309.

48. *Ibid.*, p. 311.

49. B. D. Grekov considered the Russian *obzha* as exclusively a fiscal unit, but apparently it served as a unit of measure, too; J. W. Abramovič, *op. cit.*, pp. 370-371. In Latvia, the *sokha* was used to assess feudal and state taxes and also as an agrarian measure; J. K. Zamzaris, *op. cit.*, p. 191.

50. S. Grodziski, ed., *Księgi sądowe wiejskie klucza jazowskiego, 1663–1808* [Rural judicial books of the Jazów group of estates], Warsaw, 1967, nos. 416, 457, 458, 565, 607.

51. K. Moszyński, *op. cit.*, p. 111.

52. *Ibid.*, pp. 123-124.

53. S. Grodziski, *op. cit.*, no. 416: "Józef Biryt has left his sister . . . a piece of land for a bushel."

54. K. Moszyński, *op. cit.*, p. 124.

55. S. Grzepski, *Geometria to jest miernicka nauka* [Geometry or the science of surveying], ed. H. Barycz and K. Sawicki, Wrocław, 1957, p. 109.

56. K. Sochaniewicz, "Miary roli na Podhalu w ubiegłych wiekach" [Land measures in Podhale in bygone ages], *Lud*, XXV, 1926, p. 23.

57. I. T. Baranowski, ed., *Księgi Referendarskie* [The Referendary's books], I, 1582–1602, Warsaw, 1910, p. 130.

58. *Lustracja województwa mazowieckiego 1565* [*Lustracja* of the Mazovian voivodeship], pt. I, ed. I. Gieysztorowa and A. Żaboklicka, Warsaw, 1967, p. 119.

59. J. K. Haur, *Oekonomika ziemiańska generalna* [General management of landed estates], Warsaw, 1744, p. 271.
60. *Ibid.*, pp. 274-275.
61. See Karwicki's complaint in W. Konopczyński, *Polscy pisarze polityczni XVIII w.* [Polish 18th-century political writers], Warsaw, 1966, p. 45.
62. "Genuine information about the *łan* collected by M.J.F. Niegowiecki, Mathematician and Geometrician of the Cracow Academy" (MS *ca* 1758 or 1762), *BUW*, no. 99.
63. Ł. Górnicki, *Dzieje w Koronie Polskiej*, in *Dzieła wszystkie* [History of the Polish Kingdom, in *Collected Works*], III, Warsaw, 1886, pp. 223-224.

CHAPTER 7

1. An unusual example of a commodity that we do not sell by weight today being sold in that way long ago is silk cloth, the raw material in this case being more important than workmanship; see the 1573 Cracow list of prices for various good in J. U. Niemcewicz, *Zbiór pamiętników historycznych o dawnej Polszcze . . .* [Collected sources for the history of Poland], III, Warsaw, 1822, p. 471.
2. K. Sochaniewicz ("Miary i ceny produktów rolnych na Podolu" [Measures and prices of agricultural produce in Podolia], p. 164) states that scales were unknown in the countryside in the 16th century; similarly A. Radwański, "Jakie są przymioty dobrej wagi. Jak jej fałszywość poznawać i jak za pomocą wagi fałszywej dokładnie ważyc" [Qualities of true scales. How to recognise doctored scales and to weigh accurately with same], *Nowy Kalendarz Powszechny* [New Universal Calendar] for 1836, pp. 8-10, 58-60.
3. F. Bostel, "Taryfa cen dla województwa krakowskiego z r. 1565" [Price tariff for the Cracow *voivodeship*], Arch. Kom. Hist. *PAU*, VI, 1891, p. 301.
4. Quoted after Z. Gloger, *Encyklopedia staropolska* [Encyclopaedia of Old Poland], III, Warsaw, 1958, p. 206.
5. F. Bostel, *op. cit.*, p. 301.
6. *Contra*: A. Falniowska-Gradowska, "Miary zbożowe w województwie krakowskim w XVIII w." [Grain measures in the Cracow *voivodeship* in the 18th century], *Kwartalnik Historii Kultury Materialnej*, XIII, 1965, p. 670.
7. A. Wojciechowski, "Bednarstwo" [Cooperage], in J. Burszta, ed., *Kultura ludowa Wielkopolski* [Folk culture of Greater Poland], II, Poznań, 1964, pp. 505-516.
8. F. Bostel, *op. cit.*, p. 301.
9. *Volumina Legum . . .* , vol. VII, pp. 145-146.
10. J. Burszta, C. Łuczak, eds., *Inwentarze mieszczańskie z wieku XVIII z ksiąg miejskich i grodzkich Poznania* [18th-century burghers' inventories from municipal books of Poznań], 2 vols., Poznań, 1962 and 1965,

vol. I, 1700–1758, pp. 79, 199; vol. II, 1759–1793, pp. 16, 132, 139, 161, 163, 248.

11. AGAD, KRSW 7709, cols. 31-41, 58-64.

12. Police Superintendent of Warsaw to KRSW&P, 2/14 VI 1847, AGAD, KRSW 7711, cols. 36-37.

13. Capt. Coignet's reminiscences in F. et J. Fourastié, eds., *Les écrivains témoins du peuple*, Paris, 1964, pp. 149-150.

14. Charles A. Bourdot de Richebourg, *Nouveau Coutumier Général. Corps des Coutumes Générales et Particulières de France et des provinces connues sous le nom des Gaules . . . par avocat au Parlement*, à Paris 1724, 4 vols. in 8 fascicles, f. 1, p. 531 (hereafter *NCG*, with province).

15. *NCG*, Touraine, IV, f. 2, p. 643.

16. "huit pouces et deux lignes et demi de haut et dix pouces de diamètre" (*Encyclopédie ou dictionnaire raisonné . . .*, "Boisseau," II, p. 310). Père Mercenne in "Parisienses mensure" (pre-1648) gives an even flatter Parisian bushel: "9 pouces de diamètre et 8 pouces 5 lignes de hauteur," *Archives Parlementaires*, XI, p. 473.

17. A. Falniowska-Gradowska, "Miary zbożowe . . . ," p. 670.

18. T. Furtak, *Ceny w Gdańsku w l. 1701–1815* [Prices in Gdańsk from 1701 to 1815], Lvov, 1935, p. 39.

19. This was prohibited in 1764, and yet again in 1797, implying that the ban was ineffective, cf. Furtak, *op. cit.*, p. 40.

20. Dated 1 March 1759; quoted after A. Machabey, *La métrologie dans les musées de province et sa contribution à l'histoire des poids et des mesures en France depuis le treizième siècle*, Thèse pour le doctorat d'Université, Paris, Lettres, Bibl. de la Sorbonne, W. Univ. 1959 (29), 4°, p. 192.

21. *Lustr. wielkop.* 1564, I, p. 168 (Skwirzyna), *Lustr. lubel.* 1661, p. 144 (Kazimierz), p. 145 (Karczmiska), p. 149 (Wojszyn), etc. This evidence seems to rebut A. Falniowska-Gradowska's view ("Miary zbożowe . . . ," pp. 668-669) that the general practice was to pour the grain from shoulder height and that only demands that it be poured overhead amounted to abuse.

22. *Lustr. krak.* 1789, I, p. 190.

23. *Lustr. wielkop.* 1565, pp. 168, 269; *Lustr. wielkop.* 1628, pp. 7, 116; *Lustr. lubel.* 1661, *passim*.

24. *Lustr. sand.* 1564, pp. 236, 237. Striked measures, without compressing, are also specified in *Lustr. raw.* 17th century, pp. 154, 190.

25. B. Ulanowski, ed. *Księgi sądowe wiejskie* [Rural judicial books], Cracow, 1921, II, p. 677 (year: 1753), also p. 689.

26. See A. Keckowa and W. Pałucki, eds., *Księgi Referendarii Koronnej z drugiej połowy XVIII w.* [Crown Referendary Court's books, second half of the 18th century], Warsaw, 1957, II, p. 451 (hereafter *KRK*). Only hops were to be "compressed"; *Lustr. sand.* 1789, II, pp. 16, 20, 22.

27. St. Luke 6.38.
28. "Shaking down" by the buyer was regarded as dishonest, but by the seller as generous; see P. Goubert, *Beauvais et le Beauvaisis, de 1600 à 1730. Contribution à l'histoire sociale de la France du XVIIe siècle*, Paris, 1960, p. 181.
29. Dept. of Ind. and Arts to Dept. of Gen. Admin., KRSW&P, 21 IX/ 3 X 1846, AGAD, KRSW 7711, cols. 8-16.
30. K. Kurek, *Rady dla początkujących w praktyce gospodarskiej* . . . , [Advice for beginners in management . . .], Warsaw, 1841, p. 111.
31. For examples, see B. Ulanowski, ed., *Księgi sądowe*, I, p. 587; *Lustr. sand.* 1564, p. 104.
32. *Ibid.*, pp. 79, 258; *Lustr. krak.* 1564, pp. 40, 48, 49, 91, 94, etc.
33. *Lustr. krak.* 1565, p. 94.
34. A. Wyczański, *Studia nad gospodarką starostwa korczyńskiego 1500–1660* [Studies in the economy of the Korczyn *starostship*], Warsaw, 1964, pp. 113, 221-222.
35. A. Falniowska-Gradowska ("Miary zbożowe," p. 668) disregards this point and her inferences are unlikely to be correct; similarly K. Dobrowolski, *op. cit.*, p. 565.
36. AGAD, KRSW 7711, cols. 8-16.
37. For relevant correspondence, see AGAD, KRSW 7710 and 7711.
38. Woyda to the Minister of the Interior, 15 Sept. 1815, AGAD, KRSW 7709.
39. As late as 1858 the legal standard was to apply to grain dry measures of capacity provided that the quarter and certainly not the bushel was used, *Dz. Pr. KP* [Journal of laws of the kingdom of Poland], LI, p. 367ff.
40. This, indeed, was the usual practice; see A. Wyczański, *op. cit.*, p. 167.
41. Instructions for the skipper of the Zamość estates in 1800. *Instrukcje gospodarcze dla dóbr magnackich i szlacheckich z XVII–XIX w.*, [Economic guidelines for landed estates] (hereafter *Instr. gosp.*), II, Wrocław, 1963, p. 63.
42. *Ibid.*, vol. I, Wrocław, 1958, p. 493.
43. *Ibid.*, vol. II, p. 15.
44. *Ibid.*, vol. I, p. 239.
45. *Ibid.*, vol. I, p. 401.
46. *Ibid.*, vol. I, p. 485.
47. *Ibid.*, vol. II, p. 315.
48. *Ibid.*, vol. II, pp. 447, 451.
49. *Ibid.*, vol. I, p. 89.
50. *Ibid.*, vol. II, p. 330.
51. *Ibid.*, vol. I, p. 478.
52. *Ibid.*, vol. II, p. 15.
53. *Ibid.*, vol. II, pp. 45-46.

54. *Ibid.*, vol. II, pp. 303-304, 314-315.

55. *Ibid.*, vol. I, p. 493.

56. *Ibid.*, vol. I, p. 523

57. *Ibid.*, vol. I, p. 565. Cf. J. K. Haur, *op. cit.*, pp. 15, 48.

58. *Ibid.*, vol. I, p. 385.

59. L. Musioł, *op. cit.*, pp. 69-70.

60. *Ibid.*, p. 85.

61. *Ibid.*, p. 93.

62. *Ibid.*, pp. 18, 92, 64, and *passim.*

63. J. Putek, "Śląskie miary nasypowe w Polsce," in *Z dziejów wsi polskiej* [Silesian dry measures in Poland, in History of the Polish countryside] Warsaw, 1946, p. 38, and L. Musioł, *op. cit.*, p. 38.

64. *Ibid.*

65. *Ibid.*, p. 40.

66. M. Wolański, "Śląskie miary nasypne (zbożowe) w XVIII w." [Silesian dry measures of corn in the 18th century], *Zeszyty Naukowe Uniwersytetu Wrocławskiego*, ser. A, no. 13, 1959, p. 13.

67. L. Musioł, *op. cit.*, p. 84.

68. *Ibid.*, p. 99.

69. *Ibid.*, pp. 38-39.

70. M. Wolański, *op. cit.*, pp. 8, 10.

71. L. Musioł, *op. cit.*, p. 69.

72. *Ibid.*, p. 75.

73. *Ibid.*, pp. 66-67, 91.

74. *Ibid.*, p. 82.

75. *Ibid.*, pp. 67, 69-70, and *passim.*

76. *Ibid.*, p. 100.

77. *Ibid.*, p. 96.

78. *Ibid.*, p. 103.

79. *Ibid.*, p. 14.

80. *Ibid.*, p. 40.

81. *Ibid.*, pp. 105-106.

82. *Ibid.*, p. 46.

83. C. Biernat, *op. cit.*, pp. 195-233, and Z. Binerowski, "Gdańskie miary zbożowe w XVII i XVIII wieku," *Zapiski Historyczne. Kwartalnik Poświęcony Historii Pomorza*, XXIII, 1957, pp. 59-81.

84. C. Biernat, *op. cit.*, pp. 13-14.

85. Z. Binerowski, *op. cit.*, pp. 70-71, 75.

86. *Ibid.*, pp. 67, 74.

87. C. Biernat, *op. cit.*, pp. 4-5.

88. A. Falniowska-Gradowska, "Miary zbożowe," p. 669.

89. C. Biernat, *op. cit.*, p. 5.

90. *Ibid.*, pp. 14-15.

91. Instructions for the Zamość estates skipper in 1800; *Instr. gosp.*, II, p. 63.

92. C. Biernat, *op. cit.*, p. 7.

93. Z. Binerowski, *op. cit.*, pp. 73-74; C. Biernat, *op. cit.*, pp. 4-5, 7-10, 16-17, 20-21, 32.

94. C. Biernat, *op. cit.*, pp. 15-17.

95. *Archives Départementales* of *département* Doubs, ser. L, 1679 (hereafter AD Doubs).

96. *Archives Départementales* of *département* Meurthe-et-Moselle (hereafter AD M et M), ser. L, 1040, 1560, 1280, and 2579. This author's thesis that it is the answers to the questionnaire from the parishes that are of most interest, because those sent in to Paris by cantons or *départements* arbitrarily oversimplify issues, is confirmed by B. Gille, *Les sources statistiques de l'histoire de France. Des enquêtes du XVIIᵉ siècle à 1870*, Geneva, 1964, pp. 15-16 and *passim*.

97. In addition to the striked (*rase*) and heaped (*comble*) measures, there was the practice of measuring *grain sur bord*, i.e. applying the strickle lightly, with the grain topping the rim.

98. "Lettre du Secrétaire de l'Académie des Sciences à M. le Président de l'Assemblée Nationale, de 11 novembre 1790," *AD Charente* (Angoulême), L, 1317 (and in other archives).

99. R. Vivier, "Contribution à l'étude des anciennes mesures du Département d'Indre-et-Loire au XVIIᵉ et XVIIIᵉ siècles," *Revue d'Histoire Économique et Sociale*, XIV, 1926, p. 197.

100. R. Vivier, "L'application du système métrique dans le Département d'Indre-et-Loire. 1789-1815," *Revue d'Histoire Économique et Sociale*, XVI, 1928, p. 200.

101. H. van der Wee, *op. cit* pp. 90-94.

102. M. Luzzati, *op. cit.*, p. 213.

103. J. A. Manandian, "Rimsko-bizantiyskye khlebniye myery i osnovanniye na nikh indeksy khlebniykh tsen," *Vizantiyski Vremennik*, II (XXVII), 1949, p. 63.

104. AD Doubs (Besançon), L, 1679; mentioned in 1791 by parishes Vaucluse and Provenchères.

105 L. Musioł, *op. cit.*, p. 39.

106. M. J. Tits-Dieuaide, "La conversion des mesures anciennes en mesures métriques. Note sur les mesures à grain d'Anvers, Bruges, Bruxelles, Gand, Louvain, Malines et Ypres de XVᵉ au XIXᵉ siècle," *Contribution à l'Histoire Économique et Sociale*, II, 1963, pp. 31-89; similarly in Île de France, see *Annales Historiques de la Révolution Française*, no. 179, p. 87.

107. See Goubert, *op. cit.*, p. 374.

108. AD Doubs (Besançon), L, 1679.

109. AD Charente (Angoulême), L, 590; cf. A. Machabey, *op. cit.*, p. 151.

110. Under "Boisseau"; the point is also made in *Conservatoire Nationale des Arts et Métiers. Catalogue du Musée. Section K. Poids et mesures Métrologie*, Paris, 1941, p. 66.

111. G. Fourquin, *Les campagnes de la région parisienne à la fin du Moyen Age*, Paris, 1964, p. 53.
112. *Cdd.*, Rennes, III, p. 416.
113. R. Vivier, "L'application," p. 199.
114. P. Goubert, *op. cit.*, p. 396.
115. R. Vivier, "L'application," p. 200.
116. *Cdd.*, Angers (Le May), II, pp. 667-668.
117. *Cdd.*, Angoulême (ville d'Angoulême), pp. 121-122.
118. *Ibid.* (Chassin), p. 265.
119. AD M et M (Nancy), L. 1040, 655 and 2579.
120. Archives d'États de Genève (hereafter AEG), Archives du Département de Leman (hereafter ADL), 522.
121. Ministre de l'Intérieur au Préfet du Départ . . . , 9 germinal, year XIII (a circular to all départements), as in AEG ADL, 522.
122. AD Doubs, L, 882 and AD M et M, L, 655, 1040, 2579. Question no. 13 reads: "In your view, can capacity measures of corn be substituted [replaced] without difficulty by weighing it? Consider this question with the Conseil General of parishes holding major corn markets; consult farmers, merchants, and other knowledgeable citizens, and let us know the general preference."
123. The words "because of differences in the weight of grain" are crossed out here.
124. J. Richard, "La gréneterie de Bourgogne," p. 131.
125. I. Habib, *The Agrarian System of Mughal India, 1556–1707*, Aligarh Muslim University, Asia Publishing House, 1963, p. 374.

CHAPTER 8

1. N. M. Nicolai, *Memorie, leggi ed osservazioni sulla campagne e sull'Annona di Roma*, parte terza, Roma, 1808, p. 89.
2. M. Bogucka, "Z zagadnień spekulacji i nadużyć w handlu żywnością w Gdańsku w XV–XVII w." (On speculation and abuses in food trade in Gdańsk from the 15th to the 17th centuries), *Zapiski Historyczne* [Historical Records], 1962, XXVII, no. 1, 1962, p. 16.
3. The earliest instance known to the author of a municipal attempt to conceal rising prices by means of decreeing smaller loaves was in Lyons in 1357 (see J. Richard, *La gréneterie de Bourgogne*, p. 130), when the measure of wine was reduced, too.
4. M. Bogucka, *op. cit.*, p. 15.
5. In 1458 in Geneva a similar tariff allowed for fluctuations in corn prices within the range of 1:4; J. F. Bergier, *Commerce et politique du blé*, p. 534.
6. R. Baehrel, "Épidémie et terreur. Histoire et sociologie," *Annales*, XXIII, 1951, and "La haine de classe en temps d'épidémie," *Annales ESC*, 1952. For events in Gdańsk during the scarcity of 1560, see M. Bogucka, *op. cit.*

7. A. Magier, *op. cit.*, p. 206.
8. The "permanent" tariff of 1433 in Gdańsk (preceded by one in 1416) was yet to be followed by those of 1557 and 1560; M. Bogucka, *op. cit.*, pp. 15-16.
9. S. Hoszowski, *Ceny we Lwowie w XVI i XVII w.* [Prices in Lvov in the 16th and 17th centuries], pp. 309-310.
10. *Ibid.*, p. 31.
11. B. Ulanowski, *Kilka zabytków ustawodawstwa królewskiego i wojewodziń- skiego w przedmiocie handlu i ustanawiania cen* [A few examples of royal and voivodeship legislation pertaining to trade and price control] (documents from period 1459 to 1750), Arch. Kom. Prawn. *PAU*, I, 1895, pp. 37-144.
12. A. Chmiel, *Ustawy cen dla miasta Starej Warszawy od r. 1606 do r. 1627*, [Price regulations for Old Warsaw, 1606-1627], Arch. Kom. Hist. *PAU*, VII, 1894, pp. 55-268.
13. S. Hoszowski, *Ceny we Lwówie w latach 1701-1914* [Prices in Lvov, 1701-1914], Lvov, 1934, pp. 124-125.
14. W. Adamczyk, *Ceny w Lublinie od XVI do końca XVIII w.* [Prices in Lublin in the 16th to the end of the 18th century], Lvov, 1935, p. 12.
15. S. Hoszowski, *Ceny we Lwowie w XVI i XVII w.*, pp. 31-32.
16. Quoted after S. Siegel, *Ceny w Warszawie w latach 1701–1815* [Prices in Warsaw, 1701–1815], Lvov, 1936, pp. 27-28.
17. W. Kula, *Teoria ekonomiczna ustroju feudalnego. Próba modelu*, Warsaw, 1962 [English translation: *An Economic Theory of the Feudal System*, London, 1976—TRANS.].
18. For England, see F. Woodcock, "The Price of Provisions and Some Social Consequences in Worcestershire in the XVIIIth and XIXth Century," *Journal of the Royal Statistical Society*, CVI, 1943, pp. 268-272; for Germany (Frankfurt), the tariff in 1747, M. J. Elsas, *Umriss einer Geschichte der Preise und Löhne in Deutschland*, I, p. 7, Leiden, 1936; for France, P. Raveau, *Essai sur la situation économique et l'état social en Poitou au XVIᵉ siècle*, Paris, 1931, p. 58ff.; and for Italy, G. Lombardini, *Pane e denaro a Bassano tra il 1501 a il 1799*, Venice, 1963, pp. 39-40, 42.
19. S. Hoszowski, *Ceny we Lwowie w l. 1701–1914*, p. 121.
20. E.g. E. Tomaszewski, *op. cit.*, pp. 32-33.
21. S. Hoszowski, *Ceny we Lwowie w l. 1701–1914*, p. 116; E. Tomaszewski, *op. cit.*, p. 33; for another instance, in the private town of Zaleszczyki, see W. Kula, *Szkice o manufakturach w Polsce w XVIII w.* [Essays on Polish 18th-century Manufactories], Warsaw, 1956, p. 255.
22. Cf. the tariff of 1589, B. Ulanowski, *Kilka zabytków*, p. 99; the method was repeated in all later Cracow tariffs.
23. M. Bogucka, *op. cit.*, p. 20. Bogucka sees in this phenomenon the cumulative long-term effect of speculation, but this author feels that

the speculators owe their frequent success to underlying spontaneous market trends; in periods of inflation, prices of essential goods tend to rise by the largest margin, and in modern history it is the inflationary periods that prevail.

24. P. Vilar, *op. cit.*, p. 388.
25. M. Bogucka, *op. cit.*, p. 18.

Chapter 9

1. Z. Gloger, *op. cit.*, III, p. 89; however, K. Moszyński, *op. cit.*, p. 125, does not derive these terms from *kora*.
2. L. Musioł, *op. cit.*, p. 36.
3. K. Kluk, quoted by Z. Gloger, *op. cit.*, III, p. 206.
4. H. van der Wee, *op. cit.*, II, p. 67.
5. J. F. Bergier, *Commerce et politique du blé*, p. 540. See also the catalogue of the exhibition of old weights and measures at the Musée Toulouse-Lautrec in Albi (Musée Toulouse-Lautrec. Anciens Poids et mesures, offprint from *Revue du Tarn*, 42, 1966, p. 220). The parish of Marcillac Lanville in Angoumois on floréal the 6th, year VI, in answering the questionnaire from Paris, stated: "In the courtroom of the former Marcillac judiciary there is a *boisseau* divided into halves, quarters and measures. It is a block of stone, with the dimensions employed in the former Duchy cut into the surface" (AD Charente L, 188).
6. R. Serrure, *Catalogue de la collection de poids et de mesures, par . . .* , Musée Royal d'Antiquités et d'Armures, Brussels, 1883, p. 23. In Albi, too, there had once been a standard cut in stone; A. Machabey, *Histoire générale* II: *Les premières étapes du machinisme*, p. 325.
7. A. Falniowska-Gradowska, "Miary zbożowe" [Measures of Grain], p. 669.
8. L. Musioł, *op. cit.*, p. 29.
9. *Ibid.*, p. 36.
10. M. Wolański, *op. cit.*, p. 13.
11. Cdd., Angoulême, pp. 121-122, 277-278, 282, 292-293, 307.
12. *Conserv. Nat. Catalogue du Musée.*
13. The catalogue of the metrological collection has now been published.
14. J. A. Brutails, *op. cit.*
15. R. Vivier, "Contribution," p. 184.
16. G. Bigourdan, *Système métrique de poids et mesures. Son établissement et sa propagation graduelle, avec l'histoire des opérations qui ont servi à déterminer le mètre et le kilogramme, par . . .* , Paris, 1901.
17. B. Rybakov, *op. cit.*, p. 83; J. K. Zamzaris, *op. cit.*, pp. 183-184.

Chapter 10

1. Magellan's men, however, noted something like decimal grouping in the Moluccas; A. Pigafetta, *op. cit.*, p. 203.

2. The vigesimal system is discernible in the French *quatre-vingt* and until the 17th century *six-vingt*, etc., too.

3. The Russian word *sorok* points to a quadragesimal system, which was also used in Latvia (see J. K. Zamzaris, *op. cit.*, pp. 219-220) and in parts of ancient India (see I. Habib, *op. cit.*, p. 367).

4. The sexagesimal system is generally held to have been of Babylonian origin; it was also known in ancient Greece; see W. Ciężkowska-Marciniak, *O greckich i rzymskich wagach* [Greek and Roman Weights], and *Jak ważono w starożytności* [Weighing in the ancient world]; Polish *kopa* [a common commercial term for *sixty*—Trans.] points in the same direction.

5. K. Moszyński, *op. cit.*, pp. 99, 104.

6. *Ibid.*, p. 100.

7. *Ibid.*, p. 101.

8. Archivio di Stato di Milano (hereafter ASM), Commercio, Seria nuova, cart. 229.

9. Niangoran-Bouah, *op. cit.*, mentions the following binary system—employing set prime numbers as points of departure—which operated in pre-colonial West Africa: 1-7-14-28-56-112, or 1-5-10-20-40, or 1-11-22-44-88, or 1-13-26-52, etc.

10. E.g. in Latvia, see J. K. Zamzaris, *op. cit.*, pp. 217-218.

11. *Ibid.*, pp. 218-219.

12. *Ibid.*

13. *Ibid.*, p. 220.

14. *Ibid.*, p. 215.

15. *Ibid.*, pp. 198-199.

16. J. Savary de Bruslons, *Le parfait négociant*, 1742.

17. K. Dobrowolski, *op. cit.*, p. 556; B. Rybakov, *op. cit.*, *passim*.

18. E.g. A. Plaisse, *La baronnie du Neubourg. Essai d'histoire agraire, économique et sociale*, Paris, 1961, pp. LIX, 158. In 1473, the *boisseau* in *les halles* of Beaumont-le-Roger contained 10 *pots*, and in the 18th century, 18, the size of the pots being unchanged. The *boisseau* of Argues in Normandy contained 10, then 11⅔, 12, 14, and finally 16 *pots* of constant size.

19. R. Mauny, *op. cit.*, pp. 410-419.

20. "Riduttione a sia tariffa delle diversità della misure e pesi antichi delle città, terre e luoghi delli stati di quà de' Monti del Serenissimo Carlo Emanuelle Duca di Savoia, alla equalità delle misure e pesi de S. Altezza Sereniss. nuovamente stabiliti," in Torino, apresso a Luigi Pizzamiglio, Stampator Ducale, 1613.

21. AD M et M, p. 1280.

22. AEG, ADL, p. 522.

23. B. A. Rybakov, *op. cit.*, pp. 88-91 and *passim*.

24. T. Zagrodzki, "Regularny plan miasta średniowiecznego a limitacja miernicza" [The standard plan of the medieval town and the sur-

veying limitation], *Studia Wczesnośredniowieczne* [Studies in the early Middle Ages], V, 1962, p. 1.

25. B. Baranowski, "Pośmiertna kara . . . ," p. 79.
26. M. Rojas, *Syn złodzieja*, Warsaw, 1965, p. 159 [Orig. *Hijo de Ladrón*, 1951; English translation: *Born Guilty*, 1955—TRANS.].

CHAPTER 11

1. Archivio di Stato di Torino (hereafter AST), I, *Pesi e misure*, 2.
2. H. van der Wee, *op. cit.*, p. 95.
3. Téron Ainé, *Instruction sur le système de mesures et poids uniformes pour toute la République Française . . . par . . . teneur de livres et maître d'arithmétique*, à Genève, An X, 10 prairial, p. 2.
4. A. Machabey, *op. cit.*, p. 160.
5. H. van der Wee, *op. cit.*, p. 77.
6. J. K. Zamzaris, *op. cit.*, pp. 180, 192.
7. S. Górzyński, *op. cit.*, p. 20.
8. B. A. Rybakov, *op. cit.*, p. 69 (one ell to measure cloth by and a different one for linen).
9. I. Habib, *op. cit.*, p. 376
10. *World Weights and Measures.*

CHAPTER 12

1. A study pioneered by Budé, *De Asse*, 1513. Early in the 17th century Nicolas Claude Fabri de Peiresc evinced much interest in ancient measures; see Ph. Tamizey de Barroque, ed., *Lettres de Peiresc*, 7 vols., Paris, 1888–1898, *passim*.
2. E.g. see "Poids," *Encyclopédie ou dictionnaire raisoné* . . . , à Neufchâtel, 1765, XII, pp. 849-862; cf. P. Hazard, *La crise de la conscience européenne, 1680–1715*, Paris, 1935, p. 40.
3. See Biron, "Rapport fait à la Société de Médecine de Paris, séante au Louvre. Séances des 21 et 27 pluviôse an X. Sur l'application des nouveaux poids et mesures dans les usages de la médecine, par les citoyens . . . ," Brasdor et Pelletier, publié par l'ordre de Ministre de l'Intérieur, à Paris, de l'imprimerie des sourds-muets rue de Faubourgs Saint-Jacques No. 115, an X, p. 32; found by the author with other medical writings, in Paris, Mazarine Library, côte 15.575 B.
4. [Paucton], *Métrologie ou traité des mesures, poids et monnaies des anciens peuples et des modernes*, à Paris, chez la veuve Desaint, libraire, rue du Foin, 1780, avec approbation et privilège du Roi, pp. XVI, 955.
5. Babeuf's work as *feudiste* resulted in two small publications: "Mémoire peut-être important pour les propriétaires de terres et de seigneuries ou idées sur la manutention des fiefs," 1786, and "L'Archiviste-Terriste, ou traité méthodique de l'arrangement des archives seigneuriales et de la confection et perpétuation successive des inventaires,

des titres et des terriers d'icelles, des plans domaniaux, féodaux et censuels," 1786; unfortunately, neither work has been available to me.

6. M. Dębski *Kalendarz Polski i Ruski na rok 1794* [Polish and Russian Calendar for 1794], Cracow University.

7. T. Czacki, *O litewskich i polskich prawach* [Of Lithuanian and Polish Laws], I, Warsaw, 1800, pp. 223, 289-291.

8. J. Michalski, *Z dziejów Towarzystwa Przyjaciół Nauk* [From the Past of the Society of the Friends of Learning], Warsaw, 1953, p. 148.

9. A. Sapieha, "Tablice stosunku nowych miar i wag francuskich z litewskimi i polskimi miarami i wagami," *Roczniki Tow. Warsz. Przyj. Nauk* [Tables of relations of new French weights and measures to Lithuanian and Polish weights and measures, Yearbooks of the Warsaw Society of Friends of Learning], Warsaw, 1802.

10. A. Chodkiewicz, *Tablice stosunku dawnych miar i wag francuskich i koronno-litewsko-polskich z miarami i wagami nowymi a przyjętemi we Francji,* [Tables of relations of old French to Polish-Lithuanian measures and of newly introduced French weights and measures], Warsaw, 1811.

11. J. Colberg, *Porównanie teraźniejszych i dawniejszych miar i wag w Królestwie Polskim używanych* [Comparison of old and current weights and measures used in the Kingdom of Poland], Warsaw, 1819; 2d ed., enlarged by W. Kolberg, 1838.

12. W. A. Maciejowski, "Historia dawnych polskich miar i wag w zarysie od czasów najdawniejszych aż do końca XVIII w. przedstawiona" [An outline history of Polish weights and measures from earliest times until the end of the 18th century], *Ekonomista*, 1868, 2.

13. Lubomirski, in *Encyklopedia Rolnicza* [Agricultural Encyclopaedia], IV, Warsaw, 1876.

14. M. A. Baraniecki, "O miarach prawnych i zwyczajowych w Polsce" [Statutory and customary measures in Poland], *Wszechświat*, II, 1883.

15. F. Piekosiński, "O łanach w Polsce wieków średnich" [The *łans* in medieval Poland], *Rozprawy AU*, XXI, 1882.

16. E. Stamm, "Miary długości w dawnej Polsce" [Measures of length in old Poland], *Wiadomości Służby Geograficznej*, no. 3, 1935; *Staropolskie miary* [Old Polish measures], I: *Miary długości i powierzchni* [Measures of length and area], Warsaw, 1938; and *Miary powierzchni* [Measures of area].

Chapter 13

1. For the measuring of cornfields in ancient Egypt, see Niemann, *Von Altägyptischer Technik, Beiträge zur Geschichte der Technik und Industrie*, Berlin, 1930, p. 100.

2. E.g. J. Köbel, *Geometry*, Franckfort am Mayn, 1584; S. Solski, *Geometra polski* [The Polish Surveyor], Cracow, 1683.

3. I have seen fine collections in Sweden at the Nordiska Museet of Stockholm and in the Uppsala Museum. For a drawing and description of the Polish ell of 1651, see Z. Gloger, *op. cit.*, III, p. 163.
4. The Warsaw Museum of Weights and Measures.
5. R. Serrure, *Catalogue de la collection de poids et de mesures*, Brussels, 1883. P. Burguburu, *Catalogue de poids anciens, Ville de Bordeaux*, Bordeaux, 1936; *Conserv. Nat. Catalogue du Musée*; for provincial France, see A. Machabey, *La métrologie*.
6. Hence scales rather than weights are often to be seen in museums; the former were still of value in use after a reform, whereas the latter were not and they were consigned to destruction.
7. B. A. Rybakov, *op. cit.* The same applied to old icons, i.e. the dimenions of the space the architect would allocate to them.
8. B. A. Rybakov, *op. cit.*, p. 68ff.
9. E.g. A. Falniowska-Gradowska, "Miary zbożowe" [Measures of grain], p. 666, utilizes a number of manuals of arithmetic, especially those published by the Commission for National Education.
10. J. A. Manandian, *op. cit.*, pp. 60-62.
11. Both were printed in *Trattato della Decima Fiorentina*, 4 vols. Lucca, 1756.
12. E. I. Kamyentseva, N. V. Ustiugov, *Russkaya Metrologiya*, Moscow, 1965, p. 10.
13. S. Grzepski, *op. cit.*, an annotated edition by H. Barycz and K. Sawicki, Wrocław, 1957.
14. H. van der Wee, *op. cit.*, p. 71.
15. L. V. Čerepnin, *Russkaya metrologiya. Utchebnyie posobya po vspomagatielnim istoričeskim disciplinam*, ed. A. I. Andreyev, Moscow, 1944, pp. 89-91, mentions several medieval Russian examples; and cf. V. L. Yanin in *Ocherki istorii istoricheskoi nauki v SSSR*, II, Moscow, 1960, pp. 678-680.
16. W. Beveridge, "A Statistical Crime of the 16th century," *Journal of Economic and Business History*. The "crime" was the concealment of change in a measure, which distorted the statistics of production and prices.
17. A. Gilewicz, "Studia z dziejów miar i wag w Polsce, part I: Miary pojemności i ciężaru (wagi)" [Studies in the history of weights and measures in Poland, Measures of capacity and weight], *Sprawozdania Towarzystwa Naukowego we Lwowie* [Proceedings of the Lvov Scientific Society], XVI, 1936, 3, p. 3.
18. S. Śreniowski, *op. cit.*
19. L. V. Čerepnin, *op. cit.*, p. 9.
20. M. Bloch, "Le témoignage des mesures agraires," *Annales d'Histoire Économique et Sociale*, VI, 1934.

CHAPTER 14

1. Wholesale measures sometimes differed from retail ones so as to allow for the wholesaler's profit; see M. J. Elsas, *op. cit.*, IIA, pp. 26, 45-46.

2. Pague de Fauchat, a Toulouse lawyer, writing on 3 February 1815 states: "qu'il se pratique dans ce faubourg deux espèces de mesurage, qu'on appelle d'entrée et de sortie du magasin, ou le particulier perd environ quatre hectolitres sur la vente de cente septiers de blé"; Archives Nationales (hereafter AN), F¹², 1289.

3. V. I. Šunkov, "Mery sypnykh tel v Sybiri XVII v.," in *Akademiku B. D. Grekovu ko dnyu semidyesatiletya. Sbornik Statey*, Moscow, 1952, p. 168.

4. A. Gilewicz, "Wczesnosłowiańskie miary i wagi," in *Słownik Starożytności Słowiańskich* [Early Slavonic Weights and Measures, in The Dictionary of Slavonic Antiquities], pp. v, 209.

5. M. J. Tits-Dieuaide, *op. cit.*, pp. 40-41.

6. See several petitions from peasants or country townsmen in AGAD, *Arch. Kamer.*, vol. 441, I. 9, and in *Uwagi praktyczne o poddanych polskich* [Practical reflections on Polish serfs], Warsaw, 1790, p. 132; cf. W. Kula, *Szkice o manufakturach*, [Essays on the manufactories], p. 359.

7. R. Mauny, *op. cit.*

8. A. Pigafetta. *op. cit.*, p. 182.

9. H. H. Wächter, "Ostpreussische Domänenwerke im 16 und 17 Jahrhundert," *Beihefte zum Jahrbuch der Albertus Universität*, Königsberg Pr. Würzburg, 1958, and the address by A. Gilewicz to the Economic History Section of the eighth General Historical Congress, Cracow, 1958.

10. A. Keckova, *op. cit.*, p. 624.

11. J. K. Zamzaris, *op. cit.*, p. 214.

12. H. van der Wee, *op. cit.*, p. 65 gives an example of the importer increasing the weight of a sack of wool.

13. M. Luzzati, *op. cit.*, p. 200; J. Heers, *Le livre de comptes de Giovanni Piccamiglio, homme d'affaires génois (1456-1459)*, Paris, 1959, p. 16; J. K. Zamzaris, *op. cit.*, p. 187; A. Falniowska-Gradowska, "Miary zbożowe" [Measures of grain], p. 679, gives instances of local measures being ousted by the spreading use of measures from large towns.

14. R. Vivier, "Contribution," p. 196.

15. Cdd., Sens, pp. 177-178.

16. Historians of measures (e.g. Falniowska-Gradowska, *op. cit.*, p. 666, or J. Putek, *op. cit.*, p. 37), in the absence of unequivocal sources, sometimes try to establish relations between measures of different towns by reference to prices; if the tendency discussed here is marked, this may result in error, since the same price of grain in the major centers of the surplus area and of the "import" area respectively may have been brought about by differences in measures.

17. Thus Falniowska-Gradowska, *op. cit.*, pp. 682-687, and in her *Świad-*

czenia poddanych na rzecz dworu w królewszczyznach województwa kra-kowskiego w drugiej połowie XVIII w. [The serfs' dues to the manor in the Crown lands of the Cracow *voivodeship* in the second half of the 18th century], Wrocław, 1964, p. 143.

18. S. Mielczarski, *Rynek zbożowy na ziemiach polskich w drugiej połowie XVI i pierwszej połowie XVIII wieku. Próba rejonizacji* [The grain market in Poland in late 16th and early 17th centuries; the quest for region-alization], Gdańsk, 1962, p. 78.
19. Examples cited after Mielczarski, *op. cit.*, pp. 81-83.
20. Cited after Mielczarski, *op. cit.*, p. 78.
21. *Lustracje rawskie, 1564, 1570*, p. 24 and *passim; Lustracje płockie, passim.*
22. H. H. Wächter, *op. cit.*, pp. III-IV (Anhang), and the map at p. IV.
23. *Lustracje podlaskie, 1570, passim.*
24. *Lustracje wielkopolskie, 1565*, p. 238.
25. K. Sochaniewicz, *Miary i ceny produktów rolnych* [Measures and prices of agricultural produce], p. 157.
26. Niangoran-Bouah, *op. cit.*

CHAPTER 15

1. *Annales*, VI, 1934, p. 280.
2. H. Navel, *op. cit.*, is inclined to take this view, which is also emphasized by A. Machabey, *Histoire générale des techniques*, II: *Les premières étapes du machinisme*, p. 313.
3. H. Navel, *op. cit.*
4. *Ibid.*: he found marked on the wall of a castle in Normandy the foot of 1589 which was approximately the same as the foot obligatory from 1667.
5. A. Gilewicz, *Studia z dziejów miar* [Studies in the history of measures], p. 5.
6. C. Verlinden, *Dokumenten voor de Geschiedenis van Prijzen en Louen* . . . , Bruges, 1959, p. 3. For the debate on the Belgian municipal measures, see M. J. Tits-Dieuaide, *op. cit.*, pp. 33, 57, 60, 64.
7. G. Parenti, *Prime ricerche sulla rivoluzione dei prezzi in Firenze*, Florence, 1939, p. 73; and M. Luzzati, *op. cit.*, pp. 210-211.
8. Z. Binerowski, *op. cit.*, p. 64.
9. L. Musioł, *op. cit.*, p. 12.
10. *Ibid.*, p. 57.
11. Y. Bezard, *La vie rurale dans le sud de la région parisienne de 1450 à 1560*, Paris, 1929, p. 36.
12. G. Fourquin, *op. cit.*, p. 52.
13. A. Machabey, *La métrologie*, p. 427.
14. R. Mauny, *op. cit.*
15. M. Legendre, *Survivance des mesures traditionelles en Tunisie*, Paris, 1958.

16. G. S. Aurora," Economy of a Tribal Village," *Economic Weekly*, 26 Sept. 1964.

17. *Conserv. Nat. Catalogue du Musée*, p. 17.

18. H. van der Wee, *op. cit.*, II, pp. 67, 75, 80, 87-89.

CHAPTER 16

1. H. Van der Wee, *op.cit.*, II, P. 65.

2. *Ibid.*, p. 95.

3. *Ibid.*, p. 73.

4. Hence difficulties in standardizing measures of cloth; *ibid.*, p. 95.

5. R. Mauny, *op.cit.*

6. H. van der Wee, *op.cit.*, II, p. 68; J. W. Thompson, *op.cit.*, II, pp. 596-597.

7. S. Hoszowski, *op.cit.*, p. 59.

8. W. Adamczyk, *Ceny w Warszawie w XVI i XVII w.* [Prices in Warsaw in the 16th and 17th centuries], Lvov, 1938, p. 29.

9. This was usual in Poland and took place in France too; see J. A. Brutails, *op.cit.*, P. 6.

10. For medieval attempts to unify measures in France, England, and Germany, see J. W. Thompson, *op.cit.*, II, pp. 596-597.

11. C. G. Coulton, *Medieval Village, Manor and Monastery*, New York, 1960, p. 45.

12. It was no mere coincidence that the *Geometria Culmensis* appeared at that time; *Ein agronomischer Traktat aus der Zeit des Hochmeister Ulrich von Jungingen*, ed. H. Mendthal, Leipzig, 1886.

13. J. L. Zamzaris, *op.cit.*, p. 187.

14. L. Musioł, *op.cit.*, p. 46.

15. E. Hamilton, *American Treasure and the Price Revolution in Spain, 1501–1650*, Cambridge, Mass., 1934, pp. 153-158.

16. *Ibid.*, p. 165. For Portuguese efforts to standardize measures in Brazil, see F. Mauro, *Le Portugal et l'Atlantique au XVIIᵉ siècle, 1570–1670*, Paris, 1960, pp. 225, 228, 264; and for attempts to establish the Paris weights and measures in the French Antilles, see J. Rennard, *Baas, Blénac ou les Antilles françaises au XVIIᵉ siècle*, Fort-de-France, 1935.

17. *Ibid.*, p. 159. For similar problems in Siberia, see V. I. Šunkov, *op.cit.*, p. 167. For the disappearance of the municipal standards of Niort with the departure of the persecuted Huguenots, see A. Machabey, *Revue de Métrologie Pratique et Légale*, 1952, no. 4, pp. 165-180.

18. For a fuller discussion, see chapter 21, below.

19. *Instruzione sulle misure e sui pesi che si usano nella Repubblica Cisalpina, publicata per ordine del Comitato Governativo*, Milano, anno X, 1801.

20. *Riduttione o sia tariffa delle diversità delle misure . . .*

21. Cited after S. G. Strumilin, *op.cit.*, p. 8.

22. V. I. Šunkov, *op.cit.*, p. 106. For a modern attempt to standardize

measures in Latvia in order to standardize the dues paid by peasants, see J. K. Zamzaris, *op.cit.*, p. 190.

23. S. Hoszowski, *Ceny we Lwowie w XVI i XVII w.* [Prices in Lvov in the 16th and 17th centuries], p. 60.

24. M. J. Tits-Dieuaide, *op.cit.*, pp. 65-66.

25. L. Musioł, *op.cit.*, p. 47. M. Wolański, *op.cit.*, pp. 11-12.

26. A. Huščava, "A pozsonyi méró fejlódése az 1588. Évi törvényes ren-dezésig" (after French summary, *Evolution du muid, (mérö) de Bratislava jusqu'aux dispositions légales de 1588*), in *Fejezesele a magyar mérésiigy torteneseböl*, Budapest, 1959, pp. 25-44, and his "Metrologický vyvin v obdobi rozkladu feudalneho zriadenia na Slovensku v r. 1790–1848," *Historické Stúdie*, VII, 1961, pp. 189-200; A. Paulinyi, "Soupis mer z r. 1775, a zavedenie Yednotney mery uhla v r. 1776–1780 v stredoslovenskej banskey, oblasti," *Historické Stúdie* (Bratislava), XI, 1966, pp. 263-281.

27. A. Huščava, *Metrologický vyvin* . . .

28. L. Musioł, *op.cit.*, pp. 48-50, 57.

29. *Sto let gosudarstvennoi sluzby mer i vesov v SSSR*, Komitet po delam mer i izmiritelnych priborov v SSSR, Moscow, 1945, pp. 21-29. L. V. Čerepnin, *op.cit.*, pp. 76-84. *Programma kursa "Russkaia metrologia" dla Moskovskogo gosudarstvennogo istoriko-arkhivnogo Instituta*, Moscow, 1954, p. 7.

30. M. Luzzati, *op.cit.*, pp. 217-218.

CHAPTER 17

1. A. Favre, *Les origines du système métrique*, Paris, 1951; L. D. Isakov, *Na vse vremena, dla vsekh narodov. Očerki po istorii metričeskoi sistemy*, Pet-rograd, 1923; cf. Also K. Kowalczewska, "Historia systemu metry-cznego (1791-1921)" [History of the Metric System], and W. Kas-perowicz, "Metryczny system miar" [Metric System of Measures], both in *Przegląd Techniczny* [Technical Review], special issue, "System me-tryczny miar. Stotrzydziestolecie, 1791-1921" [One Hundred and Thirty Years of the Metric System], Warsaw, 1921.

2. Z. Binerowski, *op.cit.*, M. Wolański, *op.cit.*

3. J. P. Eckermann, *Rozmowy z Goethem* [Conversations with Goethe], II, Polish translation, Warsaw 1960, p. 68.

4. The metric system was seen thus by Mendeleyev at the first congress of Russian naturalists in Petersburg in 1867; see his *Sotchinenya*, XXII, p. 27, Moscow, 1950.

5. Throughout this book the antithesis between pre-metric "represen-tational" measures and metric measures of convention is consistently maintained. It is unfortunate that the term "conventional" is applied by some writers to the former measures, implying some sort of a primeval "social contract"; there had been, of course, no such con-tract, and pre-metric measures were by no means fixed "arbitrarily,"

313

as H. T. Wade would have it (in "Weights and Measures," *Encyclopaedia of Social Sciences*, XV, pp. 389-392, New York, 1949), but determined by the needs of life and work. It was the basic unit of the metric system that was largely arbitrary and wholly dependent on convention.

6. K. Marx, *Kapitał*, I, Warsaw, 1951, p. 76ff.

CHAPTER 18

1. E.g. at the towns of Skwirzyna and Pobiedziska and the village of Krasne in Greater Poland; *Lustr. 1565*, I, pp. 168, 269, and 44, respectively; similarly in the Lublin region in the town of Kazimierz and villages of Karczmiska and Woyszyn, *Lustr. 1661*, pp. 144, 145, and 149.

2. E.g. in the Sandomierz region in the villages of Myronyce and Byelany, *Lustr. 1564*, pp. 236, 237; Similarly in the Rawa region, villages of Krobino and Jaktorów, pp. 190, 154.

3. Greater Poland *Lustr.*, II, p. 161.

4. Sandomierz *Lust. 1564*, p. 104 (villages of Wolya Mała and Wolya Wielka).

5. As at the village of Wargoczyn, Sandomierz *Lustr. 1564*, pp. 222-223.

6. S. Kutrzeba, ed., *Materiały do dziejów robocizny w Polsce w XVI wieku* [Source materials for the history of the *corvée* in Poland in the 16th century], 1913.

7. S. Kuraś, ed., *Ordynacje i ustawy wiejskie z archiwów metropolitarnego i kapitulnego w Krakowie, 1451–1689* [Rural ordinances and prescripts in the metropolitan and capitular archives in Cracow], Cracow, 1960, p. 42.

8. See I. T. Baranowski, ed., *Księgi referendarskie* [Records of the Referendary's Court], I, 1582–1602, Warsaw, 1910, p. 19.

9. Rawa *Lustr. 1661*, pp. 154, 190.

10. *Ibid.*, pp. 167-168 (the town of Mszczonów).

11. As at Rokitno, Greater Poland *Lustr. 1628*, I, p. 22, or at Pyzdry, *ibid.*, p. 161.

12. *Ibid.*, p. 22.

13. *Ibid.*, p. 161.

14. Z. Ćwiek, *Z dziejów wsi koronnej w XVII wieku* [Crown villages in the 17th century], Warsaw, 1966, p. 165.

15. *KRK* [Records of the Crown of Referendary's Court], I, p. 320; the year was 1777; cf. J. Leskiewicz and J. Michalski, eds., *Supliki chłopskie XVIII w. z archiwum prymasa Michała Poniatowskiego* [Peasants' supplications in the archives of primate M. P.], Warsaw, 1954, pp. 151, 154, 156.

16. *KRK*, II, p. 265.

17. *Ibid.*, I, p. 258.

18. *Ibid.*, II, p. 231.

19. *Ibid.*, I, p. 447.
20. *Supliki chłopskie*, p. 319.
21. *Ibid.*, p. 334.
22. *Ibid.*, p. 450.
23. *Ibid.*, p. 457.
24. Cracow *Lustr. 1789*, I, pp. 130, 131-132.
25. *Ibid.*, p. 188.
26. This, as well as Ćwiek, *op.cit.*, and the Referendary's decrees of 1582 and 1602 (Baranowski, *op.cit.*, pp. 114, 128) show that Falniowska-Gradowska's view in her "Miary zbożowe" [Measures of grain] is incorrect and that undoubtedly pouring grain from shoulder height was deemed unfair.
27. Cracow *Lustr. 1789*, I, p. 190.
28. *Ibid.*, pp. 196-197.
29. Sandomierz *Lustr. 1789*, II, pp. 96-97, 106.
30. *Ibid.*, p. 100.
31. *Ibid.*, p. 143.
32. *Ibid.*, p. 162.
33. *Supliki chłopskie*, p. 496.
34. 1787; *KRK*, II, pp. 619-620.
35. 1784; *ibid.*, p. 407.
36. 1777; *ibid.*, I, p. 315.
37. *Supliki chłopskie*, pp. 308, 311.
38. Sandomierz *Lustr. 1789*, II, p. 97.
39. *Ibid.*, p. 142.
40. *Supliki chłopskie*, p. 23, 72.
41. *Ibid.*, p. 76. For the inspector accepting a similar complaint from the villagers at Kompin, *ibid.*, pp. 92, 94.
42. *Ibid.*, pp. 245, 210.
43. Examples in G. Labuda, ed., *Inwentarze starostw puckiego i kościerskiego z XVII w.* [Inventories of the Kościerz and Puck districts in the 17th century], *Fontes*, Toruń, 1954; S. Pawlik, ed., *Polskie instruktarze ekonomiczne z końca XVII i z XVIII w.* [Polish manuals of husbandry in late 17th and 18th centuries], I, Cracow, 1915, p. 263.
44. *KRK*, II, pp. 677, 707.
45. 1785; *ibid.*, II, p. 460.
46. *Supliki chłopskie*, p. 312.
47. *Ibid.*, pp. 334, 396.
48. *Ibid.*, p. 466.
49. *Ibid.*, pp. 373, 375, And elsewhere; also, *Uwagi praktyczne o poddanych polskich względem ich wolności i niewoli* [Practical comments relating to the freedom and servitude of Polish serfs], Warsaw, 1790, p. 116.
50. E.g. Sandomierz *Lustr. 1789*, II, pp. 19, 66, 71, 76.
51. *Supliki chłopskie*, p. 325.
52. *Ibid.*, p. 375.

53. 1786; S. Pawlik, ed., *op.cit.*, I, p. 276.

54. 1785; *KRK*, II, p. 475.

55. For examples, see S. Kutrzeba and A. Mańkowski, eds., *Polskie ustawy wiejskie XV–XVIII w.* [Polish rural statutes from the 15th to the 18th century), XI, 1938, p. 401; S. Pawlik, ed., *op.cit.*, I, pp. 229, 283; W. Konopczyński, ed., *Diariusze sejmowe z XVIII w.* [18th-century Seym daily records], Warsaw, 1937, III, p. 70; and B. Baranowski et al., eds., *Instrukcje gospodarcze . . . z XVII–XIX w.* [Estate management manuals . . . of the 17th to the 19th century], Wrocław, 1959–1963, I, pp. 113, 365ff.

56. S. Pawlik, ed., *op.cit.*, I, pp. 284-285.

57. *Ibid.*, I, p. 114.

58. For a dismissal of an innkeeper for using false measures, see B. Ulanowski, ed., *Księgi sądowe* [Judicial books], II, p. 36.

59. J. Osiński, *Opisanie polskich żelaza fabryk . . .* [A description of Polish ironworks], Warsaw, 1782, p. 77.

60. B. Ulanowski, ed., *op.cit.*, I, p. 304.

61. *Ibid.*, pp. 764-765.

62. Of many source examples, see A. Vetulani, ed., *Księgi sądowe wiejskie klucza łąckiego* [Rural judicial books of the Łącz complex of estates], I, 1526-1730, Wrocław, 1962, p. 212.

63. A. Wyczański, *op.cit.*, p. 122.

64. Cracow *Lustratio 1789*, I, pp. 129-135.

65. B. Baranowski et al., eds., *op.cit.*, I, p. 89.

66. *Ibid.*, II, p. 330.

67. *Ibid.*, I, p. 485.

68. *Ibid.*, I, p. 493.

69. *Ibid.*, I, p. 667.

70. *Ibid.*, II, p. 15.

71. *Ibid.*, I, p. 386.

72. *Ibid.*, I, p. 405.

73. *Ibid.*, II, p. 14.

74. A. Gostomski, *Gospodarstwo (1588)* [The landed estate], ed. S. Inglot, Wrocław, 1951, pp. 106-108.

75. B. Baranowski et al., eds., *op.cit.*, I, pp. 523, 534, 535, 550, 596, 602, 605, 606, 621, 667.

76. A. Jabłonowska, *Ustawy powszechne dla rządców dóbr moich . . .* [General guidelines for the administrators of my estates . . .], Siemiatycze, 1785, Warsaw, 1786 and 1787; B. Baranowski et al., eds., *op.cit.*, I, pp. 113, 365, and the request to the steward at Zwierzyniec dated 11 May 1798 in vol. II.

77. A. Jabłonowska, *op.cit.*, VI, p. 9.

78. B. Baranowski et al., eds. *op.cit.*, I, pp. 447, 478; vol. II, pp. 18, 303-304, 314-315.

79. J. Leskiewiczowa, "W sprawie publikacji instruktarzy gospodarczych"

[On the publication of manuals of husbandry], *Kwartalnik Historii Kultury Materialnej*, IX, 1961, p. 807.

80. B. Baranowski et al., eds., *op.cit.*, I, pp. 365-366.
81. A. Gostomski, *op.cit.*, p. 112 (italics added—W.K.).

Chapter 19

1. W. Kula, *Teoria ekonomiczna* ... , pp. 90-93; [English version: *An Economic Theory of the Feudal System*, London, 1976—Trans.].
2. *Volumina legum* ... *1732-1782*, II, pp. 49-50; (ed. J. Ohryzko, Petersburg, 1859).
3. Proclamation by the deputy *voivode* of Przemyśl, dated 30 March 1666, in *Akta grodzkie i ziemskie z czasów Rzeczypospolitej Polskiej z archiwum t.zw. bernardyńskiego we Lwowie* [Town and country records of the Polish Republic, from the so-called Bernardine archives in Lvov], 25 vols., Lvov, 1868–1935, XXI, pp. 500-501; hereafter cited as *AGZ*.
4. *Lauda sejmików w Sądowej Wiszni* [Local laws passed by the dietines at Sądowa Wisznia], 6 June 1667, *AGZ*, XXI, p. 453; hereafter cited as *Laudum wisz.*
5. Przemyśl vehmgericht, 16 February 1669, *AGZ*, XXI, pp. 492-496.
6. Regulation of the vehmgerichts at Przemyśl, 15 January 1674, *AGZ*, XXII, p. 14.
7. *Laudum wisz.*, 16 August 1674, *ibid.*, p. 27.
8. *Ibid.*, 1 June 1677, p. 86.
9. *Ibid.*, 24 May 1683, pp. 163-164.
10. *Ibid.*, 10 July 1685, p. 176.
11. *Ibid.*, p. 183.
12. *Ibid.*, 16 December 1687, p. 202, similarly pp. 214-215, and yet again in 1689 (5 December), p. 229.
13. E.g. *AGZ*, XX, p. 415 (1638), and XXI, pp. 492-496 (1669).
14. *Ibid.*, XXII, p. 647.
15. Laudum wisz., 19 September 1713, *AGZ*, XXII, pp. 532-533.
16. *Laudum sejmiku lwowskiego gospodarskiego* [Lvov laudum], 6 November 1764, in *AGZ*, XXIII, p. 486.
17. *Laudum wisz.*, 16 August 1674, *AGZ*, XXII, p. 27.
18. *Laudum wisz.*, 1 June 1677, *ibid.*, p. 86.
19. *Laudum wisz.*, 24 May 1683, *ibid.*, pp. 163-164.
20. *Laudum wisz.*, 19 November 1693, *ibid.*, p. 268.
21. *Laudum wisz.*, 28 March 1707, *ibid.*, p. 420.
22. *Laudum wisz.*, 20 May 1707, *ibid.*, p. 426.
23. *Laudum wisz.*, 15 September 1716, *ibid.*, p. 647.
24. *Laudum wisz.*, 12 September 1730, *ibid.*, p. 727. Just this once the figure of 36 gallons is mentioned, but in 1733 it was 32 again; *AGZ*, XXIII, p. 35.
25. All the "Lvov school" investigations of prices point in this direction.
26. *AGZ*, XXI, p. 453 (6/6/1667), p. 489 (15/11/1668), p. 492 (16/2/1669);

vol. XXII, p. 176 (10/7/1685), p. 183 (10/7/1685), p. 189 (16/12/1687), p. 420 (28/3/1707), p. 426 (30/5/1707); vol. XXIII, p. 35 (14/4/1735), p. 390 (28/1/1760).

27. E.g. *AGZ*, XXI, p. 489 (15/11/1668).
28. *Ibid.*, XXII, p. 426.
29. *Ibid.*, XXIII, p. 35.
30. *Laudum wisz.*, 27 January 1638, *AGZ*, XX, p. 415.
31. F. Bostel, *op.cit.*, pp. 297-316. The phrase was repeated almost word for word in the tariff of 1573; see J. U. Niemcewicz, *op.cit.*, pp. 449-475.
32. Cracow tariffs of 1565 and 1573; *ibid.*
33. See W. Konopczyński, ed., *Diariusze. . .* , II, pp. 290-291, Warsaw, 1912.
34. J. K. Haur, *op.cit.*, p. 39; K. Kluk as cited in Z. Gloger, *op.cit.*, p. 190; see also T. Furtak, *op.cit.*, pp. 39-41.
35. Much relevant information in M. J. Elsas, *op.cit.*, I, pp. 137-161, and vol. II, *passim*.
36. C. Biernat, *op.cit.*

CHAPTER 21

1. If not exactly "decrees," such source materials clearly point to the urgency of these issues for the Crown at that time.
2. *Capitularia Regum Francorum*, I, 238, *Carol. Magn.* (in 789).
3. *Ibid.*, I, 333, *Carol. Magn.* (in 800).
4. *Ibid.*, I, 503, *Carol. Magn.* (in 813).
5. *Ibid.*, II, 182, *Carol. Calv.* (in 864).
6. Historians of weights and measures appear, however, to believe that this was so; e.g. Bigourdan, *op.cit.*, p. 2.
7. (Paucton), *Métrologie ou traité des mesures*, pp. 11-13.
8. R. Vivier, "Contribution," p. 179.
9. *Ibid.*; numerous other writers also perceive only the first two issues.
10. See R. Vivier, "Contribution," pp. 180-183, for an analysis of three successive *coutumes* with reference to the metrological rules they contain.
11. This pertained to transactions in public places, e.g. markets; see "Mesurage" in Diderot's *Encyclopaedia*. The right to control measures was a particular feudal privilege that could be leased by itself; see Y. Bezard, *op.cit.*, p. 119.
12. The Conserv. Nat. Catalogue du Musée, p. 215, is probably incorrect in stating that the seigneur as higher judge had the final say here.
13. *NCG*, Auxerre, III, 1, pp. 569-570, 593-594.
14. *NCG*, Maine, IV, 1, p. 469.
15. *NCG*, Sens, III, 1, p. 485; also pp. 507, 484, 505.
16. *NCG*, Clermont et Beauvoisis, II, 2, p. 776.
17. *NCG*, Limoges, IV, 2, p. 1149.

18. *NCG*, Orléans, III, 2, p. 808.
19. *NCG*, Pernes, I, 1, p. 389.
20. *NCG*, Bailleul, II, 2, p. 987.
21. *NCG*, Normandie, IV, 1, p. 8.
22. *NCG*, Poitou, IV, 2, p. 781.
23. *NCG*, Touraine, IV, 2, p. 602; and for Lodunois, IV, 2, p. 713.
24. *NCG*, Poitou, IV, 2, p. 781.
25. *NCG*, Touraine, IV, 2, p. 602; and for Lodunois, IV, 2, p. 713.
26. *NCG*, Anjou, IV, 1, p. 531.
27. *NCG*, Touraine, IV, 2, p. 643.
28. Jacquet, avocat au Parlement de Paris, *Traité des justices de seigneur et des droits en dépendants, conforméments à la jurisprudence actuelle des différents tribunaux du Royaume, suivi des pièces justificatives qui ont trait à la matière*, à Lyon, chez J. Breguilliat . . . , à Paris, chez Luis Cellot . . . et les frères Estienne, 1764, avec approbation et privilège du Roy (Bibl. Mazarine, A. 10.051), p. 268.
29. *Encyclopédie ou dictionnaire raisonné*, s.v. "Boisseau," II, p. 310. Savary states that in the 17th century the Paris *boisseau* was somewhat flatter; *Dictionnaire Universel de commerce, d'histoire naturelle et des arts et métiers . . . Ouvrage posthume du Sieur Jacques Savary des Bruslons, inspecteur général des Manufactures, pour le Roy, à la Douanne de Paris . . .* , à Copenhague, chez les Frères, Cl. et Ant. Philibert, 1759-1765, I, pp. 567-569.
30. *Conserv. Nat. Catalogue du Musée*, p. 69, no. 3243.
31. For an attempt by the Royal Procurator of Nantes to legalize and fix the size of the heap in the interest of the poorer buyers, dated 1 March 1759, see A. Machabey, *La Métrologie*, p. 192.
32. E. Clémenceau, *Le service des poids et mesures en France à travers les siècles*, Saint-Marcellin, Isère, 1909, pp. 89-92.
33. *Ibid.*
34. *Ibid.*, pp. 94-98.
35. *Ibid.*, pp. 99-100.
36. *Ibid.*, p. 175.
37. *Ibid.*
38. *Ibid.*, p. 104.
39. *Ibid.*, p. 107.
40. *Ibid.*, p. 85.
41. *Ibid.*, p. 106.
42. *Ibid.*, pp. 105-106.
43. *Ibid.*
44. *Ibid.*
45. *Ibid.*, p. 117.
46. *Ibid.*, p. 175.
47. *Ibid.*, p. 176.
48. *Ibid.*, p. 163.

49. *Ibid.*, p. 176.
50. *Ibid.*
51. *Ibid.*
52. *Ibid.*, p. 171.
53. (Paucton), *op.cit.*, p. 381.
54. Mersenne, *Parisienses Mensurae*, 1640, pp. 12-13. For data on striked and heaped measures, see S. Haillant, "Mesures anciennes des diverses régions vosgiennes," *Bulletin Historique et Philosophique*, Paris, 1903, pp. 9, 18ff., and table 26.
55. Mem. de l'Acad. des Sciences, 1772, II, p. 482.
56. Jacquet, *op.cit.*, p. 252.
57. If the seigneurs "n'ont pas de mesure originale qu'on nomme vulgairement étalon, leurs censitaires et justiciables étant en possession de payer les rentes qu'ils doivent à la mesure de Roy" (*Ibid.*, p. 250).
58. *Ibid.*, p. 266.
59. Vauban favored standardization of measures; see his 1696 *Description géographique de l'élection de Vezelay* . . . , in Vauban, *Projet d'une dixme royale* . . . , ed. E. Coornaert, Paris, 1933.
60. For a description of methods, see Sʳ de la Motte's 1721 application for a post of supervisor of corn measures (AN Paris, F¹², 1287).
61. R. Vivier, "Contribution," pp. 196-198.
62. F. Gerbaux, ed. Ch. Schmidt, "Procès-verbaux des Comités d'agriculture et de commerce de la Constituante, de la Législative et de la Convention," *Collection de documents inédits sur l'histoire économique de la Révolution Française*, Paris, 1906–1937, IV, p. 50.
63. B. Gille, *Les sources statistiques de l'histoire de France. Des enquêtes du XVIIᵉ siècle à 1870*, Geneva, 1964, p. 34.
64. *Ibid.*, p. 87.
65. A.R.J. Turgot, *Oeuvres de . . . et documents le concernant*, per G. Schelle, Paris, 1913–1923, V, pp. 31-33.
66. E. Clémenceau, *op.cit.*, pp.107-171.
67. A.R.J. Turgot, *op.cit.*
68. *Cdd.*, Cotentin, I, p. 799.
69. "État des mesures et réduction des boisseaux de la province d'Angoulême à la mesure et boisseaux de la ville d'Angoulême, 10 December 1790, AD Charente, L, 590. Also answers to the questionnaire of 1790, *ibid.*, L, 13091 and L. 884; cf. also "État des poids et mesures . . . en usage dans l'éntendue du district de la Rochefoucauld," *ibid.*, L, 185.
70. J. Leclére, P. Cozette, *op.cit.*, p. 9.
71. G. Fourquin, *op.cit.*, p. 49ff.
72. Y. Bezard, *op.cit.*, p. 37.
73. *Ibid.*, p. 119.
74. E. Fléchier, *Les Grands-Jours d'Auvergne*, Paris, 1964. For dishonest acts by seigneurial officials entrusted with control and marking of

measures, see *Cdd.*, Cotentin, I, pp. 195, 556-567, and *Cdd.*, Flandre, II, pp. 227-228.

75. E. Fléchier, *op.cit.*, pp. 57-58.
76. A. Young, *Voyages en France en 1787, 1788 et 1789*, ed. H. Sée, vol. II, Paris, 1931, pp. 536-541.
77. Montesquieu, *O duchu praw* (The Spirit of Laws), transl. Boy, Warsaw, 1957, bk. XXIX, chap. 18, p. 383.
78. G. Bigourdan, *op.cit.*, p. 11.
79. *Encyclopédie ou dictionnaire raisonné. . .* , X, "Mesure (Généralement)," signed D. J. (Ch^{er} de Jaucourt).
80. *Ibid.*, XII, "Poids."
81. Téron Ainé, *op.cit.*, p. 4.
82. A.R.J. Turgot, *op.cit.*, I, pp. 31-33.
83. *Ibid.*
84. *Ibid.*, II, p. 228.
85. *Cdd.*, Sens, pp. 177-178.
86. Cf. below, chap. 22, n. 246.
87. Jacquet, *op.cit.*, bk. I, chap. XXI.
88. *Ibid.*
89. *Ibid.*
90. *Ibid.* But the practice was in the seigneurs' interest, too. R. Vivier rightly emphasizes the "avidité des fermiers" and "complaisance des seigneurs"; "Contribution," pp. 185-186, also 183, 188-190.
91. Jacquet, *op.cit.*, bk. I, chap. XXI.
92. Jacquet's view here is critical of that in the decree of the Sens nobles.
93. Jacquet, *op.cit.*, bk. I, chap. XXI.
94. *Ibid.*
95. *Ibid.*
96. *Ibid.*
97. Poix de Frémonville, *La pratique universelle.*
98. M. Delamare, *Traité de la police, par . . .* , vol. II, Paris, 1722, chap. VIII, "De mesurage des grains."
99. *Ibid.*
100. *Ibid.*
101. R. Vivier, "Contribution," pp. 180-181, 185-186.
102. R. Vivier, "L'application," pp. 193, 198-199. There was also a growth of scientific interest in feudal institutions, including measures; see "Recherches sur les anciennes monnoies du Comté de Bourgogne avec quelques observations sur les poids et mesures autrefois en usage dans la même Province. Ouvrage qui a remporté le prix au jugement de l'Académie de Besançon, par un Bénédictin de la Congregation de St. Vanne, membre de plusieurs Académies," à Paris, chez Nyon . . . , à Besançon, chez les principaux libraires, 1782. Against such investigations, see the *cahier de doléances* of Neuilly (Troyes) of 1614, p. 233.

103. V. R. Mirabeau, *Philosophie rurale...*, Amsterdam, 1763, p. 157.
104. G. Bigourdan, *op.cit.*, p. 5, sees the lack of any notable merits in the Paris standards, which the authorities wished to impose nationwide, as the reason for the failure of attempts at standardization.

CHAPTER 22

1. A. de Tocqueville, *L'Ancien Régime et la Révolution*, Paris, 1964, p. 45.
2. P. Sagnac, "Les cahiers de 1789 et leur valeur," *Revue d'Histoire Moderne et Contemporaine*, 1906–1907; the same thesis, but vulgarized, in P. Gaxotte, *La Révolution Française*, Paris, 1928, p. 122.
3. *Bulletin de la Commission d'Histoire Économique*, 1906, pp. 23, 398; and 1907, p. 1.
4. The debate was now opened by H. Sée, in "La rédaction et la valeur historique des cahiers de paroisses pour les États Généraux de 1789," *Revue Historique*, CIII, 1920.
5. P. Gaxotte, *op.cit.*, pp. 120-121.
6. A. Mathiez, *Rewolucja Francuska* [The French Revolution], Warsaw, 1956, p. 43.
7. Or perhaps Gaxotte's "secret societies."
8. R. Picard, *Les cahiers de 1789 du point de vue industriel et commercial*, thèse pour le doctorat ès sciences juridiques, Faculté de Droit de l'Université de Paris, Paris, 1910, p. 27ff.
9. P. Goubert, 1789. *Les Français ont la parole*; ser. "Archives," Paris, 1964, p. 7.
10. A. Le Moy, ed., *Cahiers de doléances des corporations de la ville d'Angers et des paroisses de la sénéchaussée particulière d'Angers*, Angers, 1915.
11. P. Bois, ed., *Cahiers de doléances du Tiers-État de la sénéchaussée de Château-du-Loire*, Gap, 1960.
12. R. Picard, *op.cit.*, p. 28.
13. *Cdd.*, Angers, II, pp. 259-260.
14. The fact that the "models" were often copied precludes, in our view, quantitative analysis of the *cahiers* of the kind attempted by D. Feuerwerker in "Anatomie des 307 cahiers de doléances de 1789. Les Juifs en France," *Annales E.S.C.*, 1965, pp. 45-61.
15. M. Bouloiseau, "Notables ruraux et élections municipales dans la région rouenhaise," *Actes du Congrès des Sociétés Savantes*, 1957; and "Voeux et griefs saumorois lors des élections aux États Généraux de 1789," *ibid.*, 1962, p. 170.
16. For Franche-Comté, in AD Doubs (Besançon).
17. R. Picard, *op.cit.*, points out many crossings-out and alterations in the MSS of the cahiers, including those copied from "models."
18. R. Picard, *op.cit.*, p. 29.
19. See R. Picard, *op.cit.*, A. Favre, ed., "Les cahiers de doléances du Bugey et de la Dombes," *Bulletin d'Histoire Économique et Sociale de la Révolution Française*, 1967, p. 44.

20. E.g. *Cdd.*, Rennes (Saint-Servan), III, pp. 80-81.
21. E.g. *Cdd.*, Reims (Bazancourt-sur-Suippe), p. 254: "Qu'il n'y ait qu'un seul poid et seule mesure par tout le Royaume, et une même mesure."
22. *Cdd.*, Rennes (La-Boussac), II, p. 597: "Qu'il n'y ait à l'avenir qu'un même poids et qu'une même mesure"; this was copied by the parish of La-Chapelle-aux-Filzméens who added "pour toute la France."
23. *Cdd.*, Château-du-Loire (Montabon), p. 40, and (Flee), pp. 28-29.
24. *Cdd.*, Reims (ville de Reims, menuisiers, etc.), p. 152; and *ibid.*, (Sarcy), p. 933.
25. *Ibid.* (Trois-Puits), p. 1016, and cf. the demand by the parish of Recy-sur-Marce (*Cdd.*, Châlons-sur-Marne, p. 520).
26. *Cdd.*, Cotentin (Hauteville-près-la-mer), I, p. 352.
27. *Cdd.*, Bourges (Sancerre), p. 320, and (Saint Satur), p. 459.
28. *Cdd.*, Angoulême (Saint-André-de-Blanzac), pp. 292-293.
29. *Cdd.*, Quimper (Plozevet), p. 99.
30. *Cdd.*, Rennes (Tinténiac), III, p. 184.
31. *Cdd.*, Quimper (Beuzec-Cap-Caval), p. 204.
32. *Cdd.*, Metz (Gandrange), p. 108.
33. *Cdd.*, Bourges (Humbligny), p. 157; and *Cdd.*, Boulay (Folschwiller and Metring), p. 84.
34. J. Durand, ed., *Cahiers de doléances des paroisses du baillage de Troyes pour les États Généraux de 1614*, Paris, 1966, pp. 203, 204, 233, 299.
35. D. Ligou, "La première année de la révolution vue par un témoin. 1789–1790." Les "Bulletins" de Poncet-Delpech, député de Quercy aux États Généraux de 1789. Publications de la Faculté des Lettres et Sciences Humaines d'Alger, Paris, 1961, p. 70 (session 1 Sept. 1789). For many relevant brochures of May 1790, see file BN, 8⁰, Le[29] 626-664 at the Bibliothèque Nationale in Paris.
36. J. Jaurès, *Histoire socialiste de la Révolution Française*, I, pp. 154-155; cf. G. Lefebvre, *La grande peur de 1789*, pp. 7-8.
37. See the numerical tabulation, below.
38. *Cdd.*, Ham., p. 13.
39. *Cdd.*, Rennes (ville de Guingamp), II, p. 28.
40. *Cdd.*, Reims (Saint-Mesmes), p. 915.
41. *Ibid.*, (Rilly-la-Montaigne), pp. 850-851.
42. *Ibid.* (Bouleuse), pp. 332-333.
43. *Cdd.*, Pas-de-Calais (Vimille), II, p. 482.
44. E.g., *Cdd.*, Orléans, I, p. 102; *Cdd.*, Reims, p. 1120; *Cdd.*, Rennes, IV, p. 79.
45. Cdd., Blois (ville de Selles-sur-Cher), II, pp. 207-208.
46. *Cdd.*, Bourges (Harry), p. 153.
47. *Cdd.*, Rouen (Auzouville-sur-Ry), II, p. 337.
48. *Ibid.* (corporation des épiciers), I, p. 141.
49. *Cdd.*, Orléans (Loury), I, p. 590.

50. *Cdd.*, Cotentin (Assemblée Préliminaire du Tiers-État, Saint-Sauver-Lendelin), III, p. 146.
51. Jan Kott's phrase [the literary critic—TRANS.].
52. A. Tocqueville, *op.cit.*, p. 45.
53. Some editors supply additional information, e.g. H. Sée on the sénéchaussée de Rennes.
54. *Cdd.*, Rennes (Yvignac), III, p. 266.
55. *Ibid.* (Brain), II, p. 432.
56. *Ibid.* (Ruffigné), II, p. 393.
57. *Cdd.*, Caen, pp. 229-230.
58. *Cdd.*, Châtillon's (M. La-Neuville-aux-Larris), p. 213.
59. *Cdd.*, Rennes (Belle-Isle-en-Terre), IV, p. 195.
60. *Ibid.* (Saint-Martin-de-Janzé), I, p. 479.
61. *Cdd.*, Quimper (Pleuven), p. 359.
62. *Ibid.* (Locamand), p. 320.
63. *Ibid.* (Treffiagat), p. 218.
64. *Ibid.* (Beuzec-Cap-Caval), p. 204.
65. *Ibid.* (Esquibien), p. 129.
66. *Ibid.* (Goulien), p. 112.
67. *Ibid.* (Plozeviet), p. 99.
68. *Cdd.*, Pestivien, nb. *cahier* of 21 March 1790 in A. Bourgés, ed., *Les doléances des paysans bretons en 1789. Quelques cahiers de paroisses. Territoire de l'actuel département des Côtes-du-Nord*, Saint-Brieuc, 1953, p. 116.
69. *Cdd.*, Rennes (Tinténiac), III, p. 184.
70. *Ibid.* (Trévérien), III, p. 228.
71. *Ibid.* (Plélan), III, p. 416.
72. *Ibid.* (Yvignac), III, p. 266.
73. *Cdd.*, Rennes (Plancoet), III, p. 667.
74. *Ibid.* (Andel), III, p. 721.
75. *Cdd.*, Angers (Le-May), II, pp. 667-668.
76. *Ibid.* (Saint-Sulpice-sur-Loire), II, p. 726.
77. *Ibid.* (La-Meignanne), II, pp. 259-260.
78. *Cdd.*, Angoulême (ville d'Angoulême), pp. 121-122.
79. *Cdd.*, Quimper (Mellac), p. 218.
80. *Cdd.*, Rennes (Pommeret), III, p. 728.
81. *Cdd.*, Pas-de-Calais (Longueville), II, p. 326.
82. *Cdd.*, Rennes (Plédran), III, p. 732.
83. *Cdd.*, Beaujolais (Belleville), p. 35.
84. *Cdd.*, Pont-à-Mousson, p. 73.
85. *Cdd.*, Angoulême (ville d'Angoulême), pp. 121-122.
86. P. Sagnac, P. Caron, eds., *Les Comités de droits féodaux et de législation et l'abolition du régime seigneurial, 1789–1793*, Collection de documents inédits sur l'histoire économique de la Révolution Française, Paris, 1907, pp. 412-416.

87. *Cdd.*, Angoulême (Blanzac), p. 307.
88. *Ibid.* (Saint-André-de-Blanzac), pp. 292-293.
89. *Ibid.* (ville d'Angoulême), pp. 121-122.
90. *Cdd.*, Quimper (Primelin), p. 114.
91. *Ibid.* (Crozon), p. 226.
92. *Ibid.* (Briec), p. 189.
93. *Ibid.* (Loctudy), p. 171; for the seigneurs' irregularities; *Cdd.*, Angoulême, pp. 121-122; *Cdd.*, Cotentin, II, pp. 799-800; for demands for abolition of the seigneurs' measures, *Cdd.*, Angers, I, pp. 259, 667, 726, 739-740; *Cdd.*, Angoulême, pp. 121-122, 227-228; *Cdd.*, Bigorre, p. 397; *Cdd.*, Bretagne, p. 397; *Cdd.*, Troyes, II, pp. 136-137.
94. *Cdd.*, Rennes (Saint-Morvez), IV, p. 92.
95. *Cdd.*, Angoulême (Blanzac), p. 307.
96. *Cdd.*, Angers (Saint-André-de-la-Marche), I, p. 127.
97. *Cdd.*, Pas-de-Calais (Saint-Omer), I, p. 121, and (Tournehem), I, p. 565.
98. *Cdd.*, Angoulême (Saint-André-de-Blanzac), pp. 292-293.
99. *Cdd.*, Pas-de-Calais (Selles), II, p. 436.
100. *Cdd.*, L'Isle-Jourdan et du Vigean, BW Paris: 8⁰ Le²⁴ 328.
101. *Cdd.*, Angers (Saint-Michel-du-Bois et Chanveaux), II, pp. 739-740.
102. *Cdd.*, Reims (Saint-Pern), III, p. 238.
103. *Cdd.*, Rennes (Guichen), III, p. 469.
104. *Cdd.*, Bretagne (Pestivien), p. 115.
105. *Cdd.*, Rennes (Nouvoitou), I, p. 611.
106. *Ibid.* (Saint-Servan), III, p. 83.
107. *Cdd.*, Sézanne (Esclavolles), p. 251.
108. *Cdd.*, Pas-de-Calais (Saint-Omer), I, p. 121.
109. *Cdd.*, Rennes, IV, p. 265.
110. *Cdd.*, Blois (Saint-Laurent-de-Vatan), II, p. 341.
111. *Cdd.*, Bretagne (Plounèvez-Moedoc), pp. 135-136.
112. *Cdd.*, Angoulême (Saint-André-de-Blanzac), pp. 292-293; similarly in *ibid.* (Blanzac), p. 307.
113. *Cdd.*, Angers (Le-May), II, pp. 667-668.
114. *Cdd.*, Châtillon s/M, (Dormans), p. 146; similarly, p. 433.
115. *Cdd.*, Pas-de-Calais (Saint-Omer), I, p. 121.
116. *Cdd.*, Amont (Valleroy-Lorioz), II, p. 425; also; *ibid.* (Abbenans), I, p. 65; *ibid.* (Colombette), I, p. 377; *ibid.* (les corporations de Gray), I, p. 43.
117. *Cdd.*, Rennes (Parcé), I, pp. 227-228.
118. *Ibid.* (Dompierre-du-Chemin), I, p. 231.
119. *Ibid.* (Servon), II, p. 44.
120. *Cdd.*, Châtillon s/M. (cahier du tiers-état), p. 494; also, *Cdd.*, Châtillon s/M. (Dormans), p. 146.
121. *Cdd.*, Arques (Saint-Nicolas-de-Veules), II, p. 431.

122. *Cdd.*, Sens (Foucherolles), p. 177.
123. *Cdd.*, Quimper (Beuzec-Cap-Caval), p. 204.
124. *Cdd.*, Rennes (ville de Châteaubriant), II, p. 312.
125. *Cdd.*, Angoulême (ville d'Angoulême), pp. 121-122.
126. *Cdd.*, Châlons s/M. (Bussy-Lettrée), p. 116; cf. *ibid.* (Courtisols), p. 239, and (Souain), pp. 649-650.
127. *Cdd.*, Rennes (Saint-Pern), III, p. 238.
128. *Ibid.* (Saint-Jean-de-Béré), II, pp. 331-332, where the term is explained.
129. *Cdd.*, Troyes (Champvallon), I, pp. 542-543.
130. *Cdd.*, Rennes (Nouvoitou), I, p. 611.
131. *Cdd.*, Sézanne (Esclavolles), p. 251.
132. *Cdd.*, Reims (Villers-Marmery), p. 1111.
133. *Cdd.*, Toulouse (Ardiege), p. 125.
134. *Cdd.*, Orléans (corporation des avocats), II, p. 75.
135. *Cdd.*, Angers (Le-May), II, p. 667-668.
136. *Cdd.*, Châlons s/M. (Rouffy), p. 533.
137. *Cdd.*, Angers (La Jubaudière), II, p. 473.
138. *Cdd.*, Angoulême (ville d'Angoulême), p. 122.
139. *Cdd.*, Pas-de-Calais (Selles), II, p. 436.
140. *Cdd.*, Rennes (Saint-Servan), III, p. 83.
141. *Ibid.* (Saint-Servan), III, pp. 80-81.
142. *Ibid.* (Trézélan), IV, p. 88.
143. *Cdd.*, Amont (Port-sur-Saône), V, p. 289.
144. *Cdd.*, Rennes (Guenézan), IV, p. 149.
145. *Ibid.* (Guichen), III, p. 469.
146. *Cdd.*, Baillage de Gray, quoted after G. Bigourdan, *op.cit.*, p. 12.
147. Some examples: Châtillon-sur-Marne (cahiers du baillage), pp. 447, 457, 494; *Cdd.*, Épernay (ville), p. 70; *ibid.* (Mardeuil), p. 105; *Cdd.*, Nîmes (Poulx), II, p. 40; *Cdd.*, Caen (ville), p. 229; *Cdd.*, Varzi (ville), p. 368; *Cdd.*, Rouen (collèges), I, p. 41; *ibid.* (épiciers), I, p. 141; *ibid.*, (Cailly-Bois d'Ennebourg), II, p. 222; *ibid.*, (Cello-ville), II, p. 350; *Cdd.*, Reims (corporations) pp. 126, 186, 200; *ibid.* (Chaumuzy), p. 433; *ibid.* (Marfaux), p. 745; *ibid.* (Montbré), p. 759; *ibid.* (Saint-Martin-l'Hereux), p. 904; *ibid.* (Ville-en-Selve), p. 1056.
148. *Cdd.*, Pas-de-Calais (Annay), I, p. 147.
149. *Ibid.* (Havrincourt), I, p. 345.
150. *Cdd.*, Troyes (Germigny), II, p. 137.
151. *Cdd.*, Reims (Romigny), p. 857.
152. As they were *felt*. Social psychology is of more relevance here than "objective" appraisals.
153. *Cdd.*, Flandre-Maritime (Esquelbecq), II, 1, pp. 254-255.
154. *Cdd.*, Châtillon-sur-Marne (Verneuil Haut et Bas), p. 379; *Cdd.*, Flandre-Maritime (Ledringhem), II, 1, pp. 259-260; *Cdd.*, Reims (Rilly-la-Montaigne), pp. 850-851.
155. *Cdd.*, Autun (Rigny-sur-Arroux), p. 165; similarly *Cdd.*, Sézanne (ville

de Sézanne), p. 29, and cf. *Cdd.*, Metz (Thionville, marchands), p. 343.

157. *Cdd.*, Angoulême (ville d'Angoulême), pp. 121-122.

158. *Cdd.*, Metz (Gandrange), p. 108.

159. *Cdd.*, Rennes (Lailleu), II, p. 284; similarly, in the same judicial district, the parishes of Erbray (*ibid.*, p. 349) and Bains (p. 428)

160. Although there are many references in the *cahiers* to the *droit de minage*.

161. *Cdd.*, Sézanne (Charleville), p. 179; also applicable at Chichey, *ibid.*, p. 194, and Bannes *ibid.*, p. 76, etc.

162. *Cdd.*, Troyes (Trainel), II, p. 615.

163. *Cdd.*, Angers (cahiers des consuls), I, p. 11.

164. *Cdd.*, Nîmes (Poulx), II, p. 40; and *Ibid.* (Saint-Gervais), II, p. 223.

165. *Cdd.*, Orléans (ville d'Orléans, orfèvres), II, p. 165; cf. *Cdd.*, Quimper, p. 8.

166. *Cdd.*, Pas-de-Calais (Annay), I, p. 147.

167. *Cdd.*, Rouen (Salmonville-Le-Sauvage), II, pp. 310-311.

168. *Cdd.*, Troyes (Beugnon), I, p. 402.

169. *Cdd.*, Rouen (La-Neuville-Camp-d'Oisel), II, p. 379.

170. Quoted after G. Bigourdan, *op.cit.*, p. 12.

171. *Cdd.*, Alençon (corporation des cordonniers), p. 85.

172. *Cdd.*, Orléans (Yvoi), I, p. 199.

173. *Ibid.* (Neuvy-en-Sullias), I, p. 319.

174. *Ibid.* (Mardié), I, p. 102.

175. *Cdd.*, Nîmes (Bagnols), I, p. 100; likewise *ibid.*, (Sabran), II, p. 114, and *ibid.* (Saint-Gervais), II, p. 220.

176. *Ibid.* (Vers), II, pp. 518-519.

177. See above, chap. 2.

178. *Cdd.*, Rouen (corporation des épiciers), I, p. 141.

179. *Cdd.*, Nimes (sénéchas), II, p. 392; a highly original view.

180. *Cdd.*, Blois (L'Hôpital-Commanderie), II, p. 183; again, a unique phrase.

181. *Cdd.*, Châlons s/M. (Jonchery-sur-Suippe), pp. 359-360.

182. *Cdd.*, Épernay (Moslins-et-Morangis), p. 115; repeated *ibid.* (Morangis), p. 117, and *ibid.* (Pierry), p. 137.

183. *Cdd.*, Angers (Villeveque), II, pp. 685-686.

184. The denial of the importance of these demands by J. R. Armogathe, "Genes du cahier de doléances de la communauté de Sénéchas (Gard)," *Bulletin d'Histoire Économique*, 1968, p. 64, is quite unsubstantiated.

185. *Cdd.*, Angers (Saint-Sulpice-sur-Loire), II, p. 726.

186. *Cdd.*, Angoulême (Pérignat), pp. 227-278.

187. *Cdd.*, Civray (Villemorin et Saint Goutan), p. 351.

188. *Cdd.*, Épernay (Germaine et Vaurémond), p. 85; similarly, *ibid.* (Moslins-et-Morangis), p. 115.

189. *Cdd.*, Orléans (Loury), I, p. 590.

190. *Ibid.* (Neuville), II, p. 376.
191. *Cdd.*, Orléans (orfèvres), II, p. 165; also, Quimper (ville), p. 8.
192. *Cdd.*, Reims (Champigny), p. 405.
193. *Ibid.* (Courville), p. 548.
194. *Ibid.* (Rilly-La-Montagne), p. 853.
195. *Ibid.* (Sacy), p. 874.
196. M. C. Demay, ed., "Cahiers de doléances des villes de Cosne, Varzy, et de la paroisse de Lignorelle," *Bulletin de la Société des Sciences historiques et naturelles de l'Yonne*, 2e semestre 1886, Auxerre, 1887, p. 356—Cosne-sur-Loire.
197. *Cdd.*, Rouen (épiciers), I, p. 141.
198. Thus the demand in *Cdd.*, Rouen (Auzouville-sur-Ry), II, p. 337.
199. The attempt is described in P. Viollet, *op.cit.*, III, pp. 200-201.
200. *Cdd.*, Rouen (ville de Rouen, épiciers), I, p. 141.
201. *Ibid.* (Le-Mesnil-Durecau), II, p. 181.
202. *Cdd.*, Troyes (ville), I, p. 262.
203. M. C. Demay, ed., *op.cit.*, p. 356—Cosne-sur-Loire.
204. *Cdd.*, Troyes (Germigny), II, pp. 136-137.
205. *Cdd.*, Niort (Couture d'Argenson), p. 104.
206. *Cdd.*, Rouen (Belbeuf), II, p. 342.
207. *Cdd.*, Blois (Montils), I, p. 194; *ibid.*, (La-Marolle), II, p. 100.
208. *Cdd.*, Pas-de-Calais (Saint-Paul), I, p. 127.
209. *Cdd.*, Quimper (Mahalon-et-Cuiler), p. 133.
210. *Cdd.*, Sens (Villeneuve-la-Dondagre), p. 498.
211. *Cdd.*, Troyes (Coursan), I, p. 647.
212. *Cdd.*, Sens (Villeneuve-la-Dondagre), p. 498, and many others.
213. *Cdd.*, Rouen (Saint-Aigan-les-Rouen), II, p. 21.
214. *Cdd.*, Bigorre (Esquieze), p. 257; copied texts: *ibid.* (Esterre), p. 259, and (Sers), p. 566; and copied but modified: *Cdd.*, Bigorre (Sére), p. 552.
215. *Cdd.*, Châlon s/M. (Vatry), p. 752; also *Cdd.*, Rouen (Ry), II, p. 410.
216. *Cdd.*, Blois (Onzain), I, p. 213; the same, *ibid.* (Chaumont-sur-Loire), I, p. 217.
217. *Cdd.*, Rouen (Darnetal, Saint-Gervais-les-Rouen), II, p. 29.
218. *Ibid.* (Darnetal, Saint-Paul-les-Rouen), II, p. 41; similarly, *ibid.* (Saint-Pierre-de-Deville), II, p. 69.
219. *Cdd.*, Pas-de-Calais (Béthune), I, p. 83.
220. *Cdd.*, Rouen (Saint-Jean-de-Cardonnay), pp. 108-109.
221. *Cdd.*, Blois (Saint-Aignanhors-l'Enclos), II, p. 225.
222. *Ibid.* (ville de Selles-sur-Cher), II, pp. 207-208.
223. *Cdd.*, Rouen (Saint-Denis-de-Duclair), II, p. 80.
224. *Cdd.*, Rennes (Saint-Servan), III, p. 83.
225. *Cdd.*, Metz (Gandrange), p. 108; *Cdd.*, Pas-de-Calais (Marquion), I, p. 399; *ibid.* (Oppy), p. 447; and others.
226. *Cdd.*, Quimper (Pont-Croix), p. 92.
227. *Cdd.*, Alençon (huissiers-corporation), p. 56.

228. *Cdd.*, Rouen (Ry), II, p. 410.
229. *Cdd.*, Rennes (Saint-Mélior-des-Ondes), III, p. 118.
230. *Cdd.*, Civray (Linazay), p. 54.
231. *Cdd.*, Reims (Ormes), p. 796.
232. *Cdd.*, Rennes (Brain), II, p. 437.
233. *Cdd.*, Amont (Comberjon), I, p. 381.
234. *Cdd.*, Châlon s/M. (Champagne), p. 131; *ibid.* (Bussy-Lettrée), p. 116; *Cdd.*, Reims (ville de Reims, corporation des armuriers), p. 100; *Cdd.*, Beaujolais (Chenas), p. 48.
235. *Cdd.*, Angers (Pellouailles), II, p. 694.
236. *Cdd.*, Rouen (Darnetal), II, p. 29; *ibid.* (Darnetal-Saint-Paul-les-Rouen), II, p. 41; *ibid.* (Saint-Pierre-de-Deville), II, pp. 69-70.
237. *Cdd.*, Châlon s/M. (Margerie), p. 432.
238. *Cdd.*, Rouen (Montigny), II, p. 101; similarly and quite independently put in *ibid.* (Cailly-Fontaine-sur-Preaux), II, p. 257.
239. Cf. *Cdd.*, Cotentin, III, p. 558; *Cdd.*, Châlon s/M., p. 863; *Cdd.*, Châtillon s/M., p. 494; *Cdd.*, Serrebourgh, p. 46.
240. *Cdd.*, Troyes (*cahier* of the nobles), III, p. 173. *Archives Parlementaires*, VI, p. 78, art. 76.
241. C. L. Chassin, ed., *Les élections et les cahiers de Paris en 1789*, III, Paris, 1889, pp. 328, 347.
242. *Cdd.*, Orléans (*cahier* of the nobles), II, p. 431.
243. *Cdd.*, Longwy (*cahier* of the nobles), p. 150.
244. *Cdd.*, Orléans (*cahier* of the clergy), II, p. 418; cf. Bouloiseau, ed., "Deux cahiers de curés, normands pour les États Généraux de 1789," *Annales de Normandie*, VII, 1, p. 64. For Touraine, see R. Vivier, "L'application," p. 198.
245. *Cdd.*, Orléans (clergé), p. 53. *Cahier* of the clergy of Erreux, *Archives Parlementaires*, III, p. 202.
246. *Cdd.*, Sens (noblesse), p. 810.
247. Cf. above, chap. 21, n. 85.
248. *Cdd.*, Amont (Port-sur-Saône), II, p. 289.
249. M. C. Demay ed., *op.cit.*, p. 356—Cosne-sur-Loire.
250. J. Vidalenc points out the complete unanimity of the various social groups on the issue of standardization of measures; "Les revendications économiques et sociales de la population parisienne en 1789 d'après les cahiers de doléances," *Revue d'Histoire Économique et Sociale*, XXVII, 1948-1949, p. 282.
251. *Cdd.*, Niort (huissiers), p. 320.
252. A. de Tocqueville, *op.cit.*

CHAPTER 23

1. *Bulletin d'Histoire Économique de la Révolution*, 1912 and 1920–1921, Recueil de textes concernant la suppression des droits féodaux, Décret du 15-28 mars 1790, titre II, art. 17, 18, 19, 21, pp. 27-28.

2. *"Proposition faite à l'Assemblée Nationale . . . ,"* G. Bigourdan, *op.cit.*, *Collection générale des décrets rendus par l'Assemblée Nationale*, II, p. 370.

3. *Moniteur Universel*, no. 130, 10 May 1790.

4. F. Gerbaux, C. Schmidt, eds., *op.cit.*, I, pp. 653-654.

5. A. Tuetey, *Répertoire général des sources manuscrites de l'histoire de Paris pendant la Révolution Française*, par . . . , III, Paris, 1894, no. 1111.

6. Text in AD M et M., L, 2170, and elsewhere.

7. G. Bigourdan, *op.cit.*, pp. 21-23.

8. *Cdd.*, Rennes, III, p. 266.

9. *Ibid.*, IV, pp. 88, 149.

10. *Cdd.*, Cotentin, II, pp. 799-800.

11. P. Sagnac, P. Caron, eds., *op.cit.*, pp. 401-404, no. 179.

12. *Ibid.*, pp. 407-409, no. 182.

13. *Ibid.*, pp. 201-203, no. 85.

14. *Cdd.*, Angoulême, pp. 121-122, 277-278, 282, 292-293, 307.

15. *Ibid.*, pp. 121-122.

16. *Ibid.*

17. P. Sagnac and P. Caron, eds., *op.cit.*, pp. 412-416, no. 185.

18. *Ibid.*, 15 March 1791 (italics added—W.K.).

19. *Ibid.*

20. *Ibid.*

21. J. Godart, L. Robin, *Rapport de Messieurs . . .* , p. 41 (in the Bibl. Nat., Paris, at 8⁰ Le²⁹ 1410).

22. See AN (Paris), F¹², 1288, 1289, and AN, F¹², 210ff.

23. See reference to M. Buissart, *Archives Parlementaires*, XXV, pp. 609-610, session 5 May 1791.

24. *Archives Parlementaires*, LXX, session 1 August 1793, pp. 70-73, 112.

25. F. Gerbaux and C. Schmidt, eds., *op.cit.*, I, pp. 653-654 (17 Nov. 1790).

26. "Loi relative à l'établissement de nouvelles mesures pour les grains . . . ," 16 pp., AD Charente, L, 1317, and elsewhere.

27. G. Bigourdan, *op.cit.*, p. 34.

28. For more extreme views voiced at the National Assembly, 8 May 1790, see *Moniteur Universel*, no. 130, 10 May 1790.

29. AN, F¹², 1288.

30. *Moniteur Universel*, No. 214, 2 August 1793.

31. AN, F¹², 1289, a letter to the Minister of the Interior, 7 nivôse, year II.

32. AN, F¹², 1289. "Le chargé provisoire des fonctions du Ministre de l'Intérieur aux Administrateurs des Districts, 25 germinal an II" (printed); AD Charente, L, 1545, AD Doubs, L, 1518, and elsewhere.

33. "Le Comité d'Instruction Publique à l'administration du District . . . 8 floréal an III" (printed), AD Doubs, L, 1679.

34. G. Bigourdan, *op.cit.*, pp. 51-52.

35. Dated 17 thermidor, year III, AN, F¹², 210, p. 71.

36. C. Schmidt, *Le commerce, Instruction, recueil des textes et notes*, Commission de recherche et de publication des documents relatifs à la vie économique de la Révolution, Paris, 1912, pp. 154-155, no. 143.

37. *Moniteur Universel*, 24 and 25 germinal, year III (13 and 14 April 1795).

38. "Ministre de l'Intérieur à l'administration du Département . . . ," 23 fructidor an V, AN F^{12}, 1289.

39. G. Bigourdan, *op.cit.*, pp. 20-21.

40. "Loi relative au moyen d'établir une uniformité de poids et mesures," AD M. et M., L, 2170.

41. G. Bigourdan, *op.cit.*, p. 47.

42. Joseph-Marie Lequeno, a Jacobin member of the Convention, combined the nationalistic-patriotic and the universalistic views of the metric reform in his book *Les préjugés détruits*, published in 1792.

43. AD Charente, L, 1317. See also F. Donovan, ed., *The Thomas Jefferson Papers*, New York, 1963, p. 121.

44. Talleyrand to the government of the Cisalpine Republic, 11 floréal, year VI, ASM, Commercio, parte antica, cart. 226.

45. Cf. the ironical comment by the Amboise priest, François de Gloinec; R. Vivier, "L'application," p. 182.

46. A. Tuetey, *op.cit.*, VI, no. 2078; also F. Gerbaux and C. Schmidt, eds., *op.cit.*, II, p. 765 (session 56, 31 May 1792), and the letter by the Minister of the Interior to the chairman of the National Assembly dated 19 May 1792, AN, F^{12}, 1288.

47. *Bulletin d'Histoire Économique de la Révolution*, 1912, pp. 154-155; and G. Bigourdan, *op.cit.*, pp. 30-31.

48. Agence de Poids et Mesures à l'Agence des Transports Militaires de l'Intérieur, 23 frimaire, year IV, AN, F^{12}, 210, p. 152.

49. Agence de Poids et Mesures aux membres composant le Comité des Finances, de la Convention Nationale, 1 floréal, year III, AN, F^{12}, 210, pp. 1-4.

50. Circulaire aux Comités Civils des 48 Sections de Paris, 25 frimaire an IV, AN, F^{12}, pp. 157-158, and the Agency's letter to members of the Bureau Central de Police of 14 nivôse, year IV; *ibid.*, pp. 167-168.

51. Agency to C^{en} Audichacot, 28 frimaire, year IV, *ibid.*, p. 157.

52. AD Doubs, Besançon, L, 1679.

53. R. Vivier, "L'application," p. 192.

54. AD Charente, L, 188.

55. The same conclusion was drawn from his study of several questionnaire enquiries in 18th-century France by B. Gille, *op.cit.*, pp. 16, 87.

56. AD Charente, L, 1317 letter of 4 April 1791.

57. AD Doubs, L, 2180.

58. G. Bigourdan, *op.cit.*, p. 25. Cf. *Archives Parlementaires*, XXI, p. 323.

59. E.g. Grands Augustins, A. Tuetey, *op.cit.*, nos. 2272, 2273.

60. Circulaire du Ministre de l'Intérieur (François de Neufchâteau) aux administrations centrales du département . . . 21 fructidor an VI; C. Schmidt, *Le Commerce*, p. 306, no. 380. Cf., however, Arreté du Directoire du 29 brumaire an VII; *ibid.*, no. 388.

61. Agence Temporaire des Poids et Mesures aux Citoyens composant le District de Blaimont, 11 prairial an III; AD M. et M., L, 655.

62. For numerous examples, see AN, F^{12}, 210ff.

63. E.g., R. Vivier, "L'application," p. 204.

64. See the Agency to citizen Gindrotz, 1 brumaire, year IV, AN, F^{12}, 210, p. 116 to citizen Sicard, 4 brumaire, year IV, *ibid.*, p. 121; or to citizen La Roche, 16 nivôse, year IV, p. 169 "pourquoi donc faire une enquête des mesures incertaines, *qu'il faut même oublier telles qu'elles sont*" (emphasis added—W.K.).

65. A. Machabey, *op.cit.*, p. 238. J. Richard, "La gréneterie," p. 137, is excessively critical of the tables.

66. E.g. for Italy—AN.

67. Letter dated 11 December 1792; AN, F^{12}, 1288.

68. See F. Gerbaux, C. Schmidt, *op.cit.*, III, p. 655, nos. 48 and 56; and vol. IV, p. 242, no. 3.

69. A. Tuetey, *op.cit.*, nos. 1102, 1104, 1105, 1115 (the period 24-30 July 1793).

70. AN, F^{12}, 1288.

71. See Prefect of Police for Paris to editors of the press, 2 floréal, year XIII, AEG, Dep. Leman, 522.

72. R. Vivier, "L'application," p. 207.

73. Letter, primidi nivôse, year II, AN, F^{12}, 1288.

74. R. Vivier, "L'application," *passim*.

75. Le Ministre de l'Intérieur aux Préfets des Départements, 2 frimaire, an XI (printed), 28 pp., AD Doubs, M, 2390, and elsewhere.

76. *Ibid.*, circular dated 2 frimaire, year XI.

77. Agences des Poids et Mesures à la commission des Travaux Publiques, 14 frimaire, year IV, AN, F^{12}, 210, pp. 129-130.

78. As in Tours, R. Vivier, "L'application."

79. In Tours, R. Vivier, "L'application." In Angoulême, D. P. Albert, "Étude comparative des mesures anciennes du Département de la Charente," *Etudes Locales*, Angoulême, XIX, 1938, 177, pp. 65-68.

80. Communication from Carcassonne, 11 December 1792, AN, F^{12}, 1288.

81. See below, discussion of Napoleon's makeshift "Système Usuel."

82. F. Gerbaux and C. Schmidt, eds., *op.cit.*, IV, pp. 775-776, session 115. AN, F^{12}, 210, pp. 119-120.

84. F. Brunot, *Histoire de la langue française des origines à 1900*, IX: *La Révolution et l'Empire*, Pt. II, "Les événements, les institutions et la langue," Paris, 1937, p. 1159.

85. "Extrait d'un mémoire envoié à M. le Controleur Général par . . . Insp^r de manuf^{res} à Grenoble, le 14 juillet 1735, AN, F^{12}, 1288.
86. 3 floréal, year III, AN, F^{12}, 210, pp. 5-6.
87. 4 floréal, year III, *ibid.*, p. 8.
88. 17 thermidor, year III, *ibid.*, p. 71.
89. 14 nivôse, year IV, *ibid.*, pp. 166-167.
90. Agence des Poids et Mesures au Conseil des Anciens, 14 nivôse, year IV, AN, F^{12}, 210, pp. 166-167; similarly on the same day to the Conseil des Cinque Cents; *ibid.*, p. 167; and C^{en} Legarde, Secr^e du Directoire Exécutif, *ibid.*
91. E.g. letter dated 22 vendémiaire, year VI, AD Doubs, L, 2180.
92. Circular from the Minister of the Interior to the prefects of 18 pluviose, year XIII, AEG, ADL, 522; later examples, of 1802 and, indeed, of 1837, in B. Gille, *op.cit.*, pp. 114, 198.
93. R. Vivier, "L'application."
94. *Ibid.*; some mayors would warn local people of forthcoming controls.
95. G. Bigourdan, *op.cit.*, R. Vivier, "L'application" (reference to 18 brumaire).
97. G. Bigourdan, *op.cit.*, *Archives Parlementaires*, XLI, p. 100; and XLIII, p. 653 (session 21 May 1792).
98. A. M. Vassalli-Eandi, *Saggi del nuovo sistema metrico col rapporto delle nuove misure alle antiche misure francesi et a quelle del Piemonte, di*, edizione terza accresciuta, Turin, 1806, pp. XI-XII.
99. AN, F^{12}, 1289.
100. Circular, year VI (exact date not given); *ibid.*
101. R. Vivier, "L'application."
102. AEG, ADL, 522.
103. Ministre de l'Intérieur aux Préfets . . . 23 August 1806, AD Doubs, M, 2390.
104. Notebook of the report in AN, F^{12}, 1289.
105. Circulaire de la Commission des Travaux Publiques aux ingénieurs en chef des départements pour leur demander l'état de l'esprit publique relativement au nouveau système métrique, C. Schmidt, ed., *Le commerce*, p. 253, no. 295.
106. For the circulars, see AD Doubs (Besançon), ser. M, no. 2390, and elsewhere.
107. *Ibid.*
108. Ministre de l'intérieur aux Préfets . . . novembre 1812, AEG, ADL, 522.
109. Ministre de l'Intérieur au Préfet du Département du Leman, 22 July 1812, AEG, ADL, 522.
110. Ministre de l'Interieur aux Préfets, 14 November 1812, *ibid.*
111. De Rioccour, "Les monnais lorrains" (Seconde partie), *Memoires de la Société d'Archéologie Lorraine et du Musée Historique Lorrain*, ser. III, vol. XII, Nancy, 1884, p. 31.

112. Referred to by Vaublanc in circular to the prefects, 23 February 1816, AD Doubs, M, 2390.

113. Napoleon Bonaparte, *Proclamations, ordres du jour et bulletins de la Grande Armée*, Paris, 1964, p. 137.

114. Rapport presenté au Roy, 4 July 1814, AN, F¹², 1289.

115. Ministre de l'Intérieur aux Préfets, 23 Feb. 1816 (arrété of 21 Feb. enclosed), AD Doubs, M, 2390.

116. M. J. Guillaume, "Procès-verbaux du Comité d'Instruction Publique de la Convention Nationale, publiés et annotés par . . . ," vol. III, Table Générale, pt. II, Paris, 1957, p. 213.

117. M. F. Viaud, *Cours de poids et mesures*, I: "Législation des poids et mesures," Paris, 1933, p. 13; R. Vivier, "L'application," regards the 1812 reform as an unexpected blow to the reform when it was near completion.

118. H. Moreau, *Les récents progrès du système métrique, 1948–1954*, Paris, XXX 1955, p. 58, takes Martin's views seriously.

119. For some post-1840 survivals of traditional measures, see T. Sourbé, "Observations du Sr . . . publiciste, sur l'inexécution des lois qui régissent les poids et mesures," A Monsieur le Ministre du Commerce et des Colonies, Paris, 1881, and D. P. Albert, *op.cit.*

120. Agence . . . au Cᵉⁿˢ composant le Cᵒⁿ Militaire séante Maison de Noailles, 25 frimaire, year IV, AN, F¹², 210, pp. 155-156.

121. For attacks upon the new nomenclature in the Convention, see F. Brunot, *Histoire de la langue française*, IX, pt. II, pp. 1147-1168; and F. Gerbaux, and C. Schmidt, eds., *op.cit.*, IV, pp. 775-776.

CHAPTER 24

1. [Talleyrand], Proposition faite à l'Assemblée Nationale, sur les poids et mesures, par M. l'Évêque d'Autun . . .

2. *Archives Parlementaires de 1787 à 1860*, XV, p. 443 (session held 8 May 1790).

3. For the great diversity of weights and measures in Europe at the time (e.g. 391 different values of "the pound"), see H. T. Wade, *op.cit.*, and W. Mühe, "Mass- und Eichwesen" in *Handwörterbuch der Sozialwissenschaften*, VII, 1961, pp. 226-233.

4 All cultural borrowings involve some adaptation, and the metric system, being very rigid, does not lend itself at all well to any modifications.

5. A rare instance of formal annulment, mentioned in *Instruzione sulle misure e sui pesi che si usano nella Repubblica Cisalpina*, publicata per ordine del Comitato Governativo, Milano anno X, 1801.

6. See ASM, Commercio parte antica, c. 225, 226, and parte moderna, c. 229, 230, 231, 232, nd 391.

7. C. Beccaria Milanese, "Relazione della riduzione delle misure di lungheze all'uniformità per lo Stato di Milano," *Scrittori classici italiani*

di Economia Politica, Parte moderna, XII, Milan, 1804, pp. 243-313 (report submitted on 25 January 1780).

8. *Raccolta delle leggi proclami, ordini et avvisi publicati in Milano nell'anno VI Repubblicano*, Milano, IV, p. 217.

9. ASM, Commercio, seria nuova, c. 229.

10. See AST, Seczione I. Materie Economiche. Pesi e misure, fascicles 1-8; and Seczione II, Seziono Riunite, Archivio Sistemato. Commercio. Pesi e misure (two fascicles).

11. *Riduttione o sia tariffa delle diversità delle misure e pesi antichi delle città, terre e luoghi delli stati di quà de' Monti del Serenissimo Carlo Emanuelle duca di Savoia, alla equalità delle misure e pesi de S. Altezza Sereniss. nuovamente stabiliti*, in Torino, 1613.

12. *Parere, della Reale Accademia delle Scienze di Torino, intorno alle misure ed ai pesi* (publ. 30 April 1816 and 28 July 1816, 29 pp., private collection, Turin).

13. A. M. Vassalli-Eandi, *Saggio del nuovo sistema metrico* . . .

14. *Registro No. 2 degli Ordinati della Reale Accademia*, principiato 13 Jan. 1790, e finito le 8 piovoso anno 7-mo repubb. (27 Jan. 1799), p. 306ff., Accademia delle Scienze, Turin, MS.

15. *Mémoires de l'Académie Impériale des Sciences, Littérature et Beaux-Arts de Turin pour les années XII et XIII, Turin, an XIII*, 1805, pp. XLVIII-L; (the author was unable to find these reports).

16. *Mémoire Historique de l'Académie des Sciences, depuis 1792 jusqu'à 1805*, p. LXXVII.

17. E.g., G. Fornaseri, *L'Archivio del Departamento della Stura nell'archivio di Stato di Cuneo, 1799–1814. Inventario a cura di* . . . , Quaderni delle "Rassegna degli Archivi di Stato," II, Rome, 1960, pp. 67-68.

18. E.g. the sentence of 22 Aug. 1805 on the miller who failed to submit his measures for check; AST, I, 2.

19. G. Fornaseri, *op.cit.*

20. *Annali della Reale Accademia delle Scienze dal di 7 de agosto 1815 al di 7 di marzo 1818*, Mémoire della Reale Accademia delle Scienze di Torino, XXIII, Turin, 1818, pp. VI-VII, XIII-XIV.

21. *Parere della Reale Accademia delle Scienze di Torino intorno alle misure ed ai pesi*, Turin, 1816, 39 pp.; (copy in the Library of the Academy of Sciences, Turin, DD VI 39).

22. *Ibid.*, p. 15.

23. *La metrologia comparata ridotta a comune intelligenza o sia teorica del sistema metrico decimale* . . . , Publicazione a favore delle scuole infantili, Turin, 1847, p. 87.

24. AST, I, 2.

25. "Manifesto della Regia Camera de' Conti portante notificanza del nuovo regolamento approvato da S. M. per li pesi e misure," 29 July 1826, AST, II.

26. Notwithstanding the publication of such tables in 1824! *Riduzione*

degli antichi pesi, misure, e monete del Piemonte e delle principati città d'Europa al sistema decimale e viceversa, con apposite tavole di conti fatti dal misuratore Vicenzo Ballesio di S. Maurizio, Turin, Della Stamperia degli Eredi Botta, 1824, pp. 84, AST, II.

27. Inspezione Superiore dei Pesi e delle Misure a S. E. Monsieur le Chevalier Desambroise, Ministre des travaux publiques, de l'Agriculture et du Commerce, Turin, 15 June 1848, AST, I, 5.

28. Rapport fait au Ministre de la Marine, de l'Agriculture et du Commerce par l'Inspection Supérieure des poids et mesures sur les résultats de la vérification de 1850, Turin, 20 April 1851, AST, II.

29. See n. 27, above.

30. Inspezione Superiore dei Pesi e delle misure to the Minister, 16 May 1847, AST, I, 5.

31. "Tables de rapport des anciens poids et mesures des États de terra ferme du Royaume . . . 11 de l'Edict Royale du 11 septembre 1845," Turin, Imprimerie Royale, 1849.

32. Reports from Sardinia, 13 July 1848 and 14 Sept. 1848, AST, I, 5; see also "Ministro per gli Affari di Sardegna. Atlanta dei pesi e delle misure metriche decimali secondo il sistema introdotto nel Regno di Sardegna col Regio Editto del 1-o luglio 1844," Turin, 1844.

33. See n. 27, above.

34. Sources in Archives d'État de Genève, Archives du Département du Leman, vols. 522 and 523.

35. Téron Ainé, *op.cit.*

36. This was not appreciated by the Minister de Neufchâteaux in his communication of 4 brumaire, year XI.

37. According to Téron, *op.cit.*, there was, however, no uniform system of measures in Geneva yet.

38. Prefects of several *départements* protested; for Tours, see R. Vivier, "L'application."

39. *Recueil authentique des lois du gouvernement de la République et Canton de Genève*, II, 1816.

40. *Ibid.*, XXXVIII, 1852.

41. *Ibid.*, LXII, 1876.

42. For the historical metrological literature for Belgium and Holland, see B. H. Slicher van Bath, "Alfabetische Lijst van boeken en tijdschriftartikelen van belang voor de kennis van de oude Nederlandse en Belgische maten en gewichten," *AAG Bijdragen*, no. 11, Wageningen, 1964, pp. 210-221.

43. I am following here H. van der Wee, *op.cit.*, I, pp. 63-103, and M. Tits-Dieuaide, *op.cit.*

44. Ghisebreght, *Tables de conversion ou réduction des anciens poids et mesures de Bruxelles, Louvain, Hal, Nivelles, Diest, Tirlemont, Wavres, Grimbergen, Teralphen, Overyssque et la Hulpe*, Brussels, 1803.

45. See the letter by Talleyrand to the government of the Cisalpine Re-

public dated 11 floréal, year VI (30 April 1798), ASM Commercio, parte antica, cart. 226.

46. F. Engels noted that the proliferation of currencies, weights, and measures in pre-1870 Germany made them all unacceptable in the world market, *Teoria Przemocy* [The theory of violence] Warsaw, 1961, p. 10.

47. Participation by a country in the Convention did not necessarily mean adoption of the metric system.

48. Jefferson had succeeded in establishing decimal coinage in the United States but not decimal weights and measures.

49. C. É. Guillaume, *Les récents progrès du système métrique*. Rapport présenté à la quatrième Conférence générale des poids et mesures, à Paris, en octobre 1907, Paris, 1907; also his report, of the same title, to the 5th conference in October 1913, Paris, 1913; H. Moreau, *op.cit.*

50. *Report from Her Majesty's Representatives on the Metric System* (Parliamentary publication, unavailable to this author).

51. *Report of the Committee on Weights and Measures Legislation*, HMSO, London, 1951, 147 pp.

52. G. S. Aurora, "Economy of a Tribal Village," *Economic Weekly*, 26 September 1964; L. V. Hirsch, *Marketing in an Undeveloped Economy. The North Indian Sugar Industry*, Prentice-Hall, 1961, p. XIII.

53. R. Dumont, *op.cit.*, p. 34.

54. J. Miglioli, "Robotnicy rolni w Brazylii" [Agricultural workers in Brazil]; typescript kindly made available by the author.

55. M. Legris, "Développement agricole et paysans noirs." *Le Monde*, 10-11 Feb. 1966.

56. *Informacje o Chinach, Biuletyn* [Information about China, Bulletin], no. 8, (221), 15 Aug. 1966. India commemorates a standardizing reform in the 4th century B.C.; J. Nehru, *Odkrycie Indii* [Discovery of India], Warsaw, 1957, pp. 120-121.

57. C. S. Chen and L. M. Tseing, "La réforme des poids et mesures en Chine," in C. E. Guillaume, *Les récents progrès* (1913), pp. 110ff.

58. Chen and Tseing faithfully maintain that the basic Chinese unit of length was derived from the flute's length, and at the same time was equal to 90 grains of wheat laid end to end.

59. B. Constant, *De l'esprit de Coquête* (1813), Paris, 1947, pp. 53-54.

60. Chateaubriand, *Mémoires d'outre tombe*, pt. II, bk. 30, chap. 5 (Ambassade à Rome, 1828).

61. A. O. Makovyelski, *Istoria logiki*, Moscow, 1967, p. 46.

SOURCES AND BIBLIOGRAPHY

1. Manuscript Sources

Archive of Ancient Records in Warsaw (AGAD):
 Council of State of the Kingdom of Poland (I) — 143, 308, 309a
 Council of State of the Kingdom of Poland (III) — 904
 Secretariat of State of the Kingdom of Poland — 614/1862, 1/1820
 Governmental Commission for Revenues and Treasury — 1394, 2578
 Governmental Commission for Internal Affairs and Police — 44, 45, 5514, 5515, 7709-7718
 Commission of the Mazovian Voivodeship — 137
 Public Archive of the Potocki Family — 84

Archive of the City of Cracow:
 Free City of Cracow — V. 195

Czartoryski Collection:
 Łojko's Files — 1084, 1086

Warsaw University Library, MSS:
 Old Collection — 99

National Library, Warsaw:
 MSS — 5942, 5943

Historical Institute of the Polish Academy of Sciences, Records Department, Cracow:
 Pawiński's Files — 1, 2, 9, 13, 15, 18, 24, 29, 32, 34
 Ulanowski's Files — 18, 63, 64

Archives Nationales, Paris:
 series F^{12} — 209-211, 1287-1289, 2184
 series F^{III} — 109

Archives Départementales de la Charente (Angoulême)
 series L: 185-188, 590, 884, 1309[1] 1317, 1545
 series G: 787

Archives Départementales de Doubs (Besançon)
 series L: 882, 1080, 1142, 1343, 1518, 1548, 1679, 1799, 1841, 1859, 1869, 1900, 1940, 2021, 2038, 2094, 2121, 2156, 2180, 2857

Archives Départementales de Meurthe-et-Moselle (Nancy)
series L: 655, 839, 1040, 1280, 1282, 1560, 1869, 2170, 2579

Archives d'État de Genève
Archives du Département du Leman: 522, 523

Archivio di Stato a Torino
sezione I. Materie Economiche. Pesi e Misure: 1-8
sezione II. Archivio sistemato. Sezioni Riunite. Commercio. Pesi
e misure: 1, 2

Archivio di Stato a Milano
Commercio, Parte antica: 225, 226
Commercio, Parte moderna: 229-232, 391

Biblioteca Reale di Torino
MSS Miscellanea: 19-37, 39-62, 87-13, 160-bis-31, var. 414, misc.
83-44, 83-16

Biblioteca della Reale Accademia delle Scienze di Torino
MSS Registro 2 degli Ordini della R. Accademia

2. Bibliographies, Dictionaries, Museum Catalogues

Burguburu, P., *Catalogue des poids anciens* . . . , *Ville de Bordeaux*,
Bordeaux, 1936.
———. *Essai d'une bibliographie métrologique universelle*, Paris, 1932.
*Conservatoire Nationale des Arts et Métiers, Catalogue du Musée. Section
K. Poids et Mesures. Métrologie*, Paris, 1941, p. 222.
Ferraro, A., *Dizionario di metrologia generale. Nuova edizione aggior-
nata*, Bologna, 1959, p. 270.
Forien de Rochesnard, J., *Catalogue générale des poids*, nr.2: *Poids
d'Afrique*, Alliance Numismatique Européenne, Anvers, 1959.
Forien de Rochesnard, J., and J. Lugan, *Catalogue générale des poids*,
Alliance Numismatique Européenne, Anvers, 1959.
"Musée Toulouse-Lautrec. Anciens poids et mesures," offprint
from *Revue du Tarn*, 1966, 42, pp. 213-220.
Sahlgren, Nils, "Aldre svenska spannmalsmatt. En metrologisk stu-
die," *Nordiska museets Handlingar* 69, 1968.
Serrure, R., *Catalogue de la collection de poids et mesures, par* . . . ,
Musée Royal d'Antiquités et d'Armures, Brussels, 1883.
Slicher van Bath, B. H., "Alfabetische Lijst van boeken en Tijd-
schriftartikelen van belang voor de kennis van de oude Nederlandse en Belgische maten en gewichten," *AAG Bijdragen*,
no. 11, Wageningen, 1964, pp. 210-221.

Vlajinac, M., *Rečnik našikh starikh miera y toku viekova*, Sveska 1 and 2, Belgrade 1961 and 1964, p. 338.

World Weights and Measures. Handbook for Statisticians, United Nations Statistical Papers, series M, no. 21, rev. 1, New York, 1966, p. 138.

Zupko, Ronald E., *A Dictionary of English Weights and Measures from Anglo-Saxon Times to the Nineteenth Century*, University of Wisconsin Press, 1969.

3. Early Printed Works (down to ca. 1815)

"Annali della Reale Accademia delle Scienze dal di 7 de agosto 1817 al di 7 di marzo 1818," *Mémoire della Reale Accademia delle Scienze di Torino*, vol. XXIII, Torino, della stamperia Reale, 1818.

Babeuf, *Mémoire peut-être important pour les propriétaires de terres et de seigneuries ou idées sur la manutention des fiefs*, Paris, 1786.

———. *L'Archiviste-Terriste ou traité méthodique de l'arrangement des archives seigneuriales et de la confection et perpétuation successive des inventaires, des titres et des terriers d'icelles, des plans domaniaux féodaux et censuels*, Paris, 1786.

Balducci Pegolotti, Francesco, "Libro di divisamenti di Pesi e di Misure per il Mercatanti" (written ca. 1330), in *Trattato della Decima Fiorentina*, Lucca, 1766.

Biron, *Rapport fait à la Société de Médecine de Paris, séante au Louvre. Séances des 21 et 27 pluviôse an X. Sur l'application des nouveaux poids et mesures dans les usages de la médecine, par les citoyens . . . Brasdor et Pelletier, publié par l'ordre du Ministre de l'Intérieur*, à Paris, de l'imprimerie des sourds-muets, rue et Faubourgs Saint Jacques no. 115, an X, p. 32.

Bonnay, de, *Rapport fait au nom de Comité d'Agriculture et de Commerce sur l'uniformité à établir dans les poids et mesures, par M. le Marquis . . . , député du Nivernois, membre de ce Comité et opinion de M. Bureaux de Pusy sur le même sujet, Imprimé ensemble par ordre de l'Assemblée Nationale du 6 mai 1790, séance du soir*, à Paris, de l'Imprimerie Nationale, 1790.

Borgo da Luca, *Summa di aritmetica*, 1494.

Boudé, *De Asse*, 1513.

Bourdot de Richebourg, Charles A., *Nouveau Coutumier Général. Corps de Coutumes Générales et Particulières de France et des provinces connues sous le nom des Gaules . . . par . . . , avocat au Parlement*, à Paris, 1724, 4 vols. in 8 fascicles.

Chaudois, J.J.H., *Tables de comparaisons des poids et mesures metriques avec poids et mesures, anciennement en usage dans les communes du département de la Haute-Saône, publiées par ordre de M Hilaire Préfet de la Haute-Saône, Baron de l'Empire, chevalier de la Légion d'Honneur, Redigées par . . . contrôleur des contributions directes . . .* , à Vesoul, 1810, p. 102.

Chodkiewicz, A., *Tablice stosunku dawnych miar i wag francuskich i koronno-litewsko-polskich z miarami i wagami nowymi a przyjętymi we Francji* [Tables of relations between former French weights and measures and those of Poland-Lithuania and the weights and measures newly adopted in France], Warsaw, 1811.

Ciscar, Don Gabriel, *Memoria elemental sobre los nuevos pesos y medidas decimales fondados en la naturaleza*, Madrid, en la Imprenta Reale, 1800, p. 62.

Czacki, T., *O litewskich i polskich prawach* [Lithuanian and Polish laws], vol. I, Warsaw, 1800.

Delamare, *Traité de la police, par . . .* , vol. II, Paris, 1722.

Dębski, Maciej, Kalendarz Polski i Ruski na rok 1794 przez . . . filozofii doktora, jeometrę JKM-ci ułożony [Polish and Ruthenian Calendar for 1794, prepared by . . . Ph.D., royal surveyor] Cracow Academy Printers.

Dictionnaire Universel de commerce, d'histoire naturelle et des arts et métiers . . . , Ouvrage posthume du Sieur Jacques Savary des Bruslons, inspecteur général des Manufactures, pour le Roy, à la Douane de Paris . . . , à Copenhague, chez les Frères Cl. et Ant. Philibert, 1759-1765.

Dino F., *Libro di mercanzie e misure*, 4 vols., Firenze, 1481.

Encyclopédie ou dictionnaire raisonné des sciences, des arts et des métiers (Diderot et al.), à Neufchâtel, 1751-1765.

Edict wegen allgemeiner Regulierung des Maasses und Gewichts in der Provinz Südpreussen, de dato Berlin den 31 Januar 1796 (in German and Polish).

Estienne Charles, *L'agriculture et Maison Rustique de MM., et Jean Liebault, docteur en medecine . . .* , à Lyon, 1618.

Franiatte J. B., *Tableau des anciens poids et mesures en usage dans le ci-devant Haute Auvergne comparés aux poids et mesures du nouveau système métrique*, Riom, 1802 (a copy in the Library Arch. Départ. de Cantal, cote II. 10150).

Franz der Zweite, *Sr. k.k. Majestät . . . politische Gesetze und Verordnungen für österreichischen, böhmischen und galizischen Erbländer*, vol. 16 (1 I–31 XII 1801), Vienna, 1815.

Ghisebreght, *Tables de conversion ou réduction des anciens poids et me-*

sures de Bruxelles, Louvain, Hal, Nivelles, Diest, Tirlemont, Wavres, Grimbergen, Teralphen, Overussque et La Hulpe, Brussels, 1803.

Godart J., Robin L., *Rapport de Messieurs . . . , commissaires civils, envoyés par le Roy dans le Département du Lot. en exécution du Décret de l'Assemblé Nationale du 13 Décembre 1790, remis ou Roy le 6 avril par . . . en présence de M. Duport, ministre de la Justice, et présenté par lui à Sa Majesté. Imprimé par ordre de l'Assemblée Nationale à Paris, de l'Imprimerie Nationale 1791*, p. 139, Bibl. Nat. 8° Le^{29} 1410.

Haur J. K., *Oekonomika ziemiańska generalna*, Warsaw, 1744. [General management of landed estates].

Herburt J., *Statuta Regni Poloniac ab . . . digesta*, Cracow, 1563.

Hersenne, *Parisienses Mensurae*, 1640.

Instituto Ligure. Tableau du Département de Gênes, 1811.

Instruction sur les mesures déduites de la grandeur de la terre uniformes pour toute la République et sur les calculs relatifs à leur division décimale, par la Commission temporaire des poids et mesures républicaines en exécution des Décrets de la Convention Nationale à Nancy . . . an II-e de la République une et indivisible, pp. XXXII, 224 (with numerous tables).

Instrukcja do wymiaru zboża laską korcową zwaną przez JW Tadeusza Dembowskiego Ministra Skarbu dnia 21 maja 1809 aprobowaną [Instructions for measuring grain with a bushel rod, approved by the Treasury in 1809], Warsaw, The Piarists' Printing Works, 1810.

Instruzione sulle misure e sui pesi che si usano nella Repubblica Cisalpina, publicata per ordine del Comitato Governativo, Milano, anno X, 1801, pp. XIV, 128.

Jabłonowska, A., *Ustawy powszechne dla rządców dóbr i instruktarz ekonomiczny* [General guidelines for administrators of estates and a manual of husbandry], Siemiatycze, 1785, Warsaw, 1786 and 1787.

Jacquet, avocat au Parlement de Paris, *Traité des justices de seigneur et des droits en dépendants, conformément à la jurisprudence actuelle des différents tribunaux du Royaume, suivi des pièces justificatives qui ont trait à la matiere*, à Lyon, chez Luis Cellot . . . et les frères Estienne, 1764, avec approbation et privilège du Roy (Libr. Mazarine, A.10.051).

Konstytucje i przywileje na Walnych Sejmach Koronnych od roku 1550 aż do roku 1569 uchwalone [Statutes and privileges passed by the Great Seym from 1550 to 1569], Cracow, printed by Mikołaj Schafenberger.

Köbel, J., *Geometry*, Franckfort am Mayn, 1584.

"Krótka wiadomość o nowych miarach francuskich i dawniejszych polskich" [A short note on the new French and old Polish measures], *Pamiętnik Lwowski*, V, 1817, pp. 329-336.

Łempicki, I., "O porównaniu miar i wag zagranicznych z krajowymi końcem wyjaśnienia dzieł o gorzelnictwie . . ." [A comparison of foreign and our measures, to clarify works on distilling], *Pamiętnik Lwowski*, IV, 1819, pp. 113-119.

Mémoires de l'Académie Impériale des Sciences, Littérature et Beaux-Arts de Turin pour les années XII et XIII, Turin, an XIII, 1805.

Mémoire historique de l'Académie des Sciences depuis 1792 jusqu'à 1805.

Mirabeau, de, marquis Victor Riquetti, *Philosophie rurale . . .* , Amsterdam, 1763.

————. *L'Ami des hommes ou Traité de Population*, Avignon, 1756.

Moniteur Universel, 1790–1795.

Nicolai N. M., *Memorie, leggi et osservazioni sulla campagna e nell'Annona di Roma*, parte terza, Rome, 1808.

Osiński J., *Opisanie polskich żelaza fabryk . . .* , Warsaw, 1782.

Ostrowski Tomasz X. S. P., *Prawo cywilne albo szczególne narodu polskiego . . . ułożone przez . . .* , vol. II, w Warszawie, w drukarni JKM i Rzeczypospolitej u XX Scholarum Piarum, 1784.

Parere della Reale Accademia delle Scienze di Torino intorno alle misure ed ai pesi. Della Stamperia Reale, Torino 1816, p. 39. (Bibl. Akademii Nauk w Turynie, klocek sygn. DD VI 39).

Pasi Bartholomeo, *Tariffa dei Pesi e Misure*, Venice, 1540.

[Paucton], *Métrologie ou traité des mesures, poids et monnaies des anciens peuples et des modernes*, à Paris, chez la veuve Desaint, libraire rue du Foin, 1780, avec approbation et privilège du Roi, pp. XVI, 955.

Poix de Frémonville, *La pratique universelle pour la rénovation des terriers et des droits seigneuriaux . . . par Edme de la . . . , bailly des villes et marquisat de la Palisse, comissaire aux droits seigneuriaux*, à Paris, chez Morel ainé et Gissay, 1746, avec l'approbation et privilège du Roy.

Postanowienia dla miasta JKM Osiecka przez Komisję Dobrego porządku dekretem sądów królewskich utwierdzone, w Warszawie 1785.

Raccolta delle leggi, proclami ed avisi publicati a Milano nell'anno VI Repubblicano, Milan, an VI.

Recherches sur les anciennes monnoies du Comté de Bourgogne avec quelques observations sur les poids et mesures autrefois en usage dans la même Province. Ouvrage qui a remporté le prix au jugement de l'Académie de Besançon, par un Bénédictin de la Congregation de St.

Vanne, membre de plusieurs Académies, à Paris, chez Nyon . . . , à Besançon, chez les principaux libraires, 1782, p. 221.

Riduttione o sia tariffa delle diversità delle misure e pesi antichi delle città, terre e luoghi delli stati di quà de' Monti del Serenissimo Carlo Emanuelle duca di Savoia, alla equalità delle misure e pesi de S. Altezza Sereniss. nuovamente stabiliti, in Torino, apresso a Luigi Pizzamiglio, Stampator Ducale, 1613.

Roland, *Traité par abrégé sur le changement des mesures, poids, monnoies, papier-monnoie et Eres ordonnés par le Gouvernement, depuis 1790 jusqu'à présent . . . , par le S^r . . . géomètre, ex-administrateur de Département du Doubs*, à Besançon, de l'imprimerie de Cl.-Fr. Mourgeon, 1810, 144.

Sapieha, A., *Tablice stosunku nowych miar i wag francuskich z litewskimi i polskimi miarami i wagami*, Roczniki Tow. Warsz. Przyj. Nauk, Warsaw, 1802.

Schelenius, J., *Cursus mathematici, IV Theil, darinnen ist Geodesia . . .* , Reval, 1665.

Skrzetuski, Wincenty X., S. P., *Prawo polityczne narodu polskiego, przez . . .* , t. II, w Warszawie, w drukarni JKM i Rzeczypospolitej u XX scholarum Piarum, 1784.

Solski, S., *Geometra polski*, Cracow, 1683.

Swinden van M., *Lettre de . . . professeur de physique et Membre de la Commission des poids et mesures de l'Institut National de France, à son collègue Vassali-Eandi*, Amsterdam, juillet 1802.

Tableau des poids et mesures du Département de Montenotte, Savone, 1811.

Tables de comparaison entre les mesures anciennes du Département de la Charente, et celles qui les remplacent dans le nouveau système métrique . . . , Rédigées par la Commission temporaire des poids et mesures du même Département . . . , à Angoulême, an X.

Tables de comparaison entre les mesures anciennes et celles qui les remplacent dans le nouveau système métrique, avec leur explication et leur usage, pour le département de la Haute-Garonne, Toulouse, an X.

[Talleyrand], *Proposition faite à l'Assemblée Nationale, sur les poids et mesures, par M. l'Éveque d'Autun*, à Paris, de l'Imprimerie Nationale, 1790.

Tariffe particolari di tutte le diversità de Pesi e Misure vecchie de Stati di S. A. Serenissima di qua da Monti per far con facilità e enqvisitezza tutte le riduttioni de Cotastri o registri et altre misure e pesi antichi d'essi Stati nelli nuovi et uquali, e per unità, che si possono desirata, in Torino, apresso a Luigi Piezamiglio, stampator Ducale, 1513.

Taripha de pesi e misure di Miser Bartholomeo Di Paxi da Venezia, per Albertin da Lizona Vercellose, Venice 1503.

Téron Ainé, *Instruction sur le système de mesures et poids uniformes pour toute la République Française . . . , par . . . , teneur de livres et maître d'arithmetique*, à Genève, an X, 10 prairial.

Tournon, le Comte de, *Études statistiques sur Rome et la partie occidentale des États Romains, contenant une description topographique et les recherches sur la population, l'agriculture, les manufactures, le commerce, le gouvernement, les établissements publiques, et une notice sur les travaux exécutés par l'administration française par . . . , Pair de France, Grand-Officier de la Légion d'Honneur, préfet de Rome de 1810 à 1814*, 2 vols., Paris, 1832.

Uwagi praktyczne o poddanych polskich względem ich wolności i niewoli [Practical comments on the liberty and slavery of Polish serfs], Warsaw, 1790.

Uzzano, da Giovanni, "Libro di Gabelle, pesi e misure di più e diversi luoghi" (written in 1440), in *Trattato della Decima Fiorentina*, Lucca, 1766.

Vassalli-Eandi, A. M., *Saggio del nuovo sistema metrico col rapporto delle nuove misure alle antiche misure francesi ed a quelle del Piemonte di . . . , professore di fisica cell Imp. Università di Torino*, edizione terza accresciuta, Presso i Fratelli Pomba, Turin, 1806, pp. XII, 296.

Venturi, *Rapporto della Commissione di commercio al Gran Consiglio sopra il nuovo campione di misura lineare con annotazioni del Cittadino . . . , rappresentante del popolo*, Milano, dalla Tipografia Nazionale, anno VI, p. 99.

Zebranie diariuszów trzech walnych sejmów [Collected diaries of three Great Sejms], *Convocationis, Electionis et Coronationis, 1764*, Warsaw, 1765.

4. EDITED SOURCES: DOCUMENTS AND AUTHORS

4.1. Inventories [Lustracje, sing. Lustracja—TRANS.] *of Royal Domains, 16th to 18th centuries*

LESSER POLAND

Lustracja województwa KRAKOWSKIEGO 1564 [Cracow voivodeship, TRANS.], pt. I, ed. J. Małecki, Warsaw, 1962, PWN.

Lustracja województwa KRAKOWSKIEGO 1564, pt. II, ed. J. Małecki, Warsaw, 1964, PWN.

Lustracja województwa KRAKOWSKIEGO 1789, pt. I, krakowski, proszowicki and książski districts, eds. A. Falniowska-Gradowska and I. Rychlikowa, PAN, Cracow, Collection of the Commission for Historical Studies no. 4, Wrocław-Warsaw-Cracow, 1962, Ossoliński Institute.

Lustracja województwa KRAKOWSKIEGO 1789, pt. II, lelowski district, and the kłobuckie and brzeźnickie *starostships*, ed. A. Falniowska-Gradowska and I. Rychlikowa, PAN, Cracow, Collection of the Commission for Historical Studies no. 7, Wrocław-Warsaw-Cracow, 1963, Ossoliński Institute.

Lustracja województwa LUBELSKIEGO 1565 [Lublin voivodeship— TRANS.], ed. A. Wyczański, Wrocław-Warsaw, 1959, Ossoliński Institute.

Lustracja województwa LUBELSKIEGO 1661, ed. H. Oprawko and K. Schuster, Warsaw, 1962, PWN.

Lustracja województwa SANDOMIERSKIEGO 1564–1565 [Sandomierz voivodeship—TRANS.], ed. W. Ochmański, Wrocław-Warsaw-Cracow, 1963, Kielce Scientific Society, Ossoliński Institute.

Lustracja województwa SANDOMIERSKIEGO 1789, pt. I, sandomierski and opoczyński districts and stężycka *ziemia*, ed. H. Madurowicz-Urbańska, PAN, Cracow, Collection of the Commission for Historical Studies no. 10, Wrocław-Warsaw-Cracow, 1964, Ossoliński Institute.

Lustracja województwa SANDOMIERSKIEGO 1789, pt. II, radomski district, ed. H. Madurowicz-Urbańska, PAN, Cracow, Collection of the Commission for Historical Studies no. 12, Wrocław-Warsaw-Cracow, 1967, Ossoliński Institute.

MAZOVIA

Lustracja województwa MAZOWIECKIEGO 1565, [Mazovia voivodeship—TRANS.] pt. I, ed. I. Gieysztorowa and A. Żaboklicka, Warsaw, 1967, PWN.

Lustracja województwa MAZOWIECKIEGO 1565, pt. II, ed. I. Gieysztorowa and A. Żaboklicka, Warsaw, 1968, PWN.

Lustracje województwa PŁOCKIEGO 1565–1789 [Płock voivodeship— TRANS.], ed. A. Sucheni-Grabowska and S. M. Szacherska, Warsaw, 1964, PWN.

Lustracja województwa RAWSKIEGO 1564 and 1570 [Rawa voivodeship—TRANS.], ed. Z. Kędzierska, Warsaw, 1959, PWN.

Lustracje województwa RAWSKIEGO XVII w. [17th century—TRANS.], ed. Z. Kędzierska, Wrocław-Warsaw-Cracow, 1965, Ossoliński Institute.

PODLASIE

Lustracja województwa PODLASKIEGO 1570 and 1576 [Podlasie voivodeship—TRANS.], ed. J. Topolski and J. Wiśniewski, Wrocław-Warsaw, 1959, Ossoliński Institute.

ROYAL PRUSSIA

Lustracja województwa MALBORSKIEGO i CHEŁMIŃSKIEGO 1565 [Malbork and Chełmno voivodeships—TRANS.] ed. S. Hoszowski, Gdańsk Scientific Society, Department I of Social and Humanistic Studies, Gdańsk, 1961.
Lustracja województwa MALBORSKIEGO i CHEŁMIŃSKIEGO 1570, ed. S. Hoszowski, Gdańsk Scientific Society, Department I of Social and Humanistic Studies, Gdańsk, 1962.
Lustracja województwa POMORSKIEGO 1565 [Pomeranian voivodeship—TRANS.], ed. S. Hoszowski, Gdańsk Scientific Society, Department I of Social and Humanistic Studies, Gdańsk, 1961.
Lustracja województwa PRUS KRÓLEWSKICH z roku 1624 z fragmentami lustracji 1615 r. [voivodeships of Royal Prussia—TRANS.], ed. S. Hoszowski, Gdańsk Scientific Society, Department I of Social and Humanistic Studies, Gdańsk, 1967.

GREATER POLAND

Lustracja województwa WIELKOPOLSKICH i KUJAWSKICH 1564–1565 [voivodeships of Greater Poland and Cuiavia—TRANS.] pt. I ed. A. Tomczak, C. Ohryzko-Włodarska, J. Włodarczyk, Bydgoszcz Scientific Society, Bydgoszcz, 1961, PWN.
Lustracja województwa WIELKOPOLSKICH i KUJAWSKICH 1564–1565, pt. II, ed. A. Tomczak, C. Ohryzko-Włodarska, J. Włodarczyk, Bydgoszcz, 1963, Bydgoszcz Scientific Society.
Lustracja województwa WIELKOPOLSKICH i KUJAWSKICH 1628–1632, pt. I, Poznań and Kalisz voivodeships, ·ed. Z. Guldon, Wrocław-Warsaw-Cracow, 1967, Ossoliński Institute.
Lustracja województwa WIELKOPOLSKICH i KUJAWSKICH 1628–1632, pt. II, Leczyca, Brześć-Cuiavia and Inowrocław voivodeships and dobrzyńska *ziemia*, ed. Z. Guldon, Bydgoszcz, 1967, Bydgoszcz Scientific Society.

4.2. Resolutions (Lauda) and Records of Dietines

Lauda Sejmików Ziemi DOBRZYŃSKIEJ [Dobrzyn], ed. F. Kulczycki, PAU, Cracow, 1887.
Lauda Sejmikowe HALICKIE [Halicz], ed. W. Hejnosz, Akta Grodzkie

i Ziemskie (AGZ) z Archiwum Ziemskiego we Lwowie, vol. XXIV: 1575–1695, vol. XXV: 1696–1772, Lvov, 1931–1935.

Akta sejmikowe województwa KRAKOWSKIEGO [Cracow], ed. S. Kutrzeba (vol. I) and A. Przyboś (vols. II–IV), vol. I: 1572–1620, vol. II: 1621–1660, vol. III: 1661–1673, vol. IV: 1674–1680, PAU & IH PAN, Cracow-Wrocław, 1932–1963.

Lauda i instrukcje. Dzieje Ziemi KUJAWSKIEJ [Cuiavia], ed. A. Pawiński, vol. II-IV, Warsaw, 1888.

Akta Sejmikowe województwa POZNAŃSKIEGO i KALISKIEGO [Poznań and Kalisz], ed. W. Dworzaczek, vol. I, pts. 1 and 2, 1572–1632, PTPN, Poznań, 1957–1962.

Lauda sejmikowe WISZEŃSKIE, LWOWSKIE, PRZEMYSKIE i SANOCKIE [Wisznia, Lvov, Przemyśl and Sanok], ed. A. Prochaska, Akta Grodzkie i Ziemskie (AGZ) z Archiwum Ziemskiego we Lwowie, vol. XX: 1572–1648, vol. XXI: 1648–1673, vol. XXII: 1673–1732, vol. XXIII: 1731–1772, Lvov, 1909–1928.

4.3 Books [Księgi] of Rural Judicial Records

Księgi sądowe wiejskie, ed. B. Ulanowski, vol. I–II, Starodawne Prawa Polskiego Pomniki, vol. XI-XII, PAU, Cracow, 1921.

Księgi sądowe wiejskie klucza JAZOWSKIEGO, 1663–1808, ed. S. Grodziski, Wrocław, 1967.

Księga sądowa kresu KLIMKOWSKIEGO, 1600–1762, ed. L. Łysiak, IHPAN, Wrocław, 1965.

Księgi sądowe wiejskie klucza ŁĄCKIEGO, 1526–1811, ed. A. Vetulani, IH PAN, 2 vols., Wrocław, 1962–1963.

4.4 Manuals for Landed Estates

Polskie instruktarze ekonomiczne z końca XVII i XVIII w., ed. S. Pawlik, 2 vols., PAU, Cracow, 1915–1929.

Polskie ustawy wiejskie XV–XVIII w., ed. S. Kutrzeba and A. Mańkowski, Arch. Kom. Praw. PAU, XI, 1938.

Instrukcje gospodarcze dla dóbr magnackich i szlacheckich z XVII–XIX w., ed. B. Baranowski, J. Bartyś, A. Keckowa and J. Leskiewicz, 2 vols., IHPAN, Wrocław, 1959–1963.

4.5. Cahiers de Doléances

Cahiers de doléances des corps et corporations de la ville d'ALENÇON pour les États Généraux de 1789, ed. R. Jouanne, Alençon, 1929.

Cahiers de doléances du bailliage d'AMONT, ed. G. and L. Abencour, vol. I, Besançon 1918, vol. II, Auxerre, 1927.

Cahiers de doléances des corporations de la ville d'ANGERS et des paroisses de la sénéchaussée particulière d'Angers pour les États Généraux de 1789, ed. A. Le Moy, 2 vols., Angers, 1915–1916.

Cahiers de doléances de la sénéchaussée d'ANGOULÊME et du siège royal de COGNAC pour les États Généraux de 1789, ed. P. Boissonnade, Paris, 1907.

Cahiers de doléances du bailliage d'ARQUES (secondaire de Caudebec) pour les États Généraux de 1789, ed. E. Le Pasquier, 2 vols., Lille, 1922.

Cahiers de doléances des paroisses et communautés du bailliage d'AUTUN pour les États Généraux de 1789, ed. A. de Charmasse, Autun, 1895.

Cahiers des currés et des communautés ecclésiastiques du bailliage d'AUX-ERRE pour les États Généraux de 1789, ed. Ch. Porée, Auxerre, 1927.

Cahiers de doléances du BEAUJOLAIS pour les États Généraux de 1789, publiés par le Conseil Général du Rhône à l'occasion du 150-e anniversaire de la Révolution Française, ed. C. Faure, Lyon, 1939.

Cahiers de doléances de la sénéchaussée de BIGORRE pour les États Généraux de 1789, ed. G. Balencie, 2 vols., Tarbes, 1925–1926.

Cahiers de doléances du bailliage de BLOIS et du bailliage secondaire de ROMORANTIN pour les États Généraux de 1789, ed. F. Lesueur and A. Cauchie, 2 vols., Blois 1908.

Cahiers de doléances . . . du bailliage de BOULEY et du BOUZONVILLE, ed. N. Dorvaux and P. Lesprand, Quellen zur Lothringischen Geschichte, vol. IX, Metz, 1908.

Cahiers de doléances du bailliage de BOURGES et du bailliage secondaire de VIERZON et d'HENRICHEMONT pour les États Généraux de 1789, ed. A. Gandilhon, Bourges, 1910.

"*Les cahiers de doléances du BUGEY et de la DOMBES,*" ed. A. Favre, *Bulletin d'Histoire Économique et Sociale de la Révolution Française*, 1967, pp. 43-50.

Les cahiers d'observations et doléances du Tiers-État de la ville de CAEN en 1789, ed. F. Mouzlot, Paris, 1912.

Cahiers de doléances de la sénéchaussée de CAHORS pour les États Généraux de 1789, ed. V. Fourastié, Cahors, 1908.

Cahiers de doléances. Département de la Marne, vol. I: *Bailliage de CHÂ-LONS-sur-MARNE*, ed. G. Laurent, Épernay, 1906.

Cahiers de doléances du Tiers-État de la sénéchaussé de CHÂTEAU-du-LOIR, ed. P. Bois, Gap, 1960.

Cahiers de doléances. Département de la Marne, vol. III: *Bailliage de CHÂTILLON-sur-MARNE*, ed. G. Laurent, Épernay, 1911.

Cahiers de doléances de la sénéchaussé de CIVRAY pour les États Généraux de 1789, ed. P. Boissonnade et L. Cathelineau, Niort, 1925.

Cahiers de doléances des villes de COSNE, VARZY et de la paroisse de LIGNORELLE, ed. M. C. Demay, *Bulletin de la Société des Sciences Historiques et Naturelles de l'Yonne*, 2° semestre 1886, Auxerre, 1887.

Cahiers de doléances du bailliage de COTENTIN (Coutance et secondaire) pour les États Généraux de 1789, ed. E. Bridery, 3 vols. Paris, 1907, 1908, 1914.

Les doléances des paysans bretons en 1789. Quelques cahiers de paroisses. Territoire de l'actuel département du CÔTES-du-NORD, ed. A. Bourgés, Sainte-Brieuc, 1953.

Mémoire contenant les doléances, remonstrances et réclamations du pays souverain de DONESAN à l'occasion des États Généraux de 1789, ed. F. Pasquier, *Extrait du Bulletin de la Société Ariégeois des Sciences, Lettres et Arts*, Foix, 1890

Cahiers de doléancees des communes du bailliage d'ÉPERNAY en 1789, ed. P. Pélicier, Châlons-sur-Marne, 1900.

Les cahiers de la FLANDRE-MARITIME en 1789, ed. A. de Saint-Léger and Ph. Sagnac, 2 vols., Dunquerque—Paris, 1908-1910.

Cahiers de doléances, remonstrances et demandes du Tiers-État de la ville et du baillage de HAM, ed. A. Bernot, Amiens, 1883.

Cahiers de doléances du bailliage de HAVRE (secondaire de Caudebec), ed. E. le Parquier, Épinal, 1929.

Les cahiers de doléances de l'ISLE-JOURDAIN et du VIGEAN en 1789, ed. P. Boissonade (offprint, n.d.)

Cahiers de doléances des bailliages de LONGUYON, de LONGWY et de VILLERS-la-MONTAGNE pour les États Généraux de 1789, ed. P. d'Arbois de Jubainville, Nancy, 1952.

Cahiers de doléances de la sénéchaussée de MARSEILLE pour les États Généraux de 1789, ed. J. Fournier, Marseille, 1908.

Cahiers de doléances des bailliages des Généralités de METZ et de NANCY pour les États Généraux de 1789, Département de la Moselle, vol. I: *THIONVILLE*, ed., N. Dorvaux et P. Lesprand, Bar-le-Duc, 1922.

Cahiers de doléances du bailliage de MIRECOURT, ed. E. Martin, Épinal, 1928.

Cahiers de doléances en 1789 (NEVERS), Extrait du Bulletin de la Société Nivernaise des Lettres, Sciences et Arts, Nevers, 1896.

Cahiers de doléances de la sénéchaussé de NÎMES pour les États Généraux de 1789, ed. E. Bligny-Bondurand, 2 vols., Nîmes, 1908-1909.

Cahiers de doléances des sénéchaussées de NIORT et de SAINT-MAIXENT et des communautés et corporations de Niort et Saint-Maixent pour les États Généraux de 1789, ed. L. Cathelineau, Niort, 1912.

Cahiers de doléances du Tiers-État de la ville de NOGARO, capitale du bas Armagnac, ed. E. Dellas, Auch, 1897.

Deux cahiers de curés, NORMANDS pour les États Généraux de 1789, ed. M. Bouloiseau, *Annales de Normandie*, VII, 1.

Cahiers de doléances du bailliage d'ORLÉANS pour les États Généraux de 1789, ed. C. Bloch, 2 vols. Orléans 1906-1907.

Cahiers de l'ordre du clergé du bailliage d'ORLÉANS pour la convocation des États Généraux de 1789, ed. A. Couzet, Orléans, 1889.

Les élections et les cahiers de Paris en 1789, ed. Ch. L. Chassin, vol. III, Paris, 1889.

Cahiers de doléances de 1789 dans le département du PAS-de-CALAIS, ed. H. Loriquet, 2 vols., Arras, 1891.

Cahiers de doléances de PONT-à-MOUSSON, ed. Z. E. Harsany, Paris, 1946.

Cahiers de doléances des sénéchaussées QUIMPER et de CONCARNEAU pour les États Généraux de 1789, ed. J. Savina et D. Bernard, 2 vols., Rennes, 1927.

Cahiers de doléances pour les États Généraux de 1789. Département de la Marne. Bailliage de REIMS, ed. G. Laurent, Reims, 1930.

Cahiers de doléances de la sénéchaussé de RENNES pour les États Généraux de 1789, ed. H. Sée and A. Lesort, 4 vols., Rennes, 1909-1912.

Cahiers de doléances du Tiers-État du pays et jugerie de RIVIÈRE-VERDUN pour les États Généraux de 1789, ed. D. Ligou, Gap, 1961.

Cahiers de doléances du Tiers-État du bailliage de ROUEN, ed. M. Bouloiseau, 2 vols. Paris 1957-1960.

Cahiers de doléances de la colonie de SAINT-DOMINGUE pour les États Généraux de 1789, ed. B. Mauriel, Paris, 1933.

Cahiers de doléances des Prévotés bailliagères de SARREBOURG et de PHALSBOURG et du bailliage de LIXHEIM pour les États Généraux de 1789, ed. P. Lesprand et L. Bour, Metz, 1938.

Cahiers de doléances du bailliage de SENS pour les États Généraux de 1789, ed. Ch. Porée, Auxerre, 1908.

Cahiers de doléances de SÉZANNE et CHÂTILLON-sur-MARNE, ed. G. Laurent, Épernay, 1909.

Cahiers paroissiaux des sénéchaussées de TOULOUSE et de Comminges en 1789, ed. E. Pasquier and Fr. Galabert, Toulouse, 1928.

Cahiers de doléances du bailliage de TROYES (principal et secondaires) et du bailliage de BAR-sur-SEINE pour les États Généraux de 1789, ed. J. J. Vernier, 3 vols., Troyes, 1909-1911.

Cahiers de doléances des paroisses du bailliage de TROYES pour les États Généraux de 1614, ed. Y. Durant, Paris, 1966.

Cahiers de doléances pour Varzi, ed. C. Domay, Auxerre, 1886.

Cahiers de doléances de VERDUN, ed. P. d'Arbois et Jubainville (offprint, n.d.).

Deux cahiers de la noblesse, 1649-1651. Problèmes de stratification sociale eds. R. Mousnier, J. P. Labatut, Y. Durand, Paris, 1965.

4.6 Miscellaneous

Akta grodzkie i ziemskie z czasów Rzeczypospolitej Polskiej z archiwum tzw. bernardyńskiego we Lwowie, vols. 20-25, Lvov, 1909-1935.

Baranowski, I. T., ed., *Księgi referendarskie*, vol. I: 1582-1602, Warsaw Scientific Society, Historical Commission, no. 3, Warsaw, 1910.

Beccaria, Cesare Milanese, *Relazione della reduzione delle misure di lunghezze all'uniformità per lo Stato di Milano*, Scrittori classici italiani di economia politica. Parte moderna, vol. XII, Milan, 1804, pp. 243-313.

Biblia Święta to jest całe Pismo Święte Starego i Nowego Testamentu, British and Foreign Bible Society, Warsaw, 1958.

Bostel, F., *Taryfa cen dla województwa krakowskiego z r. 1565*, Arch. Kom. Hist. PAU, vol. VI, 1891.

Burszta, J. and Łuczak, C., eds., *Inwentarze mieszczańskie z wieku XVIII z ksiąg miejskich i grodzkich Poznania*, vol. I: 1700-1758, vol. II: 1759-1793, Poznań, 1962 and 1965.

Petri Mariae Calandri Compendium de agrorum corporumque dimensione (ca. 1596), ed. G. V. Soderini, in *I due trattati della agricoltura*, Bologna, 1902.

Chateaubriand, *Mémoires d'outre-tombe*, pt. II, book 30, chap. 5.

Chmiel, A., *Ustawy cen dla miasta Starej Warszawy od r. 1606 do r. 1627*, Arch. Kom. Hist. PAU, vol. VII, 1894.

Chmielowski, B., *Nowe Ateny*, Warsaw, 1966.

Chomętowski, W., ed., *Diariusz sejmu piotrowskiego RP 1565, poprzedzony Kroniką 1559-1562, objaśnił . . . , wydał Władysław hr. Krasiński*, Biblioteka Ordynacji X Krasickich, 1868.

Coignet, Capitain, *Les cahiers du . . .* , in *Les écrivains témoins du peuple*, Présentation de F. et J. Fourastié, Paris, 1964.

Condorcet, A. N., *Szkic obrazu postępu ducha ludzkiego przez dzieje*, BKF, Warsaw, 1957.

Constant, B., *De l'esprit de Conquête* (1813), coll. Le Jardin du Luxembourg, Paris, 1947.

Deresiewicz, J., ed., *Materiały do dziejów chłopa wielkopolskiego w II połowie XVIII w.*, vol. III, Wrocław, 1957.

Donovan, F., ed., *The Thomas Jefferson Papers*, New York, 1963.

Firdausi, *Opowieść o miłości Zala i Rudabe*, ed. F. Machalski, Wrocław, 1961.

Flawiusz, Jozef, *Starożytności żydowskie*, Warsaw, 1962.

Fléchier, E., *Les Grands-Jours d'Auvergne*, Paris, 1964.

Gerbaux, F. and Schmidt, Ch., eds., *Procès-verbaux des Comités d'agriculture et de commerce de la Constituante, de la Législative et de la Convention*, Collection de documents inédits sur l'histoire économique de la Révolution Française, 4 vols., Paris, 1906-1937.

Gostomski, Anzelm, *Gospodarstwo (1588)*, ed. S. Inglot, Wrocław, 1951.

Górski, J. and Lipiński, E., eds., *Merkantylistyczna myśl ekonomiczna w Polsce XVI i XVII wieku*, Warsaw, 1958.

Grzepski, Stanisław, *Geometria to jest miernicka nauka*, ed. H. Barycz and K. Sawicki, Wrocław, 1957.

Guillaume, M. J., ed., *Procès-verbaux du Comité d'Instruction Publique de la Convention Nationale, publiés et annotés par . . .* , vol. III: Table Générale, fasc. 2, Paris, 1957.

Handelsman, M., ed., *Żywot chłopa polskiego na początku XIX stulecia*, Warsaw, 1907.

Hezjod, *Prace i dnie*, Wrocław, 1952.

Kato, *O gospodarstwie wiejskim*, transl. S. Łoś, Wrocław, 1956.

Keckowa, A., Pałucki, W., eds., *Księgi Referendarii Koronnej z drugiej połowy XVIII w.*, 2 vols., Warsaw, 1955 and 1957.

Kluk, Krzysztof, *O rolnictwie, zbożach, łąkach, chmielnikach . . .* (1779), ed. S. Inglot, Wrocław, 1954.

Konopczyński, W. ed., *Diariusze sejmowe z XVIII w.*, 3 vols., Warsaw, 1911-1937.

Kraushar, A., "Projekt polski z r. 1814 w sprawie ujednostajnienia w Europie miar, wag, monet i wymiarów odległości," from MS, *Miscellanea Historyczne*, III, Lvov, 1904.

Kuraś, St., ed., *Ordynacje i ustawy wiejskie z archiwów metropolitalnego i kapitulnego w Krakowie, 1451-1689*, Cracow, 1960.

Kutrzeba, St., ed., *Materiały do dziejów robocizny w Polsce w XVI wieku*, Arch. Kom. Prawn., PAU, vol. IX, 1913, pp. 1-198.

Labuda, G., ed., "Inwentarze starostw puckiego i kościerskiego z XVII w.," *Fontes*, 39, Toruń, 1954.

Leskiewicz, J., Michalski J., eds., *Supliki chłopskie XVIII w. z archiwum prymasa Michała Poniatowskiego*, Warsaw, 1954.

Magier, A., *Estetyka miasta stołecznego Warszawy*, Wrocław, 1963.

Mahomet, *Le Coran*, transl. by E. Monet, Paris, 1958.

Materiały do dziejów Sejmu Czteroletniego, eds. J. Michalski, E. Rostworowski, J. Woliński, 5 vols., Wrocław, 1955-1960.

Matuszewicz, M., *Pamiętniki . . . 1714-1765*, ed. A. Pawiński, Warsaw, 1876.

Mendthal, H., ed., *Ein agronomischer Traktat aus der Zeit des Hochmeister Ulrich von Jungingen*, Leipzig, 1886.

Montaigne, *Essays*, Paris, 1950.

Nax, Jan Ferdynand, *Wybór pism*, ed. W. Sierpiński, Warsaw, 1956.

Niemcewicz, J. U., *Zbiór pamiętników historycznych o dawnej Polszcze . . .*, vol. III, Warsaw, 1822.

Peirsec de Nicolas Claude Febri, *Lettres de . . .*, ed. Ph. Tamizey de Barroque, 7 vols., Paris, 1888-1898.

Pigafetta, Antonio, *Premier voyage autour du monde par Magellan, 1519-1522*, ed. L. Peillard, Paris, 1964.

Pigafetta, Filippo and Lopez Duarte, *Description du Royaume de Congo et des contrées environnantes par . . .*, ed. W. Bal, Publications de l'Université Lovanium de Léopoldville, Louvain-Paris, 1963.

Rome de Jean-François, *Brève relation de la fondation de la mission des frères mineurs capucins . . . au Royaume de Congo par . . .*, Rome, 1648, ed. F. Bontinck, Publications de l'Université Lovanium de Léopoldville, Louvain-Paris, 1964.

Sadowski, Z., ed., *Rozprawy o pieniądzu w Polsce pierwszej połowy XVII wieku*, Warsaw, 1959.

Sagnac, Ph. and Caron, P., eds., *Les comités des droits féodaux et de législation et l'abolition du régime seigneurial (1789-1793)*, Collection de documents inédits sur l'histoire économique de la Révolution Française, Paris, 1907.

Schmidt, Ch., ed., *Le commerce. Instruction, recueil des textes et notes*, Commission de recherche et de publication des documents relatifs à la vie économique de la Révolution, Paris, 1912.

Tuetey, Alexandre, Répertoire général des sources manuscrites de l'histoire de Paris pendant la Révolution Française, par . . . , vols. III, VI, IX, Paris, 1894, 1902, 1910.

Turgot, A.R.J., *Oeuvres de . . . et documents le concernant . . . , par G. Schelle*, 5 vols., Paris, 1913-1923.

Ulanowski, B., Kilka zabytków ustawodawstwa królewskiego i wojewodzińskiego w przedmiocie handlu i ustanawiania cen (documents of period 1459-1750), Arch. Kom. Prawn. PAU, vol. I, 1895.

Vauban, *Projet d'une dixme royale* . . . , ed. E. Coornaert, Paris, 1933.

Volumina legum. Przedruk zbioru praw staraniem XX Pijarów w Warszawie od roku 1732 do roku 1782 wydanego, ed. J. Ohryzko, vol. I-VIII, Petersburg, 1859, vol. IX, Kom. Hist. AU Cracow, 1889, vol. X, Poznań, 1952.

Young, A., *Voyages en France en 1787, 1788 et 1789*, ed. H. Sée, vol. II, Paris, 1931.

"Zbiór praw, dowodów i uwag . . . dla objaśnienia zaszczytów stanowi miejskiemu ex iuribus municipalibus służących . . . ," in *Materiały do Dziejów Sejmu Czteroletniego*, vol. III, Wrocław, 1960.

5. Works on Historical Metrology

Abramovič, G. W., "Neskolko izyskanii iz oblasti russkoi metrologii XV-XVI v. (Korobia, kopna, obzha)," *Problemy Istočnikovedeniya*, XI, 1963, pp. 364-390.

Albert, D. P., "Étude comparative des mesures anciennes du Département de la Charente," *Études locales*, Angoulême, XIX, 1938, 117.

Alberti, H. J. von, *Mass und Gewicht. Geschichtliche und tabellarische Darstellungen von des Anfängen bis zur Gegenwart*, Berlin, 1957, pp. 580.

Bakka, S., *Zamiana miar i wag polskich na rossyjskie i rossyjskikh na polskie* [Polish and Russian weights and measures in terms of each other], Opracował i ułożył . . . , Warsaw, 1849.

[Ballesio Vincenzo], *Riduzione degli antichi pesi, misure e monete del Piemonte e delle principali città d'Europa al sistema decimale e viceversa, con opposite tavole di conti fatti dal misuratore* . . . , de S. Maurizio, Torino, dalla Stampetia degli Eredi Botta, 1824.

Baraniecki, A., "O miarach prawnych i zwyczajowych w Polsce" [Legal and customary measures in Poland], *Wszechświat*, II, 1883, pp. 678-679, 694-696, 717-719.

Baranowski, B., "Pośmiertna kara za 'złą miarę' w wierzeniach ludowych [Posthumous punishment for 'bad' measures in folk beliefs], *Łódzkie Studia Etnograficzne*, VII, 1965, pp. 73-85.

Bazavalle, R., "Zur Geschichte des Judenburger Masses," *Zeitschrift des historischen Verein für Steiermark*, XXVI, 1931, pp. 190-191.

————. "Zur Geschichte der Grazer Masses," *Zeitschrift des historischen Verein für Steiermark*, XXV, 1929, pp. 47-78.

Belov, G. von, *Die Verwaltung des Mass- und Gewichtswesens im Mittelalter. Antwort an Herrn Schmoller*, Münster-Regensburg, 1893.

Berriman, A. E., *Historical Metrology. A new Analysis of the Archeological and the Historical Evidence Relating to Weights and Measures*, London, 1953.

Beveridge, W., "A Statistical Crime of the XVIth Century," *Journal of Economic and Business History*, I, 1929, 4.

Biernat, C., "Stanowisko Rady Gdańskiej wobec nadużyć mierników zbożowych w XVII i XVIII w." [Gdańsk Municipal Council's Attitude towards the Abuses by Grain Measurers in 17th & 18th Centuries], *Roczniki Dziejów Społecznych i Gospodarczych*, XV, 1953, pp. 195-223.

Bigourdan, G., *Système métrique des poids et mesures. Son éstablissement et sa propagation graduelle, avec l'histoire des opérations qui ont servi à déterminer le mètre et le kilogramme, par . . .* , Paris, 1901.

Bigwood, G., "Notes sur les mesures à blé dans les anciens Pays-Bas. Contribution à la métrologie belge," *Annales de la Société Archéologique de Bruxelles*, 19, 1905, pp. 5-55.

Binerowski, Z., "Gdańskie miary zbożowe w XVII i XVIII wieku," *Zapiski Historyczne. Kwartalnik Poświęcony Historii Pomorza*; XXIII, 1957, 1-3, pp. 59-81.

Bloch, M., "Le témoignage des mesures agraires," *Annales d'Histoire Économique et Sociale*; VI, 1934.

[Borghino Barnardo], *Nuovo trattato d'instruzione sull'applicazione del Sistema Metrico-Decimale al commercio in generale e ad ogni ramo d'amministrazione, sequito . . . , opera utile ad ogni ceto di persone dal Liquidatore . . . , professore d'aritmetica*, Turin, 1853.

Broc, A. F., *Nouveau Code des Poids et Mesures contenant les lois, décrets, ordonnances, circulaires et arrêtés ministeriels, dispositions pénales et jurisprudence de la Cour de Cassation, traité méthodique du système métrique, contenant 60 tables de conversion et un tableau des rapports des anciennes mesures locales des principales villes de France avec les mesures nouvelles, suivi de considérations sur les améliorations à apporter au système métrique et à son application, terminé par une table générale et analitiques des matières. Ouvrage d'une utilité indispensable pour les préfets, sous-préfets, maires, administrateurs, directeurs et percepteurs des contributions, contrôleurs et vérificateurs des poids et mesures, commissaires et officiers de police, notaires, avocats, propriétaires, négotians, marchands en détail, par M. . . . , avocat à la*

Cour Royale de Paris, et P.—C. Lavenas, auteur du Nouveau Code et Manuel pratique des Huissiers . . . , Paris, à la Librairie du Commerce, chez Renard, libraire, rue Sainte-Anne, n. 71, 1834, pp. 624 (Warsaw, Bibl. SGPiS: 64.486).

Brutails, J. A., *Recherches sur l'équivalence des anciennes mesures de la Gironde*, Bordeaux 1912.

Bryant, J., *Weight and Measures Department Annual Report* (Nairobi), 1964, p. 9, tab., Government of Kenya (Library of African Studies, Warsaw University, 5871).

[Bugoni], *Pesi e misure. Origine e vantaggi del sistema metrico decimale, risposta alle obbiezioni che si fanno contro questo sistema*, Piacenza, 1839.

Bulletin d'Histoire Économique de la Révolution, 1912 and 1920-1921, Recueil de textes concernant la suppression des droits féodaux.

Burguburu, P., "La livre carnassière, ancienne livre de boucherie," *Bulletin du Comité des Travaux Historique, Section Sciences Économiques et Sociales*, 1939.

———. *Métrologie des Basses-Pyrénées*, Bayonne, 1924.

Ciężkowska-Marciniak, W., *Jak ważono w starożytnej Grecji? [How were things weighed in ancient Greece?]*, National Museum, Warsaw, 1957.

———. "O greckich i rzymskich wagach i miarach" [Greek and Roman weights and measures], *Meander*, 1956, 1-2, pp. 40-56.

Ciszewski, S., "Początki miernictwa i pierwotne wagi" [Beginnings of surveying and primitive weights], (proof copy, 1934).

Clémenceau, E., *Le service des poids et mesures en France à travers des siècles*, Saint-Marcellin, Isère, 1909.

Čerepnin, L. V., *Russkaya metrologiya. Utchebnyie posobya po vspo-magatielnim istoričeskim disciplinam*, general editor A. I. Andreyev, vol. IV, Moscow, 1944.

Falniowska-Gradowska, A., "Miary zbożowe w województwie krakowskim w XVIII w.," *Kwartalnik Historii Kultury Materialnej*, XIII, 1965, 4, pp. 663-688.

Favre, A., *Les origines du système métrique*, PUF, Paris, 1951.

Fourcault, N. *Evaluation de poids et mesures anciennement en usage dans la province de Franche-Comté ou au Comté de Bourgogne, par* . . . , Besançon, 1872.

Gavazzi, Milovan, "Slovenska myere za priedivo i tkivo prema seksakezimalnom sistemu," *Slavia*, III, pp. 655-672.

Gilewicz, A., "Studia z dziejów miar i wag w Polsce," pt. I: "Miary pojemności i ciężaru (wagi)" [Studies in history of Polish weights and measures; measures of capacity and mass (weight)], offprint, *Sprawozdania Tow. Naukowego we Lwowie* (Lvov), XVI, 1936, 3, pp. 3-10.

―――. "Wczesnosłowiańskie miary i wagi," in *Słownik Starożytnosci Słowiańskiej* [Early Slavonic weights and measures, in: A Dictionary of Slavonic Antiquity], pp. 204-211.

Glamann, K., "Om kapitelstakst og kornmal," *Historisk Tidsskrift*, 11, IV, 4.

Górska-Gołaska, K., *Pomiary gruntowe w Wielkopolsce, 1763-1861* [Land surveys in Greater Poland], Studia nad źródłami kartograficznymi Wielkopolski z okresu reform agrarnych, Wrocław, 1965.

Górzyński, S., *Z dziejów jednostek miar w dawnej Polsce* [Studies in former Polish units of measures], Główny Urząd Miar, Warsaw, 1948.

Guillaume, Ch. Éd., *Les récents progrès du système metrique*. Rapport présenté à la quatrième Conférence générale des poids et mesures, à Paris en octobre 1907, Paris, 1907.

―――. *Les récents progrès du système metrique*. Rapport présenté à la cinquième Conférence générale des poids et mesures, à Paris, en octobre 1913, Paris, 1913.

Gumowski, M., "Najstarsze systemy wag w Polsce" [Oldest Polish systems of weights], *Studia Wczesnośredniowieczne*, II, 1953, pp. 19-36.

Haillant, S., "Mesures anciennes des diverses régions vosgiennes," *Bulletin Historique et Philosophique*, Paris, 1903.

Hallock, W., and Wade, T. H., *Outlines of the Evolution of Weights and Measures and the Metric System*, New York, 1906.

Hassler, F. R., *Comparison of Weights and Measures of Length and Capacity*, Reported to the Senate of the United States by the Treasury Department in 1832, and Made by . . . , 22 Congress, 1st Session, Washington, 1832, no. 299.

Hilliger, B., "Studien zu mittelalterlichen Massen und Gewichte. Kölner Mark and Karolinger Pfund," *Historische Vierteljahrschrift*, 1900, 2, pp. 161-215.

Hultsch, F., *Griechische und römische Metrologie*, 2d ed., Berlin, 1882.

Huščava, A., "Kuta. Stara miera konopi a lanu," *Slovenský Národopis*, II, Bratislava, 1954, 1-2, pp. 7-11.

―――. "K deyinam naystaršikh dlżkovykh mier na Slovensku," *Slovenský Národopis*, V, Bratislava, 1957, 3-4, pp. 292-306.

―――. "Metrologický vyvin v obdobi rozkladu feudalneho zriadenia na Slovensku v r. 1790-1848," *Historické Stúdie*, VII, Bratislava, 1961, pp. 189-200.

Huščava, A., "A pozsonyi méró fejlódése az 1588. Évi törvényes rendezésig"; French summary: Evolution du muid (mérö) de Bratislava jusqu'aux dispositions légales de 1588, in *Fejezesele a magyar mérésügy torteneseböl*, Budapest, 1959, pp. 25-44.

Isakov, L. D., *Na vse vremena, dla vsekh narodov. Očerki po istorii metričeskoi sistemy*, Petrograd, 1923.

Janin, V. L., "Metrologia," in *Očerki istorii istoričeskoi nauki v SSSR*, vol. II, Moscow, 1960.

Jarry-Geroult, R. A., "Économie rurale: sur la signification de la métrologie rurale traditionelle," *Comptes rendus des séances de l'Académie des Sciences*, vol. 234, 1952, pp. 1197-1199.

Kamientseva, E. I., and Ustiugov, N. V., *Russkaya Metrologiya*, Moscow, 1965.

Keckowa, A., " 'Bałwany' wielickie. Z badań nad historią techniki górnictwa solnego w Polsce w XVII-XVIII wieku" [Wieliczka salt-blocks: researches into history of salt-mining in Poland], *Kwartalnik Historii Kultury Materialnej*, VI, 1958, 4, pp. 620-639.

Kolberg (Kołobrzeg), Janusz, *Porównanie miar i wag teraźniejszych i dawniejszych w Królestwie Polskim używanych z zagranicznymi, przez . . . , wydanie wtóre, przerobił i powiększył Wilhelm Kolberg* [Comparison of weights and measures, present and past, of the Kingdom of Poland, with foreign ones . . . , 2d ed. revised and enlarged by Wilhelm Kolberg], Warsaw, 1838.

Kowalczewska, Z., Kasperowicz, W., "System metryczny miar . . . 1791-1921" [The metric system of measures], offprint from *Przegląd Techniczny*, Warsaw, 1921.

Leclére, J., and Cozette, P., "Les mesures anciennes en usage dans le canton de Noyon," *Bulletin Historique et Philosophique*, Paris, 1903.

Lederer, E., *Régi magyar Örmértékek*, Sazadok, 1923-1924.

Legendre, M., *Survivance des mesures traditionelles en Tunisie*, Paris, 1958.

Luzzati, M., "Note di metrologia pisana," *Bolletino Storico Pisano*, XXXI-XXXII, 1962-1963, pp. 121-220.

Machabey, A., *La métrologie dans les musées de province et sa contribution à l'histoire des poids et des mesures en France depuis le treizième siècle*, Thèse pour le doctorat d'Université, Paris, Lettres, Bibl. de la Sorbonne, W. Univ., 1959 (29), 4°.

Maciejowski, W. A., "Historia dawnych polskich miar i wag w zarysie od czasów najdawniejszych aż do końca XVIII wieku przedstawiona" [Historical sketch of Polish weights and measures from earliest times until the end of the 18th century], *Ekonomista*, 1868.

Maity, S. K., "Land Measurement in Gupta India," *Journal of Social and Economic History of the Orient*, I, 1957/1958, pp. 98-107.

Malhomme, L., "Spostrzeżenia nad wagami i monetami w Niemczech na wzór metrycznych francuskich zaprowadzonemi i wskazanie sposobu poprawienia ich, tudzież zastosowania miar długości, powierzchni i objętości do miar metrycznych francuskich [Proposal for a better comprehensive metrication in Germany], *Roczniki Gospodarstwa Krajowego*, XXXII, pp. 1-24.

————. "O zastosowaniu miar, wag i pieniędzy rosyjskich do metrycznych" [Russian weights, measures and money and metrication], *Roczniki Gospodarstwa Krajowego*, XXXII.

Manuele del Tempo per Economia ovvero, Almanacco Gregoriano ad uso famigliare . . . , Invenzione utilissima del Regioniere Pietro Toselli Milanese . . . , Milan, 1838.

Mauricet, A., *Des anciennes mesures de capacité et de superficie dans les départements du Morbihan, du Finistère et des Côtes-du-Nord*, Vannes, 1893.

Mazzi, A., *Il Sextarius Pergami. Saggio di ricerche metrologiche*, Bergamo, 1877.

Meitzen, A., *Volkshufe und Königshufe*, in (*Festgabe für Georg Héussen*), Tübingen, 1889.

La metrologia comparata ridotta a comune intelligenze ossia teorica del sistema metrico decimale, applicata all'uso pratico con quadri comparativi ed illustrativi da un membro dell'Accademia I. et. R. del Georgofili di Firenze, Publicazione a favore delle scuole infantili, Turin, 1847.

Ministero per gli Affari di Sardegna. Atlante dei pesi e delle misure metriche decimali secondo il sistema introdotto nel Regno di Sardegna col Regio Editto del 1° Luglio 1844, Turin, 1844.

[Molinari, Vincenzo], *Trattato di metrologia universale ossia pesi, misure e moneta di Tutti gli Stati del Mondo, col rapporto ai pesi, alle misure ed alle monete di Napoli e del Piemonte col metodo di riduzione del sistema di Napoli alle decimale e viceversa, e colla giunta di 19 tavole di confronto . . . lavoro di . . . esequito su parziali trattati di metrologia e sulle informazione di esperti negozianti*, Naples 1862, 2nd ed. Naples, 1963.

Moreau H., *Les récents progrès du système métrique 1948-1954. Rapport présenté à la dixième conférence générale des poids et mesures réunie à Paris*, en 1954, Paris, 1955.

Morineau M., "Jauges et méthodes de jauge anciennes et modernes," *Cahiers des Annales*, 24, Paris, 1966, pp. 119.

Mühe W., *Mass- und Eichwesen* [w:] *Handwörterbuch der Sozialwissenschaften* vol. VII, 1961, pp. 226-233.

Musioł, L., *Dawne miary zboża na Górnym Śląsku. Przyczynek do metrologii śląskiej* [Former measures of corn in Upper Silesia], mimeo publ. by author, Opole, 1963.

Navel, H., *Recherches sur les anciennes mesures agraires normandes, Acres, vergées et perches*, Caen, 1932.

Niangoran-Bouah, *Weights for the Weighing of Gold. One of the Aspects of African Philosophical and Scientific Thought before Colonisation.* First International Congress of Africanists, 1962, University of Ghana, Accra (mimeo).

Ogrizko, Z. A., "K voprosu o yedinitsakh izmerenya zemelnykh ploščadei v XVIII v.," *Problemy Istočnikovedeniya*, IX, 1957.

Paculski, K., Skrzypek, A., Żukowski, T., "Miary na ziemiach polskich w drugiej połowie XVIII i w XIX wieku jako odbicie zjawisk gospodarczych" [Measures in lands of Poland in later 18th and 19th centuries as reflection of economic life], *Kwartalnik Historii Kultury Materialnej*, XV, 1967, 2, pp. 357-364.

Pańkowski, W., *Zarys Polskiego prawa o miarach*, Biblioteka "Poradnika Przedsiębiorcy" [Outline of Polish law regarding measures; Library of Businessman's Manuals], series A, vol. I, Poznań, 1938.

Paulinyi, A., "Soupis mer z r. 1775 a zavedenie yednotney mery uhla v r. 1776-1780 v stredoslovenskey banskey oblasti," *Historické Štúdie*, XI, Bratislava, 1966, pp. 263-281.

Paulme, D., "Systèmes pondéraux et monétaires," *Revue Scientifique*, 5, 1942.

Petruševski, T., *Metrologiya ili opisaney mer, vesov, monet i vremyačislenya*, S. Petersburg, 1831.

Piekosiński, F., "O łanach w Polsce wieków średnich" [The *łans* in medieval Poland], *Rozpr. AU*, XXI, 1882.

Programma kursa "Russkaya metrologiya" dla Moskovskogo gosudarstvennogo istoriko-arkhivnogo Instituta, Moscow, 1954, p. 10.

Putek, J., "Śląskie miary nasypowe w Polsce," in *Z dziejów wsi polskiej* [Silesian dry measures in Poland, in Studies in history of Polish village], Warsaw, 1964.

[Quantin, M.], *Tableaux des poids et mesures légaux et usuels, précédé de recherches sur les poids et mesures en usage dans toutes les communes du département en 1789, par . . . , archiviste du Département de l'Yonne*, in Auxerres, 1839.

Rauszer, Z., "Jednostki miar używane w Królestwie Kongresowym w dobie poprzedzającej wydanie Dekretu o miarach z dn. 8 lutego 1919" [Units of measures in the Congress Kingdom of Poland prior to the Decree on measures of 8 Feb. 1919], *Miesięcznik Statystyczny*, I, 1920, pp. 76-87.

––––––. "Pierwsze dziesięciolecie polskiej administracji miar i narzędzi mierniczych" [First decade of Polish administration of measures and measuring instruments], *Przegląd Techniczny*, 1929, pp. 67-68, 179-192.

Report of the Committee on Weights and Measures Legislation, HMSO, London, 1951.

Richard, J., "Arpentage et chasse aux loups. Une définition de la lieu de Bourgogne au XVᵉ s.," *Annales de Bourgogne*, XXXV, 1963, 4, pp. 239-250.

––––––. "La gréneterie de Bourgogne et les mesures à grain dans le duché de Bourgogne," *Mémoirs de la Société pour l'histoire du droit du Duché et des institutions des anciens pays bourgignons, comtois et romands*," X, 1944-1945, pp. 117-145.

Riduzione degli antichi pesi, misure e monete del Piemonte e delle principali città d'Europa al sistema decimale e viceversa, con apposite tavole di conti fatti dal misuratore Vincenzo Ballesio di S. Maurizio, Turin, Stamperia degli Eredi Botta, 1824.

Riduzione di pesi nazionali e stranieri. Pesi nazionali e stranieri dichiarati e ridotti de P.F.R. (Pietro F. Rocca—signature on the inside of title page), Genoa, Stamperia Casamara.

Riocour, de, *Les monnais lorrains (seconde partie)*. Mémoires de la Société d'Archéologie Lorraine et du Musée Historique Lorrain, ser. III, vol. XII, Nancy, 1884.

Rybakov, B. A. "Russkiye sistemy mer dliny XI-XV vekov. Iz istorii narodnykh znanii," *Sovyetskaya Etnografiya*, 1949, 1, pp. 67-91.

Sartori, P., "Zählen, messen, wägen," *Am-Urquell. Monatschrift für Volkskunde*, VI, 1895, pp. 9-12, 58-60, 87-88, 111-113.

Schmoller, G., "Die Verwaltung des Mass- und Gewichtswesens im Mittelalter," *Jahrbuch für Gesetzgebung, Verwaltung und Volkswirtschaft* (Schmollers Jahrbuch), XVII, 1892.

Sedlaček, A. "Pamieti a doklady o staročeskykh mirakh a vahakh," *Rozpravy ČAV, Ulmeni třida* I, Prague, 1923.

Sochaniewicz, K., "Miary i ceny produktów rolnych na Podolu w XVI w." [Measures and prices of agricultural produce in Podolia in 16th century], *Lud*, XXVIII, 1929, pp. 145-166.

––––––, "Miary roli na Podhalu w ubiegłych wiekach" [Measures of cultivable land in the Tatra region in past ages], *Lud*, XXV, 1926, pp. 19-37.

Sourbé, T., *Observations du S^r . . . publiciste, sur l'inexécution des lois qui régissent les poids et mesures. À Monsieur le Ministre du Commerce et des Colonies*, Paris, 1881.

Stamm, E., "Miary długości w dawnej Polsce,"*Wiadomości Służby Geograficznej*, Warsaw, 1935, 3.

———. "Miary powierzchni w dawnej Polsce," *PAU*, Rozpr. Wydz. Hist.-Filoz., XLV, 1936.

———. *Staropolskie miary*, vol. I: *Miary długości i powierzchni*, Warsaw, 1938.

Sto let gosudarstvennoj służby mer i vesov v SSSR, Komitet po delam mer i izmeritelnykh priborov v SSSR, Moscow, 1945.

Streltsin, S., "Contribution à l'histoire des poids et des mesures en Ethiopie," *Rocznik Orientalistyczny*, XXVIII, 1965, 2, pp. 75-85.

Strumilin, S. G., *O merach feodalnoi Rosii* in *Voprosy istorii narodnogo khoziaystva SSSR*, Moscow, 1957, pp. 7-32.

Szewczyk, J., "Włóka. Pojęcie i termin na tle innych średniowiecznych jednostek pomiaru ziemi," *Prace Geograficzne*, no. 67, Warsaw, 1968.

Šunkov, V. I., "Mery sypnych tel w Sybiri XVII v.," in *Akademiku B. D. Grekovu ko dniu semidesiatiletiia. Sbornik Statej*, Moscow, 1952, pp. 166-171.

Śreniowski, S., "Uwagi o łanach w ustroju folwarczno-pańszczyźnianym wsi polskiej," *Kwartalnik Historii Kultury Materialnej*, III, 1955.

Tables du rapport des anciens poids et mesures des États de terraferme du Royaume [de Piémont—WK] *avec les poids et mesures du système métrique décimal, dressées par la Commission des Poids et Mesures et publiées par le Ministère d'Agriculture et Commerce aux termes de l'art. 11 de l'Édict Royale du 11 septembre 1845*, Turin, Imprimerie Royale, 1849.

Tablice zamiany miar i wag rosyjskich na polskie i nawzajem, w Komitecie miar i wag ułożone, a z mocy art. 7 postanowienia Rady Administracyjnej Królestwa z dnia 2/14 marca 1848 przez Komisję Rządową Spraw Wewnętrznych i Duchownych do powszechnego użytku wydane, Warsaw, 1849.

Tavole di confronto delle misure Piacentine colle misure del nuovo sistema metrico, con appendici . . . , calcolate de G. V., seconda ed., corretta ed accresciuta, Piacenza, 1826.

Thaa, von, *Mass und Gewicht* in *Österreichisches Staatswörterbuch. Handbuch des gesamten österreichischen Rechtes*, herausgegeben von Ernst Mischler und Josef Ulbrich, vol. III, Vienna, 1907, pp. 532-537.

Tits-Dieuaide, M. J., "La Conversion des mesures anciennes en mesures métriques. Note sur les mesures à grain d'Anvers, Bruges, Bruxelles, Gand, Louvain, Malines et Ypres du XV-e au XIX-e siècle," *Contributions à l'Histoire Économique et Sociale*, II, Brussels, 1963, pp. 31-89.

Viaud, F., *Cours de poids et mesures*, vol. I: *Législation des poids et mesures*, Paris, 1933.

———. *L'Intervention de l'État dans le domaine de la métrologie. Mesures, régulation, automatisme*, vol. 30, no.10, Oct. 1965, pp. 75-81.

Viedebantt, O., "Altes und ältestes Weg- und Längenmass," *Zeitschrift für Etnographie*, 45, 1913.

———. *Forschungen zur Metrologie des Altertums*, Leipzig, 1917.

Vieweg, R., "Some Aspects in the History of Metrology," *U.A.E. Standardization Bulletin*, III, Cairo, 1964, 2, pp. 1-16.

———. *Mass und messen in kulturgeschichtlicher Sicht*, Franz Steiner Verlag, Wiesbaden, 1962.

———. *Mass und messen in Geschichte und Gegenwart*, Cologne and Opladen, 1958.

Vivier, R., "L'application du système métrique dans le Département d'Indre-et-Loire. 1789-1815," *Revue d'Histoire Économique et Sociale* XVI, 1928, pp. 182-229.

———. "Contribution à l'étude des anciennes mesures du Département d'Indre-et-Loire au XVII-e et XVIII-e siècles," *Revue d'Histoire Économique et Sociale*, XIV, 1926, pp. 179-199.

Wade, H. T., *Weights and Measures* in Encyclopaedia of the Social Sciences (Seligman), XV, New York, 1949, pp. 389-392.

Wichmann, T., "Les vielles mesures de longueur des peuples finno-ougriens," *Journal de la Société Finno-Ougrienne*, 42, 1928, pp. 13-27.

Witkowski, J., "Miary długości metryczne i stosunek ich do miar angielskich i dawnych polskich," *Przegląd Techniczny*, XLI, 1903, pp. 81-85.

Wolański, M., "Śląskie miary nasypowe (zbożowe) w XVIII w." *Zeszyty Naukowe Uniwersytetu Wrocławskiego*, seria A, no. 13, 1959, pp. 3-43.

Zagrodzki, T., "Regularny plan miasta średniowiecznego a limitacja miernicza," *Studia Wczesnośredniowieczne*, V, 1962, 1, 102.

Zamzaris, J. K., "Metrologia Latvii v period feodalnoi rozdroblennostii razvitogo feodalisma (XIII-XVI vv.)," *Problemy-Istočnikovedenia*, IV Moscow, 1955, pp. 177-222.

6. Works of Relevance to Metrology

Adamczyk, W., *Ceny w Lublinie od XVI do końca XVIII w.* [Prices in Lublin from the 16th to the end of the 18th century], Lvov, 1935.

———. *Ceny w Warszawie w XVI i XVII w.* [Prices in Warsaw in the 16th and 17th centuries), Lvov, 1938.

Appadorai, A., *Economic Conditions in Southern India, 1000-1500 A.D.*, University of Madras, 1936.

Askenazy, S., "Trybun Gminu," w *Dwa Stulecia* [A tribune of the people, in Two Centuries], Warsaw, 1910.

Aurora, G. S., "Economy of a Tribal Village," *Economic Weekly*, 26 Sept. 1964.

Biot, *Pismo podręczne dla budującego drogi żelazne, albo wykład zasad ogólnych sztuki budowania drogi żelaznej przez . . . jednego z członków zarzadzających wykonaniem robót drogi żelaznej od St. Etienne do Lyonu* [A manual for railroadbuilders, or general principles of railroad building . . . by a member of the management of the building of the railroad from St. Étienne to Lyon], transl. R.K.W. Górski, Warsaw, 1842.

Baehrel, R., "Épidémie et terreur. Histoire et sociologie," *Annales Historiques de la Révolution Française*, XXIII, 1951, pp. 115-146.

———. "La haine de classe en temps d'épidémie," *Annales ESC*, 1952, pp. 351-360.

Baranowski, B., *Pożegnanie z diabłem i czarownicą* [A farewell to the devil and the witch], Łódź, 1965.

Baranowski, L. T., "Dobra puławskie miedzy I-szym a III-cim rozbiorem," w *Wieś i folwark* [The Puławy estates between the 1st and 3rd partition, in The village and the demesne farm], Warsaw, 1914.

Bataille, M., "Viollet-le-Duc, jardinier des pierres," *Archéologie*, 1965, 6, pp. 50-56.

Belov, von, G., *Die Entstehung der deutschen Stadtgemeinde*, Düsseldorf, 1889.

———. *Der Ursprung der deutschen Stadtverwaltung*, Düsseldorf, 1892.

Bergier, J. F., "Commerce et politique du blé à Genève aux XV-e et XVI-e siècles," *Revue Suisse d'Histoire*, XIV, 1964, 4, pp. 521-550.

———. *Genève et l'économie européenne de la Renaissance*, Paris, 1963.

Beveridge, W., *Prices and Wages in England*, vol. I, London, 1940.

Bezard, Y., *La vie rurale dans le sud de la région parisienne de 1450 à 1560*, Paris, 1929.

Bloch, R., *Etruskowie* [The Etruscans], Warsaw, 1966.

Bobińska, C., "Pewne kwestie chłopskiego użytkowania gruntu i walka o ziemię," w *Studia z dziejów wsi małopolskiej w drugiej połowie XVIII w.* [Some issues in the use of land by the peasants and the struggle for land, in Studies in the history of the village in Lesser Poland in the second half of the 18th century], Warsaw, 1957.

Bogucka, M., "Z zagadnień spekulacji i nadużyć w handlu żywnością w Gdańsku w XV-XVII w.," *Zapiski Historyczne*, XXVII, 1962, 1, pp. 7-22 [Aspects of speculation and profiteering in the food trade in Gdańsk in the 15th to the 17th century].

Bouloiseau, M., "Notables ruraux et élections municipales dans la région rounnaise," *Actes du Congrès des Sociétés Savantes*, 1957.

———. "Voeux et griefs saumurois lors des élections aux États-Généraux de 1789," *Actes du Congrès des Sociétés Savantes*, 1962.

Brunot, F., *Histoire de la langue française des origines à 1900*, vol. IX: *La Révolution et l'Empire*, pt. II: *Les événements, les institutions et la langue*, Paris, 1937.

Brückner, A., *Encyklopedia staropolska* [Encyclopaedia of Old Poland], Warsaw, 1939.

Bulletin de la Commission d'Histoire Économique, 1906-1907.

Burszta, J., ed., *Kultura Ludowa Wielkopolski* [The Popular Culture of Greater Poland], vol. II, Poznań, 1964.

Collection Générale des décrets rendus par L'Assemblée Nationale, vol. II.

Confino, M., *Domaines et seigneurs en Russie vers la fin du XVIIIe siècle. Étude de structures agraires et de mentalités économiques*, Collection Historique de l'Institut d'Études Slaves, vol. XVIII, Paris, 1963.

Coulton, C. G., *Medieval Village, Manor and Monastery*, New York, 1960.

Czarnowski, S., *Kultura*, w *Dzieła* (Culture, in Collected Works), vol. I, Warsaw, 1958.

Ćwiek, Z., *Z dziejów wsi koronnej XVII wieku* [Aspects of the history of villages on royal demesnes in the 17th century], Warsaw, 1966.

Dobrowolski, K., "Dzieje wsi Niedźwiedzia w powiacie limanowskim do schyłku dawnej Rzeczypospolitej," w *Studia z historii społecznej gospodarczej poświęcone Franciszkowi Bujakowi* [The history of the village of Niedźwiedź in the Limanowa district until the decline of the old Commonwealth of Poland, in Studies in social and economic history in honour of Franciszek Bujak], Lvov, 1931.

Dumont, R., *Terres vivantes. Voyage d'un agronome autour du monde*, Paris, 1961.

Elsas, M. J., *Umriss einer Geschichte der Preise und LX Löhne in Deutschland*, vols. I-II, Leiden, 1936-1949.

Engels, F., *Teoria przemocy* [The theory of violence], Warsaw, 1961.
———. *Wojna chłopska w Niemczech* [Peasants' war in Germany], Warsaw, 1950.

Ereciński, T., Prawo przemysłowe miasta Poznania w XVIII w. [Industrial law of the city of Poznań in the 18th century], Poznań, 1934.

Estreicher, T., and Mościcki, H., "Aleksander Chodkiewicz," in *Polski Słownik Biograficzny* [Polish biographical dictionary], vol. III.

Falniowska-Gradowska, A., *Świadczenia poddanych na rzecz dworu w królewszczyznach województwa krakowskiego w drugiej połowie XVIII wieku* [Serfs' dues owing to the manor in the royal estates of the Cracow voivodeship in the second half of the 18th century], Wrocław, 1964.

Feuerwerker, D., "Anatomie des 307 cahiers de doléances de 1789. Les Juifs en France, *Annales ESC*, 1965.

Fornaseri, G., "L'Archivio del Departamento della Stura nell'archivio di Stato di Cuneo, 1799-1814. Inventario a cura di . . . ," *Quaderni delle "Rassegna degli Archivi di Stato,"* vol. II, Rome, 1960.

Fourastié, F. and J., eds., *Les écrivains témoins du peuple*, Paris, 1964.

Fourquin, G., *Les campaignes de la région parisienne à la fin du Moyen Age*, Paris, 1964.

Furtak, T., *Ceny w Gdańsku w l. 1701-1819* [Prices in Gdańsk from 1701 to 1819], Lvov, 1935.

Gaxotte, P., *La Révolution Française*, Paris, 1928.

Geslin de Bourgogne, J., and Barthelemy, A. de, *Anciens évêchés de Bretagne*, vol. II, pp. CCIII, CCXI-CCXXV.

Gille, B., *Les sources statistiques de l'histoire de France. Des enquêtes du XVIIe siècle à 1870*, Geneva, 1964.

Gloger, Z., *Encyklopedia staropolska* [Encyclopaedia of Old Poland], Warsaw, 1958.

Goubert, P., *Beauvais et les Beauvaisis de 1600 à 1730. Contribution à l'histoire sociale de la France du XVIIe siècle*, Paris, 1960.
———. 1789. *Les Français ont la parole*, series "Archives," Paris, 1964.

Górkiewicz, M., *Ceny w Krakowie w latach 1796-1914* [Prices in Cracow from 1796 to 1914], Poznań, 1950.

Górnicki, Ł., *Dzieje w Koronie Polskiej, w Dzieła wszystkie* [History of the kingdom of Poland, in: Collected works], vol. III, Warsaw, 1886.

Grynwaser, H., *Przywódcy i "burzyciele" włościan, w Pisma* [Leaders and ringleaders of countryfolk, in: Works], vol. II, Wrocław, 1951.

Habib, I., *The Agrarian System of Mughal India, 1556-1707*, Aligarh Muslim Asia Publishing House, 1963.

Hamilton, E., *American Treasure and the Price Revolution in Spain, 1501-1650.*

Hauser, H., *Recherches et documents sur l'histoire des prix en France de 1500 à 1800*, Paris, 1936.

Hazard, P., *La crise de la conscience européenne 1680-1715*, Paris, 1935.

Heers, J., *Le livre de comptes de Giovanni Piccamiglio, homme d'affaires génois (1456-1459)*, Paris, 1959.

Hoffman-Krayer, E., "Bächtold-Stäubli, H.," in *Handwörterbuch des deutschen Aberglaubens*, V, Berlin, 1932-1933, p. 1852, *s.v. Mass, messen.*

Hołubowicz, W., *Opole w wiekach X-XII* [Opole in the 10th to 12th centuries], Katowice, 1956.

Hoszowski, S., *Ceny we Lwowie w XVI i XVII wieku* [Prices in Lvov in the 16th and 17th centuries), Lvov, 1928.

———. *Ceny we Lwowie w latach 1701-1914* [Prices in Lvov from 1701 to 1914], Lvov, 1934.

Instrukcja do wyciągania intraty z dóbr i szacowania placów po miastach, w Warszawie, nakładem Komisji Rządowej Przychodów i Skarbu, druk N. Glücksberga, 1828 [Manual of extracting revenues from estates and valuing urban sites, published in Warsaw by the Government Commission for Revenues and Treasury, printed by N. Glücksberg, 1828].

Janáček, J., "Rudolfinské drahotni rády," *Rozpravy ČSAV*, 1957.

Kolendo, J., *Postęp techniczny a problem siły roboczej w rolnictwie starożytnej Italii* [Technical progress and the problem of manpower in agriculture of ancient Italy], Wrocław-Warsaw, 1968.

Konopczyński, W., *Polscy pisarze polityczni XVIII w.* [Polish 18th-century political writers], Warsaw, 1966.

Kostrzewski, J., *Kultura prapolska* [Primeval Polish culture], Poznań, 1947.

Kötschke, R., *Allgemeine Wirtschaftsgeschichte des Mittelalters*, Jena, 1924.

Kraushar, A., *Towarzystwo Królewskie Przyjaciół Nauk, 1800-1832* [The Royal Society of Friends of the Sciences], Warsaw, 1902.

Kula, W., *Problemy i metody historii gospodarczej* [Problems and methods of economic history], Warsaw, 1963.

———. *Szkice o manufakturach w Polsce w XVIII w.* [Essays on 18th century manufactories in Poland], Warsaw, 1956.

Kula, W., *Teoria ekonomiczna ustroju feudalnego. Próba modelu*, Warsaw, 1962 [English translation by L. Garner: *an Economic Theory of the Feudal System—towards a model of the Polish economy 1500-1800*, London, 1976—Trans.]

Ladurie, le Roy, E., *Les paysans de Languedoc*, Paris, 1966.

Lafargue, P., *Badania nad pochodzeniem pojęcia sprawiedliwości i pojęcia dobra*, w *Pisma wybrane*, vol. I [Enquiries into the origins of the concept of justice and the concept of the good, in Selected works], Warsaw, 1961.

Legris, M., "Développement agricole et paysans noirs," *Le Monde*, 10-11 Feb. 1966.

Leskiewiczowa, J., "W sprawie publikacji instruktarzy gospodarczych" [On the publication of economic manuals], *Kwartalnik Historii Kultury Materialnej*, IX, 1961.

Lombardini, G., *Pane e denaro a Bassano tra il 1501 al 1799*, Venice, 1963.

Machabey, A., "Les sources," *Revue de Métrologie Pratique et Légale*, 1952, 4.

———. "Les techniques de mesure," in *Histoire générale des techniques*, vol. II: *Les premières étapes du machinisme*, PUF, Paris, 1965.

Marx, K., *Kapitał*, vol. I, Warsaw, 1951.

Mathiez, A., *Rewolucja Francuska* [The French Revolution], Warsaw, 1956.

Mauny, R., *Tableau géographique de l'Ouest Africaine au Moyen Age d'après les sources écrites, la tradition et l'archéologie*, Dakar, IFAN, 1961.

Mauro, F., *Le Portugal et l'Atlantique au XVII-e siècle, 1570-1670*, Paris, 1960.

Mendeleyev, D. I., *Sotchinenya*, vol. 22, Moscow, 1950.

Michalski, J., *Z dziejów Towarzystwa Przyjaciół Nauk* [Studies in the history of the Society of Friends of the Sciences], Warsaw, 1953.

Michelet, J., *Histoire de la Révolution Française*, Paris, 1952.

Mielczarski, S., *Rynek zbożowy na ziemiach polskich w drugiej połowie XVI i pierwszej połowie XVII wieku. Próba rejonizacji* [The grain market in the lands of Poland in the second half of the 16th and first half of the 17th century; an attempt at regionalization], Gdańsk, 1962.

Montesquieu, *O duchu praw* [The spirit of laws], Warsaw, 1957, bk. XXIX, chap. 18.

Moszyński, K., *Kultura ludowa Słowian* [Popular culture of the Slavs], vol. II, pt. I, Cracow, 1934.

Niemann, *Von altägyptishcer Technik. Beiträge zur Geschichte der Technik und Industrie*, Berlin, 1930.

Okolski, A., *Wykład prawa administracyjnego obowiązującego w Królestwie Polskim* (The administrative law of the Kingdom of Poland), 3 vols., Warsaw, 1884.

Orsini-Rosenberg, S., *Geneza i rozwój folwarku pańszczyźnianego w dobrach katedry gnieźnieńskiej w XVI w.* [The genesis and development of the corvée-based manorial demesne in the estates of the Gniezno cathedral in the 16th century], Poznań, 1925.

Parenti, G., *Prime ricerche sulla rivoluzione dei prezzi in Firenze*, Florence, 1939.

Pazdro, Z., *Organizacja i praktyka żydowskich sądów podwojewodzińskich w okresie 1740-1772 na podstawie lwowskich materiałów archiwalnych* [Organization and practice of the Jewish voivodeship courts as evidenced by the archives at Lvov], Lvov, 1903.

Pelc, J., *Ceny w Krakowie w latach 1369-1600* [Prices in Cracow from 1369 to 1600], Lvov, 1935.

Picard, R., *Les cahiers de 1789 au point de vue industriel et commercial. Thèse pour le doctorat ès sciences juridiques*, Faculté de Droit l'Université de Paris, 1910.

Plaisse, A., *La baronnie de Neubourg. Essai d'histoire agraire, économique et sociale*, Paris, 1961.

Prato, G., *La vita economica in Piemonte a mezzo il secolo XVIII*, Turin, 1908 (photostat version 1966 used).

Prus, B., *Kroniki* [Chronicles], vol. IV, Warsaw, 1955.

Raveau, P., *Essai sur la situation économique et l'état social en Poitou au XVIe siècle*, Paris, 1931.

Recueil authentique des lois du gouvernement de la République et Canton de Genève, vol. II, 1816.

Rennand, J., *Baas, Blénac ou les Antilles françaises au XVIIe siècle*, Fort-de-France, 1935.

Rolbiecki, J. G., *Prawo przemysłowe miasta Wschowy w XVIII w.* [Industrial law of the town of Wschowa in the 18th century], Poznań, 1951.

Rutkowski, J., *Historia gospodarcza Polski (do 1864 r.)* [An economic history of Poland until 1864], Warsaw, 1953.

———. "Studia nad położeniem Słościan w Polsce w XVIII w." w *Studia z dziejów wsi polskiej XVI-XVIII w.* [The condition of the Polish countryfolk in the 18th century, in Studies on the Polish countryside from the 16th to the 18th century], Warsaw, 1956.

Rybarski, R., *Handel i polityka handlowa Polski w XVI stuleciu* [Poland's trade and commercial policy in the 16th century], 2 vols., Poznań, 1929.

Rychlikowa, I., *Studia nad towarową produkcją wielkiej własności w Małopolsce w latach 1764-1805*, cz.I, "Towarowa gospodarka zbożowa," Wrocław, 1966.

Sagnac, P., "Les cahiers de 1789 et leur valeur," *Revue d'Histoire Moderne et Contemporaine*, 1906-1907.

Sée, H., "La redaction et la valeur historique des cahiers de paroisses pour les États Généraux de 1789," *Revue Historique*, CIII, 1920.

Siegel, S., *Ceny w Warszawie w latach 1701-1815* [Prices in Warsaw from 1701 to 1815], Lvov, 1936.

————. *Ceny w Warszawie w latach 1816-1914* [Prices in Warsaw from 1816 to 1914], Poznań, 1949.

Skrzypek, M., "J.-M. Lequino," *Euhemer*, 1965, 5.

Smoleński, W., *Komisja Boni Ordinis warszawska, 1765-1789* [Warsaw's Commission of law and order, 1765-1789], Warsaw, 1913.

Spirkin, A., *Pochodzenie świadomości* [The origins of consciousness], Warsaw, 1966.

Szelągowski, A., *Pieniądz i przewrót cen w XVI i w XVII w. w Polsce* [Money and the price revolution in the 16th and 17th century in Poland], Lvov, 1902.

Tannenbaum, F., *Une philosophie du travail. Le syndicalisme*, Paris, 1957.

Thompson, J. W., *Economic and Social History of Europe in the later Middle Ages, 1300-1530*, New York, 1958.

————. *Economic and Social History of the Middle Ages, 300-1300*, vol. II, New York, 1959.

Tocqueville, de A., *L'Ancien Régime et la Révolution*, Paris, 1964.

Tomaszewski, E., *Ceny w Krakowie w latach 1601-1795* [Prices in Cracow from 1601 to 1795], Lvov, 1934.

Topolski, J., *Gospodarstwo wiejskie w dobrach arcybiskupstwa gnieźnieńskiego od XVI do XVIII wieku* [The farmstead in the estates of the archbishopric of Gniezno from the 16th to the 18th century], Poznań, 1958.

Verlinden, Ch., *Dokumenten voor de Geschiedenis van Prijzen en Lonen . . .* , Bruges, 1959.

Vidaleno, J., "Les revendications économiques et sociales de la population parisienne en 1789 d'après les cahiers de doléances," *Revue d'Histoire Economique et Sociale*, XXVII, 1948-1949.

Vilar, P., *La Catalogne dans l'Espagne moderne*, vol. II, Paris, 1962.

Wächter, H. H., "Ostpreussische Domänenvorwerke im 16 und 17

Jahrhundert," *Beihefte zum Jahrbuch der Albertus Universität*, Königsberg/Pr., Würzburg, 1958.

Warschauer, A., "Handel, Gewerbe und Verkehr," in *Das Jahr 1793. Urkunden und Aktenstücke zur Geschichte der Organisation Süd-preussens. Veröffentlichungen der Historischen Gesellschaft für die Provinz Posen*, vol. III, Posen, 1895.

Wee, van der, H., *The Growth of the Antwerp Market and the European Economy*, vol. I, The Hague, 1963.

Wolff, Ph., *Commerce et marchands de Toulouse (vers 1350-vers 1450)*, Paris, 1954.

Woodcock, F., "The Price of Provisions and some Social Consequences in Worcestershire in the XVIIIth and XIXth Century," *Journal of the Royal Statistical Society*, CVI, 1943.

Wyczański, A., *Studia nad gospodarką starostwa korczyńskiego 1500-1660* [Studies in the economy of the starostship of Korczyn, 1500-1660], Warsaw, 1964.

INDEX

absolute monarchy, 175-177, 210-216
absolutist reform, 115
Africa: binary system in, 306n9; indigenous measures of, 5, 7, 10-11, 25; inertia of measures in, 112; king's measure in, 18; metrological reform in, 284; variations in measures in, 89, 104, 109-110
agrarian measures. *See* land measures
agronomy, and historical metrology, 90-91
Alexander the Great, 115
Alexander II (tsar of Russia), 157
Alfonso X (king of Spain), 116
Alfonso XI (king of Spain), 116
alienation, of measure, 122-123
"analogous" measures, 259, 260-261, 262
Ananias of Shiraz, 96
Angola, variations based on value in, 89
anthropometric measures, 4-5, 24-28
Antwerp: grain measure in, 65-66; metrological reform in, 278
Arago (astronomer), 243
Arbogast (at 1793 Paris Convention), 238-239
Archangel Michael, 10, 14
Archimedes, 243
architecture: anthropometric measure in, 27-28; inferring measure from, 95
Architos (Greek philosopher), 287
Arginus, Phidon, 14
Arnold, S., 93
arepennis, 5
arpent, 5, 29, 30, 31
arpent royal, 240
arquebus shot, 294n5
artefacts, inferring measure from, 95-96
Ashanti: king's measure in, 18; measures of gold by, 5, 11
aune: attempts to standardize, 172, 177; division of, 260, 277; grievances about, 191

Austria, metrological reform in, 117, 118, 278, 279

Babeuf, Gracchus, 91
bakers: dishonest measure by, 15; variable bread measure by, 71-78
baker's dozen, 84
Balbe (ambassador from Piedmont), 271, 272
Balducci, Francesco Peglotti, 96
banalités, 206-207, 218-219
bar, 6
Baraniecki, M., 93
Baranowski, B., 16
Barcelona, bread measure in, 78
barrel, 84
basket, 6
Batavian Republic, metrological reform in, 277-278
Bathory, Stephen (king of Poland), 130
Beccaria, Cesare, 269, 270, 271, 273
Belgium: Carolingian reform in, 163; grain measure in, 65-66; inertia of measures in, 111; metrological reform in, 277-278; variations based on value in, 89
Bezard, Y., 111
Bible: authority and measure in, 18; and historical metrology, 90; origin of weights and measures in, 3; on pouring of dry measure, 49; realistic vs. symbolic measures in, 9-10
bichetée, 33
bichot, 85
Biernacki, Ludwik, 51-52
binary system, 306n9
Bloch, M., 111
boatload, 6
body parts: as measures, 4-5, 24-28; superstitions about measuring, 13
Bogucka, M., 72
Bois, P., 187
boisseau: accessibility of measure for, 197-198; codification of, 46, 166-167; dishonest practices with, 175,

375

Library of Congress Cataloging-in-Publication Data

Kula, Witold.
Measures and men.

Translation of: Miary i ludzie.
Bibliography: p.   Includes index.
1. Weights and measures—History.   I. Title.
QC83.K813 1985   530.8'09   85-42690
ISBN 0-691-05446-0